Introduction to Ecological Biochemistry

FOURTH EDITION

J.B. Harborne
University of Reading, UK

ACADEMIC PRESS
Harcourt Brace & Co., Publishers
London • San Diego • New York
Boston • Sydney • Tokyo • Toronto

ACADEMIC PRESS LIMITED
24/28 Oval Road
LONDON NW1 7DX

United States Edition published by
ACADEMIC PRESS INC.
San Diego, CA 92101

First Edition published 1977
Second Edition 1982
Third Edition 1988
Third Edition reprinted 1989, 1992, 1997

This book is printed on acid-free paper

A catalogue record for this book is available from the British Library

ISBN: 0-12-324685-7
0-12-324686-5 (pbk)

Typeset by Alden Multimedia, Northampton
Printed in Great Britain by The University Press, Cambridge

Contents

3. Plant Toxins and Their Effects on Animals

4. Hormonal Interactions Between Plants and Animals

5. Insect Feeding Preferences

6. Feeding Preferences of Vertebrates, Including Man

7. The Co-evolutionary Arms Race: Plant Defence and Animal Response

8. Animal Pheromones and Defence Substances

9. Biochemical Interactions Between Higher Plants

10. Higher Plant–Lower Plant Interactions: Phytoalexins and Phytotoxins

Foreword

Science, says François Jacob, attempts to confront the possible with the actual, and by so doing must inevitably renounce a unified world concept. Jeffrey Harborne puts this to rights in one corner of the biological and biochemical fields, and that is perhaps the main reason why one is so wholeheartedly delighted with his book. We have all been waiting impatiently for a synthesis fusing plant/insect relationships with their ecological biochemistry, but in addition to the welcome unification of the underlying theme, the subject matter is presented with felicitous directness and simplicity. We are given a masterly *überblick*, so that with a sigh of appreciation and relief we know for the first time not only where we are, but also the lines along which we should now proceed. Although the complicated and tangled subject matter is presented in a straightforward fashion, with perspicacious sifting and appraisal of the evidence, without frills and in a logical and scientific manner, there is still a romantic undertone which reveals the author as a fine and enthusiastic naturalist as well as a laboratory investigator.

Curiously enough, one of the first attempts to link the fields of entomology and plant chemistry in the modern sense was initiated by field naturalists (British Empire builders in fact)—their imagination fired by the evolutionary implications of the theory of warning coloration and mimicry among butterflies. It was their infectious enthusiasm and ebullient writing, now often dismissed as "anecdotal", which first encouraged me to investigate the chemistry of Lepidoptera/plant relationships. The observations of C. F. M. Swynnerton were particularly enthralling and it was tragic that the vast piles of notes he had accumulated, by virtue of his drive and tireless energy, could not be satisfactorily decoded when he died suddenly in the prime of life. In 1915 he published an account of feeding experiments (in Africa) with butterflies as prey and woodhoopoes, hornbills and babblers as predators, which showed without doubt that Danaids, unlike their mimics, were toxic species. He also demonstrated unequivocally that taste and odour, as well as the chemical ingredients producing emesis in his captive birds, were deterrent qualities possessed by these butterflies. His description of the manner in which a parent bird, who had swallowed the nauseating insect, tried to dissuade its young from eating it was wholly delightful and telling. We are awaiting further research into bird vision and behaviour in this field.

Simultaneously, E. B. Poulton, who encouraged Swynnerton in his experimentation and astute observations, and had an encyclopaedic knowledge of the field, came to the conclusion after evaluating the theories of Haase, Wallace, Müller, Meldola and Slater that these authors' speculations were justified: aposematic butterflies and moths, the larvae of which feed on poisonous plants, can derive protective toxins form their foliage and serve as models for innocuous species. He began to agitate for chemical proof of these theories. It was necessary, perhaps, that such proof should be provided by a combined biochemist and botanist, but no such individual materialized until 50 years later when T. Reichstein, whose knowledge of the chemistry of Asclepiad and Aristolochic plants was enormous, turned his attention to the Danaids and Papilios which fed upon them.

In the meantime, however, Jane van Z. Brower had made a great step forward, with an impeccable series of laboratory experiments proving that mimicry really worked. Further important observations of hers (later emphasized and elaborated in a series of joint papers with Lincoln Brower) were that mimics as well as models are somewhat distasteful to predators, and that the disagreeable experience produced by the plant toxins stored by a butterfly model leave a lasting impression on the predators. However, it was T. Reichstein's matchless and trustworthy chemistry which suddenly floodlit the scene and linked the herbivores and their host plants in a biochemical synthesis.

In a sense this scene had already been set by Gottfried Frankel's intuitive interpretation of the protective role of secondary plant substances, which in turn inspired Ehrlich and Raven to produce their memorable paper on the co-evolution of plants and butterflies. The field naturalists who provided the basic observations which made their synthesis possible, in particular Sevastopulo, van Someren and Carcasson, must not be forgotten. It is a curious facet of modern science that these authors actually experience difficulty in publishing their invaluable lists of food-plant records. Also, all of those interested in Danaid ecological chemistry were fortunate in having to hand the knowledge accumulated and published by F. A. Urquhart on the general biology, life-style and migrating habits of the Monarch.

We are now entering on a period when undreamed of subtle adjustments between plant and insect become apparent—thus the ovipositing female Monarch selects for preference those species of *Asclepias* as food plants, from which the larvae can best assimilate and store the toxic secondary plant substances, the mimics of certain Danaids not only resemble the models in appearance but secrete chemicals which mimic the cardioactivity of cardenolides.

Reichstein's identification of ten cardiac glycosides in both *D. plexippus* and its food plant seemed to be the spark which ignited a small conflagration, for up to 1966 there was really relatively little interest in the co-evolution of plant and insect biochemistry. Such an explosion of interest—*embarras de richesse*—leaves the scene in some sort of confusion, and we can now be grateful to Jeffrey Harborne for drawing the scattered pieces of the jigsaw together, thereby achieving a synthesis, richly flavoured with his own original ideas and lucid interpretations.

One of the most stimulating aspects of this book is the many doors he deliberately leaves ajar. It is all too easy when reviewing an intricate field to give a student new to the area the feeling that everything is now known about the subject. This book has exactly the reverse effect on the reader: a dozen new ideas spring to mind at the end of each chapter. Although grateful for an integrated and unified survey, particularly because of its intricacies and scattered literature, one is stimulated to take the next step forward and push hard against the nearest door.

Ashton Wold
Peterborough, England
July, 1977

MIRIAM ROTHSCHILD

Preface

The last two decades have witnessed the growth of a new inter-disciplinary subject, variously termed ecological biochemistry, chemical ecology or phytochemical ecology, which is concerned with the biochemistry of plant and animal interactions. Its development has been due in no small measure to the increasingly successful identifications of organic molecules in microquantities, following the application of modern chemical techniques to biological systems. It has also been due to the awareness of ecologists that chemical substances and particularly secondary metabolites such as alkaloids, tannins and terpenoids have a significant role in the complex interactions occurring between animal and animal, animal and plant or plant and plant in the natural environment. A further stimulation has been the possible applications of such new information in the control of insect pests and of microbial diseases in crop plants and in the conservation of natural communities. The present text is intended as an introduction to these new developments in biochemistry that have so enormously expanded our knowledge of plant and animal ecology.

This book is based on a course taught by the author over a number of years, both at Reading and at other universities. It has been planned so that it is suitable for second or third year university teaching in Departments of Botany, Biochemistry and Biological Sciences. Two general points may be emphasized about the use of this text for university teaching. First, each chapter is intended to be self-contained, so that there should be no problem if the order is rearranged to meet the requirements of a particular course. Second, in the bibliography of each chapter, the major books and review articles have been separated from the other references with the intention that these might form a reading list for students. It is also hoped that the book will be of more general value as a simple introduction to a new subject area.

In preparing this fourth edition, every effort has been made to bring it up to date with the latest developments in the subject. New sections have been provided on the cost of chemical defence and on the release of predator-attracting volatiles. The section on increased synthesis of toxins has been rewritten. A summary has been added to Chapter 3 and new references have been included in many places.

The author is grateful to Dr Miriam Rothschild, FRS for her introductory foreword. By her own pioneering experiments with aposematic insects and equally her encouragements of other scientists, Dr Rothschild has contributed more than anyone else to this new subject and this book owes much to her example, The author is also grateful to many friends and colleagues who have provided him with reprints of their work in this field. He also thanks Miss Valerie Norris, his secretary, for her assistance in revising the text. Finally, he has pleasure in acknowledging his debt to the editorial staff of the publishers for their enthusiasm and expertise in handling this venture.

This edition, like the third, is dedicated to the memory of Professor Tony Swain,

1922–1987, a close friend and colleague, and one of the major pioneers in the development of the subject of ecological biochemistry.

March 1993

Jeffrey B. Harborne
University of Reading

1 The Plant and Its Biochemical Adaptation to the Environment

I. Introduction

The marriage between such diverse disciplines as ecology and biochemistry may seem at first a curious alliance. Ecology is largely observational, is concerned with interactions between living organisms in their natural habitats and is carried out in the field. By contrast, biochemistry is experimental, is concerned with interactions at the molecular level and is carried out at the laboratory bench. Nevertheless, these two distinctive disciplines have cross-fertilized in recent years with astonishing success and a whole new area of scientific endeavour has opened up as a result. Ecological biochemistry is only one of a variety of phrases that have been employed to describe these exciting developments.

To the ecologist, knowledge of biochemistry has illuminated to a remarkable degree the complex interactions and co-evolutionary adaptations that occur between plant and plant, plant and animal and animal and animal. It has led to the realization, for example, that plants are functionally interdependent with respect to their animal herbivores and form what are termed 'plant defense guilds' (Atsatt and O'Dowd, 1976). Similarly, to the biochemist, studies in ecology have provided for the first time a rational and satisfying explanation for at least a part of the enormous proliferation of secondary metabolism that is observed in plants. Much of the purpose of the synthesis of complex molecules of terpenoids, alkaloids and phenolics lies in their use as defence agents in the plant's fight for survival against animal depredation.

The aim, therefore, of the present text is to provide an account for the student reader of the explosive development in ecological biochemistry that has occurred in the last two decades. The various chapters deal in turn with the plant and its interactions with animals and with other plants, while animal–animal interactions are considered in some detail in Chapter 8. It should be emphasized at this point that the biochemistry of many interactions has been deliberately simplified here in order to present a coherent story. It must be recognized that a given interaction between a plant host species and its animal predator species can be very subtle and complex and certain aspects of such an interaction may require many years of study before all is revealed. Furthermore, a third organism may have a controlling influence on a plant–animal interaction, such as a parasitoid of the animal or a microbial infection of the plant.

The term plant is generally used throughout this book to refer to higher plants and mainly to angiosperms, gymnosperms and ferns. Fungi, bacteria and viruses will usually be referred to as micro-organisms; other groups of plants will rarely be mentioned—i.e. algae, mosses and liverworts—largely because their ecological biochemistry has not yet been studied in much detail.

The selection of animals mentioned in this text is restricted to those taxa that have been studied experimentally and is certainly very unrepresentative of the animal kingdom as a whole. This is because plant–animal interactions in terms of feeding and defence have largely centred on the insect kingdom and only more recently have biochemical aspects of mammalian ecology been explored to any extent.

The emphasis here on the plant is due, at least in part, to the fact that plants are richer than animals in their biochemical diversity. Although secondary metabolism occurs in animals (Luckner, 1990), nevertheless, over four-fifths of all presently known natural products are of plant origin (Robinson, 1980; Swain, 1974). Some idea of the range of secondary compounds found in plants can be obtained from Table 1.1, which lists some of the major classes, together with an indication of numbers of known compounds, distribution patterns and biological activities. Many of these substances will be mentioned in more detail in subsequent chapters. The richness in secondary chemistry in plants is at least partly explicable by the simple fact that plants are rooted in the soil and cannot move; they cannot respond to the environment in ways open to animals.

In animals, secondary metabolism is associated particularly with defence and with signalling and can be very diverse in some instances (see Chapter 8). Animal metabolites do not differ generally from those produced in plants and there are many biosynthetic pathways which are common. The cyanogenic toxins linamarin and lotaustralin, for example, are accumulated by both plants and insects and they are formed in both cases from valine and isoleucine, respectively.

In this first chapter, attention is focused on biochemical adaptation. In its widest sense, this topic continues in later chapters, but here, it is taken in the narrower sense as adaptation to the physical environment. Attention is deliberately restricted to the plant kingdom, where information on biochemical adaptation is only of recent origin. Parallels between plant and animal processes will be mentioned, whenever these are relevant.

Table 1.1 Major classes of secondary plant compounds involved in plant–animal interactions

Class	Approx. number of structures	Distribution	Physiological activity
NITROGEN COMPOUNDS			
Alkaloids	10,000	Widely in angiosperms, especially in root, leaf and fruit	Many toxic and bitter tasting
Amines	100	Widely in angiosperms, often in flowers	Many repellent smelling; some hallucinogenic
Amino acids (non-protein)	400	Especially in seeds of legumes, but relatively widespread	Many toxic
Cyanogenic glycosides	40	Sporadic, especially in fruit and leaf	Poisonous (as HCN)
Glucosinolates	80	Cruciferae and ten other families	Acrid and bitter (as isothiocyanates)
TERPENOIDS			
Monoterpenes	1000	Widely, in essential oils	Pleasant smells
Sesquiterpene lactones	3000	Mainly in Compositae, but increasingly in other angiosperms	Some bitter and toxic, also allergenic
Diterpenoids	2000	Widely, especially in latex and plant resins	Some toxic
Saponins	600	In over 70 plant families	Haemolyse blood cells
Limonoids	100	Mainly in Rutaceae, Meliaceae and Simaroubaceae	Bitter tasting
Cucurbitacins	50	Mainly in Cucurbitaceae	Bitter tasting and toxic
Cardenolides	150	Especially common in Apocynaceae, Asclepiadaceae and Scrophulariaceae	Toxic and bitter
Carotenoids	600	Universal in leaf, often in flower and fruit	Coloured
PHENOLICS			
Simple phenols	200	Universal in leaf, often in other tissues as well	Anti-microbial
Flavonoids	4000	Universal in angiosperms, gymnosperms and ferns	Often coloured
Quinones	800	Widely, especially Rhamnaceae	Coloured
OTHER			
Polyacetylenes	650	Mainly in Compositae and Umbelliferae	Some toxic

Much is known about biochemical adaptation in animals and the subject is well documented in textbooks on comparative biochemistry (e.g. Baldwin, 1937; Florkin and Mason, 1960–1964). A useful reference to the subject of biochemical adaptation of animals to environmental change is that of Smellie and Pennock (1976).

Adaptation represents the ability of a living organism to fit into a changing environment, at the same time improving its chances of survival and ultimately of reproducing itself. The extensive diversity of life forms on this planet (i.e. several million species) and their presence in every type of habitat are witness to the view that living organisms indeed are morphologically and anatomically adapted to their environments. Such ideas are fundamental to the Darwinian view of nature and have been supported by much experimentation during the last century. Ideas of physiological and biochemical adaptation came later, during the 1920s and 1930s, with the experimental development of these two subjects. It is only, however, very recently that biochemical aspects have been developed sufficiently with plants to warrant their separate consideration, as in this present chapter.

Adaptation is generally considered as occurring on an extensive time-scale, involving many generations, but it can also take place during the lifetime of an individual, when it is sometimes termed acclimatization. The term adaptation is used here largely in the evolutionary sense. Biochemical adaptation is particularly closely connected to physiological adaptation and indeed it is sometimes difficult to distinguish the two. Physiological adaptation in plants will only be considered here, where appropriate, in relatively brief terms. For a comprehensive account, see Levitt (1980). Two excellent books on the subject for the student reader are Crawford (1989) and Fitter and Hay (1987).

Biochemical adaptation can operate at different levels in metabolism. It may affect the enzymes and produce amino acid substitutions in the primary sequence of protein or else alter the balance of isozymes. It may affect intermediary metabolism; an example in the case of the carbon pathway in photosynthesis is mentioned later. Finally, it may affect secondary metabolism; this is especially true of the plant's adaptation to animal feeding.

The environmental factors that plants are subject to can be broadly divided into five types:

(1) *Climatic factors.* These include temperature, light intensity, daylength, moisture and seasonal effects.
(2) *Edaphic factors.* All plants, except aquatics, epiphytes and parasites, obtain their mineral nutrition through the soil. The soil is also the source of symbiotic microbes, e.g. those required by legumes and other nitrogen-fixing plant species. Through contact with the soil, plants may have to cope with toxic heavy metals or with excess salinity. Equally, they may be subject to biochemical stress due to mineral deficiency in the soil.
(3) *Unnatural pollutants.* These are distributed through the upper atmosphere (ozone, industrial gases, gasoline fumes) or through the environment (organic pesticides) and may be potentially toxic to many plants.
(4) *Animals.* Although there is an element of symbiosis in animal feeding, herbivores are primarily hostile to plants, since they depend on them for their very existence. Many

different defence adaptations are known in plants. An element of symbiosis is also present in the case of those animals which visit plants for the purpose of pollination (see Chapter 2).

(5) *Competition from other plants.* This can be either competition between different higher plants or between different forms of plant life, e.g. higher plants and micro-organisms.

In this chapter, we are concerned only with the first three factors: the climate, the soil, and unnatural pollutants. Biochemical adaptation to animal predation and biochemical interactions between plants will be the subject matter of later chapters in this book.

II. The Biochemical Bases of Adaptation to Climate

A. General

Anatomical and morphological adaptation of plants to different climatic factors is well known and indeed its study is a major part of the science of plant ecology. Everyone is familiar with the ways that desert cacti and succulents are adapted to their parched habitats and are able to reduce moisture loss under the scorching desert sun. This is done by extending the area of soil for water uptake, by reduction of water loss through the leaf or by increased water storage in the tissues. Situations where biochemical features are involved in climatic adaptation are less often considered or discussed. Nevertheless, there is a growing awareness of the need to explore biochemical aspects, from the practical viewpoint.

In recent years, there has been much study of the hormonal control, through the sesquiterpenoid abscisic acid, of moisture loss from plants by stomatal closure. There is practical incentive here in the need to develop drought-resistant crop varieties for growing in marginal desert areas of the world. In Israel, for example, plant scientists are working on the development of agricultural crops which will grow successfully in areas of the Negev desert. Conversely, plants may have to adapt to excess moisture and something is now known of the adaptation of intermediary metabolism to the flooding of plant roots. Also, plants growing in frost conditions undergo biochemical changes in their sap constituents. Perhaps the most dramatic example of long-term biochemical adaptation to climate discovered in recent years is of the special photosynthetic pathway that tropical plants appear to have developed in response to hot and arid conditions. All these topics will be discussed in more detail in the following sections.

B. Photosynthesis in Tropical Plants

It has been apparent since the experiments of Warburg (1920) on O_2 inhibition of photosynthesis that temperate plants, when subjected to high temperatures (as on a hot summer's day), do not show the expected increase in photosynthetic rate with temperature that theoretically should occur. Their efficiency in incorporating atmospheric CO_2 into respiratory sugar is limited by carbon loss from the well-known Calvin carbon cycle

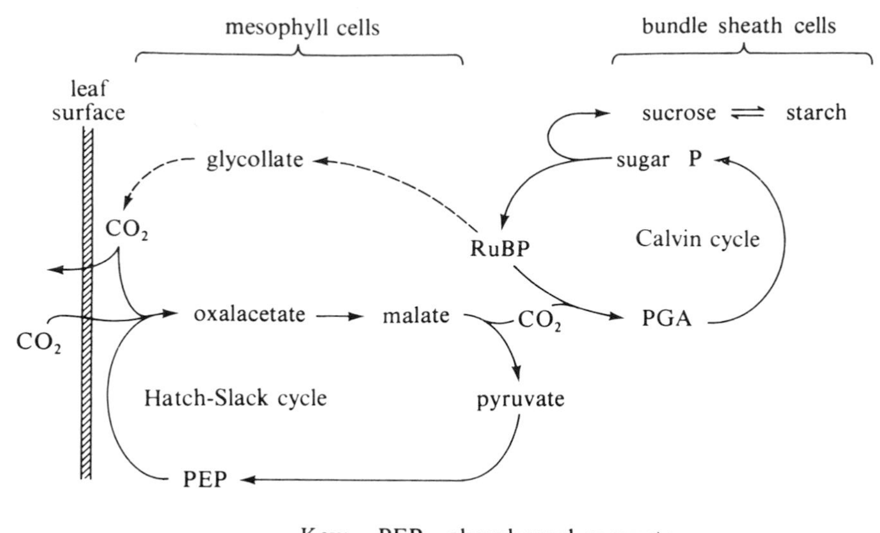

Figure 1.1 The Hatch–Slack modification of the carbon pathway of photosynthesis

via ribulose bisphosphate and glycollate (see Fig. 1.1) so that a proportion of the CO_2 originally absorbed by the plant is lost by 'photorespiration' at the leaf surface. Such losses in CO_2 conversion to sugar, due to O_2 inhibition, are not too serious since high temperatures are relatively infrequent in temperate latitudes. Such losses could, however, be much more considerable in plants growing in the tropics.

Evidence that tropical plants such as sugar-cane are able to resist inhibition by high partial pressures of O_2 and are able to photosynthesize efficiently in hot climates has only become available quite recently. These new discoveries have shown that such plants differ biochemically from temperate species (Hatch and Slack, 1970; Bjorkman and Berry, 1973). When leaves of sugar-cane are exposed to $^{14}CO_2$ for a few seconds, the first compounds labelled are not those of the Calvin cycle but are C_4 acids. Such plants have a modified pathway of carbon, which includes a new cyclic system, called the Hatch–Slack pathway, which transports CO_2 from the leaf surface to the Calvin cycle (Coombs, 1971). Plants with this so-called Hatch–Slack pathway in effect collect the CO_2 which would otherwise be lost by 'photorespiration' and 'drive it back' into the Calvin cycle to be converted to sucrose (Fig. 1.1). Such tropical plants are called C_4 plants (after the four-carbon acids involved in the Hatch–Slack pathway) to distinguish them from the more usual C_3 plants which only have the simple Calvin cycle.

This modification in biochemistry is correlated with anatomical differences; plants with the C_4 pathway have special mesophyll cells, in which the pathway is located, beside the bundle sheath cells, where the Calvin cycle operates. The anatomical differentiation of C_4 plants was recognized long before their distinctive biochemistry was elucidated, the anatomical features of such tropical species being described as the Kranz syndrome. Besides

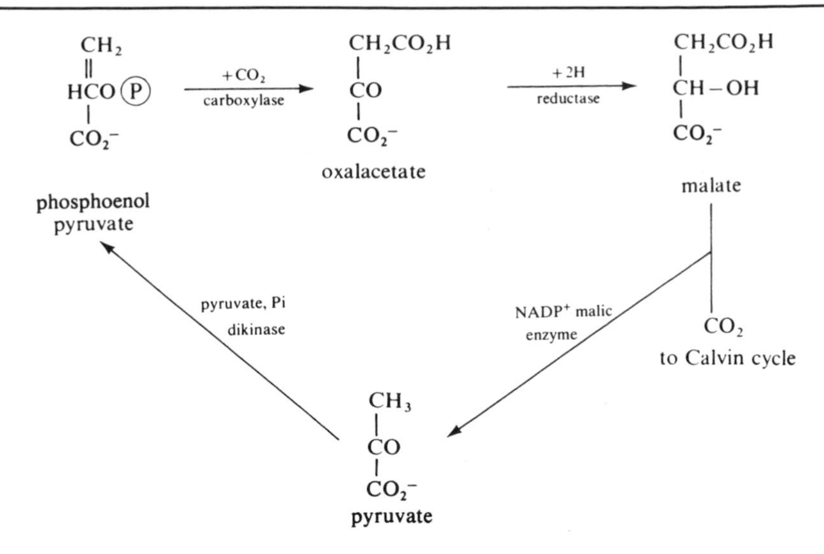

Figure 1.2 Biochemistry of the C_4 pathway in sugar-cane

recognizing C_4 plants by anatomical observations, it is possible to identify them by determining the relative uptakes of $^{13}CO_2$ and $^{12}CO_2$ by these plants. This ratio is measured in the carbon fixed by the plant as sucrose and can even be determined in dead tissue, i.e. in herbarium specimens. The relative $^{13}C/^{12}C$ ratio in C_4 plants is between −9 and − 18‰, while in C_3 plants it lies between −21 and −38‰ (Gibbs and Latzko, 1979).

The Hatch–Slack pathway was first recognized in sugar- cane, *Saccharum officinarum*, a member of the grass family, Gramineae. Subsequent investigations have shown that the majority of tropical and subtropical grasses in the sub-families Eragrostoideae, Panicoideae and Arundinoideae have this pathway (Brown, 1975). The ability to achieve optimal photosynthesis under tropical conditions is by no means restricted to grasses and at least ten other families with tropical members contain it. They include the Cyperaceae, Compositae, Euphorbiaceae, Zygophyllaceae and several families of the Centrospermae. In Cyperaceae, the distribution of C_4 plants is correlated with taxonomy; the character occurs exclusively in the tribes Cypereae and Fimbristylideae of the Cyperoideae (Raynal, 1973). In general, plants with the C_4 pathway are exclusively herbaceous and it is fairly clear that it is an advanced evolutionary condition compared to C_3 plants.

An outline of the path of carbon in plants which are adapted to tropical climates is shown in Fig. 1.1. Essentially, the purpose of the Hatch–Slack pathway is to transport CO_2 from the outside surface of the leaf to the actual site of photosynthesis in the inner chloroplast. The CO_2 is first combined with phosphoenol pyruvate (PEP) (Fig. 1.2) to give oxalacetate which is then reduced to malate, which in turn is decarboxylated to pyruvate. The CO_2 released at this stage enters the Calvin cycle by combining with ribulose bisphosphate to give phosphoglyceric acid, the first C_3 organic compound of the Calvin pathway. Most of the enzymes required in the C_4 pathway are already present in all plant species, the only really distinctive enzyme being pyruvate, Pi dikinase which regenerates

Table 1.2 Biochemical variations in the C_4 pathway of photosynthesis

Major bundle sheath (BS) decarboxylase	Energetic change	Substrate moved from mesophyll to BS	Substrate moved from BS back to mesophyll
$NADP^+$ malic enzyme	NADPH produced	Malate	Pyruvate
NAD^+ malic enzyme	NADH produced	Aspartate[a]	Alanine/pyruvate
PEP carboxykinase	ATP consumed	Aspartate[a]	PEP

[a] Formed from oxalacetate by transamination before transport and yields oxalacetate by deamination before decarboxylation.

PEP from pyruvate to complete the cycle (Fig. 1.2). However, there is evidence that other forms (or isozymes) of the enzymes common to both C_3 and C_4 plants are required to make the C_4 pathway function most efficiently. Equally important for the success of C_4 plants is the rearrangement of the cells and membranes which allow the translocation of C_4 acids and of pyruvate between the two types of cells.

Although the carboxylation of PEP via PEP carboxylase occurs uniformly in all C_4 plants, the subsequent decarboxylation of malate to yield CO_2 which is passed on (as HCO_3^-) to the Calvin cycle can vary in enzymology. Three different enzymes have been recognized (Table 1.2) and C_4 plants fall into one or other of three subgroups according to which decarboxylating enzyme is present in the bundle sheath. The scheme of Fig. 1.2 is only strictly followed in plants with the $NADP^+$ malic enzyme; the other two types differ in the actual substrates which are moved to and from the bundle sheath and mesophyll cells (Table 1.2). There are also differences in the organellar localization of the three enzymes within the bundle sheath cell, with the $NADP^+$ malic enzyme operating in the chloroplast, the NAD^+ malic enzyme in the mitochondrion and the PEP carboxykinase operating in the cytoplasm with oxalacetate instead of malate as substrate. Plants with the three different decarboxylases can be recognized anatomically as well as biochemically. All three subgroups of C_4 plants are found within the grasses, the distribution of the subgroups generally following taxonomic classification.

All available evidence indicates that species with C_4 photosynthesis have evolved from those with only the C_3 pathway; if this is true, then one would expect to find intermediate species in nature. The existence of C_3 and C_4 species within the same genus, as occurs in *Atriplex* (Chenopodiaceae), suggested to scientists that intermediates might be obtainable by simple cross-breeding experiments. However, although true hybrids were obtainable, they were found to only function photosynthetically as C_3 plants in spite of the presence of some C_4 characteristics. Continuing search among subtropical plant species has revealed several natural C_3–C_4 intermediates, the one most extensively studied being the grass *Panicum milioides* (Rathnam and Chollet, 1980). Such plants are anatomically distinctive from both C_3 and C_4 types and they have a C_4 mechanism which is operational but not as efficient as that in a true C_4 species. Finally, the view that C_4 evolved from C_3 is supported by experiments where it is possible to demonstrate the changeover from C_3 to C_4 photosynthesis in different leaves on the same C_4 plant, as for example in *Zea mays* (Crespo *et al.*, 1979).

One other point may be made about C_4 plants: it is possible that adaptation to retain photosynthetic efficiency in the tropics may have additional benefit in providing resistance to herbivores. There is evidence that herbivores, and particularly grasshoppers, avoid eating C_4 plants if given a free choice between both C_3 and C_4 plants (Caswell *et al.*, 1973; Heidorn and Joern, 1984). The reason for this may be simply due to the anatomical modifications reflected in the Kranz syndrome. Thus, the starch in C_4 plants is located further away from the leaf surface and hence is less accessible to feeders. The bundle sheaths of C_4 plants also seem to be relatively tough and in grasses at least lignin content seems to be much higher in C_4 than in C_3 plants. Finally, there is evidence that the availability of nitrogen in C_4 plants is less than that in C_3 plants. However, not all insects are affected by such factors in C_4 plants and, indeed, the lepidopteran *Paratryone melane* thrives equally well, irrespective of whether it is fed on a C_3 or a C_4 grass (Barbehenn and Bernays, 1992).

While the main environmental stress suffered by C_4 plants is high day temperatures, many plants growing in the drier parts of the tropics and subtropics may also be subjected to drought. This is true especially of desert plants, which may overcome this stress by adopting a succulent habit and storing the water that is available (see also p. 15). Many succulent plants (e.g. *Kalanchoe daigremontiana*) have been known for a long time to be idiosyncratic in their biochemistry in their penchant for storing organic acid, particularly malic, in the leaves during the night only to dissipate this acidity during the following day. Plants with this unusual behaviour are particularly common among members of the Crassulaceae and the phenomenon has become known as Crassulacean Acid Metabolism (or CAM for short). Recent research (Kluge and Ting, 1978) has now revealed that CAM plants are in fact a special type of C_4 plant in which the Hatch–Slack and Calvin cycles operate at separate times over any given 24-h period. Furthermore, CAM plants represent a unique example of a biochemical adaptation in photosynthesis which is linked to the conservation of moisture in an arid habitat.

The pathway of carbon in CAM plants, outlined in Fig. 1.3, is determined by the opening and closing of the leaf stomata. Thus, the stomata remain open at night, when water loss by transpiration is minimal, and CO_2 is taken in and combined with PEP to yield oxalacetate and then malate. This latter acid is then stored in vacuolar form until the morning. Soon after dawn, the plant then closes the stomata in order to avoid water transpiration during the heat of the day. At the same time, the second carbon cycle comes into operation and the storage malate provides the main source of CO_2 for operating the Calvin cycle, by which sugar is synthesized and then stored. Thus CAM plants are functionally similar to C_4 species, except that in CAM the sequential carboxylations of PEP and ribulose bisphosphate are separated in time, diurnally, whereas in C_4 plants they are separated spatially, i.e. anatomically. In favourable circumstances, CAM plants may exercise the C_3 photosynthetic option entirely. This and other evidence shows that CAM is a special adaptation in xerophytic plants to a harsh desert environment.

Two subgroups of CAM plants can be distinguished according to the mechanism of decarboxylation of the malate stored overnight in the vacuole. One subgroup utilizes an NADP- or NAD-linked malic enzyme, while the other uses a PEP-carboxykinase (see subgroup variation in C_4 plants, Table 1.2). Some plants which exhibit CAM appear

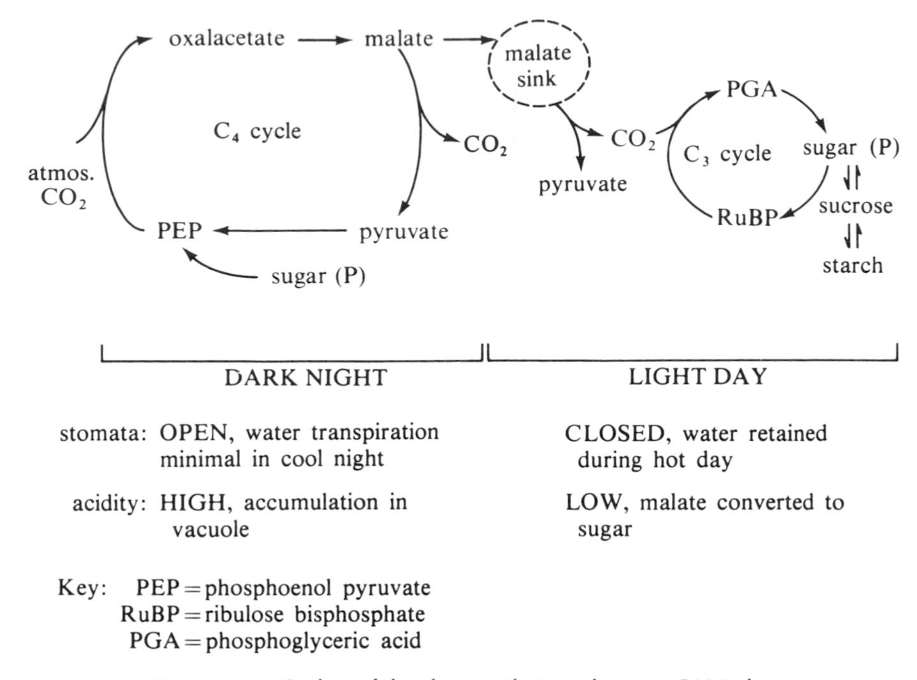

stomata: OPEN, water transpiration minimal in cool night — CLOSED, water retained during hot day

acidity: HIGH, accumulation in vacuole — LOW, malate converted to sugar

Key: PEP = phosphoenol pyruvate
RuBP = ribulose bisphosphate
PGA = phosphoglyceric acid

Figure 1.3 Outline of the photosynthetic pathway in CAM plants

to be able to shift from C_3 to CAM in response to changing environmental conditions, e.g. the removal of irrigation. How important such shifts are in natural habitats has yet to be determined but the remarkable gymnosperm *Welwitschia mirabilis,* a native of the Namibian desert of Africa, has been observed on different occasions to show C_3 metabolism, CAM and CAM cycling. The latter situation involves fluctuations in organic acid levels but little nocturnal CO_2 fixation (Ting, 1985). Overall, the CAM system of photosynthesis seems to be more flexible in its expression than the straightforward C_4 pathway.

CAM plants are as widespread in nature as C_4 plants and they have been found in members of some 25 families and 109 genera. Both specializations may be present within the same plant family but in different species. For example, in the genus *Euphorbia,* C_4 occurs in species of the North American subtropics while CAM is present in species of the African tropical zones. While these are the only two modifications in the carbon pathway so far known, relatively few plants have been thoroughly investigated biochemically and it is quite possible that other adaptive sequences in the photosynthetic carbon pathways will emerge as the result of future research.

Even within C_3 plants there may well be some adaptations of the photosynthetic enzymes to differing climates. Thus the key enzyme, ribulose bisphosphate carboxylase–oxygenase (rubisco), consists of eight large and eight small subunits and appears to have evolved over evolutionary time from a simpler form found in photosynthetic bacteria (Ellis and Gray, 1986). While the amino acid sequences of the large subunits are largely conserved, those of the small subunits are not. Variation in the small subunits could provide the basis of some thermotolerance; C_3 plants may well be adapted to growing optimally in their favoured environments through variations in these subunits. Diurnal variation in rubisco activity is now known to be controlled by a natural reversible inhibitor, 2-carboxyarabinitol-1-phosphate (Gutteridge *et al.*, 1986), and again this control system might have adaptive significance.

C. Adaptation to Freezing

Many plants and animals have the ability to resist and survive the below-zero temperatures which they may be subjected to during winter months in northern temperate and arctic regions of the world. In insects, there is evidence that freezing tolerance is achieved fairly simply by the synthesis of glycerol (see Fig. 1.4), which acts as an anti-freeze, exactly as ethylene glycol, CH_2OH—CH_2OH, does in water-cooled car engines. In higher plants, adaptation to freezing conditions seems to be more complex than this (Levitt, 1980). There is little doubt, however, that freezing tolerance is correlated with an increase of sugar content in the cell sap. Experiments show also that the ability to withstand frost can be achieved artificially by infiltrating plants with sugars.

The sugars identified in frost-resistant plants vary from plant to plant. They are often the three common sugars—glucose, fructose and sucrose—with or without oligosaccharides such as raffinose. Polyhydric alcohols such as mannitol, sorbitol and glycerol (see Fig. 1.4), which presumably could act directly as anti-freeze agents, are less frequently reported. However, all three have been found in significant amounts (up to 40% total sugars) in plants such as gardenia, apple, mountain ash and pomegranate (Sakai, 1960; see also Sakai, 1961; Sakai and Yoshida, 1968).

Whether sugars play a crucial role in resistance to freezing in plants is not yet clear. They appear to be involved in freezing tolerance in at least two ways. First, by their osmotic effect, they decrease the amount of ice formed in the vacuole. Secondly, by their metabolic effect, by being converted in the protoplasm to other constituents, they have an additional protective role. Whether these changes involve conversion of common sugars to polyhydric alcohols or whether more complex biochemistry is involved is not yet known.

glycerol	sorbitol	mannitol
	CH_2OH	CH_2OH
CH_2OH	$CHOH$	$HOCH$
$HOCH$	$HOCH$	$HOCH$
CH_2OH	$CHOH$	$CHOH$
	$CHOH$	$CHOH$
	CH_2OH	CH_2OH

Figure 1.4 Structures of polyhydric alcohols

In animals, adaptation to freezing may involve the production of both low molecular weight (e.g. glycerol) and high molecular weight cryoprotectants. For example, eight anti-freeze glycoproteins, which have the specific property of preventing ice crystals forming at temperatures below 0°C, have been isolated from the blood of antarctic fish. They range in molecular weight from 2600 to 33,700 and have a repeating sequence of alanine–alanine–threonine in their primary structure, with the disaccharide galactosyl-*N*-acetylglucosamine linked to each threonine residue. These glycoproteins may also be produced in temperate-water fish during the winter months (Knight *et al.*, 1984). However, the presence of the sugar attachments does not seem to be vital for cryoprotection, since the anti-freeze proteins of arctic fish lack sugar substituents.

Whether frost-hardy plants have a similar macromolecular mechanism of protection is not yet clear, but experiments with the tree *Nothofagus dombeyii*, which grows in the cold Chilean rain forest, suggest that this could be so. A polypeptide of molecular weight 35,000 has been isolated from cold-treated seedlings, which protected isolated thylakoid membranes from freezing damage at −20°C (Rosas *et al.*, 1986). Changes in protein content have been noted in many frost-hardy plants and it is likely that the constitutive enzymes undergo conformational and other changes in the process of cold acclimation.

One much sought-after biochemical factor in cold-hardened plants is an increase in the relative amounts of unsaturated and saturated fatty acids in the membrane lipids. According to theory, such an increase should be beneficial to the plant since it would maintain the fluidity of the membrane at the lower temperatures. Such changes in the degree of unsaturation of fatty acids have been observed in the phosphatidylglycerol fraction of chloroplast membranes. Plants with a high proportion of *cis*-unsaturated fatty acids, such as spinach, *Spinacia oleracea*, are resistant to chilling, while species like squash, *Cucurbita pepo*, with only a small proportion are not (Murata *et al.*, 1992). In cereals such as wheat and rye, freezing tolerance is correlated more closely with a decrease in the proportion of *trans*-Δ^3-hexadecenoic acid, rather than with an increase in *cis*-acids (Huner *et al.*, 1989). This protection of the chloroplast membrane allows photosynthesis to resume once the external temperature moves above freezing.

D. Adaptation to High Temperatures

The basic metabolism of living cells is generally heat sensitive, since many of the molecules involved in the intermediary pathways, e.g. ATP, are unstable at temperatures above 40°C. This is especially true of the macromolecules, the DNA of the nucleus, the RNA of the cytoplasm and the enzymes, which often become denatured and inactive with rising temperatures. And yet life can persist at high temperatures through adaptation. The most dramatic examples of this are the extreme thermophilic archaebacteria, which have been isolated from hot springs. The most thermophilic is *Pyrodictium*, which exhibits optimal growth above 80°C and will actually continue growing at temperatures up to 110°C. There has even been a claim of bacterial growth at 250°C (Baross and Deming, 1983) but this is probably not correct (see Stetter *et al.*, 1986). Even growth at the boiling point of water poses many problems to the organism, and unfortunately we know little as yet about the mechanisms present which stabilize the macromolecular components of these cells.

One type of heat adaptation that we do know more about is that of thermotolerance, or short-term adaptation to temperature stress. A wide array of organisms, ranging from bacteria to man and including the fruit fly *Drosophila melanogaster*, are recognized as responding to high temperatures by the synthesis of special heat shock (HS) proteins, with the repression of normal protein synthesis (Schlesinger *et al.*, 1982). What is astonishing about this response is its rapidity, which involves gene transcription and synthesis of new messenger RNA, and which occurs within minutes rather than hours of the heat shock. HS proteins appear to be produced whenever an organism is subjected to temperatures above those which it is normally used to and they gradually disappear when heat stress is removed.

Thermotolerance in plants is defined as the ability to survive an otherwise lethal temperature by first being exposed to a non-lethal one. Soya bean seedlings growing at 28°C will not normally survive a 2-h treatment at 45°C, but they will if they are protected by a 2-h preincubation at 40°C. A similar thermotolerance is also induced by a brief 10-min pulse at 45°C, if there is then an intermediary 2-h incubation at 28°C.

The heat shock response in plants is a fairly general one and involves the synthesis of two groups of HS protein: four of high molecular weight (68,000, 70,000, 84,000 and 92,000) and several of lower size (15,000–23,000). These HS proteins persist for some time and would appear to be primarily responsible for the ability of the plants to survive the increase in temperature. They are specially compartmented within the cell, the low molecular weight HS proteins being associated particularly with the nuclear components in the soya bean and tomato plants (Schoeffl *et al.*, 1984).

The synthesis of HS proteins has so far largely been laboratory observation and it is not yet clear how far this process is important in the field. Since plants vary in their thermotolerance, it would seem that future crop improvement might well require the selection of cultivars with increased expression of the HS protein response.

E. Adaptation to Flooding

A number of plant species are able to grow in areas, e.g. in the flat plains of glaciated river valleys, where their root systems are subjected to flooding. Indeed, some plants may spend up to half the year with their roots and lower stems immersed in water; the rest of the year the conditions are distinctly drier. Thus, they may have to regularly re-adapt their metabolism to wet and dry conditions at six-monthly intervals during the life- cycle. Species where this happens include the flag iris, *Iris pseudacorus*, and the soft rush, *Juncus effusus*. In contrast to the flag iris, the ordinary garden iris, *Iris germanica*, is intolerant to flooding. The ability to tolerate flooding may even vary within species: both tolerant and non-tolerant races of groundsel, *Senecio vulgaris*, can be distinguished.

Such plants may have to modify the respiratory pathway in their root systems to survive the change from aerobic to semi-anaerobic conditions as a result of flooding. One of the main problems in the flooded root is that the glycolytic pathway, in the absence of sufficient oxygen, will give rise to acetaldehyde via pyruvate and in the presence of high induced levels of alcohol dehydrogenase this is converted to ethanol. Ethanol is highly toxic to plant cells if it is allowed to accumulate in any quantity. In a plant such as rice, there is a sufficient flow of water in the paddy field to carry the ethanol away. However, in the thick roots of many woody plants, some adaptation to the problem posed by ethanol poisoning may be essential for survival. Indeed non-tolerant plant species may be killed by ethanol accumulating in this way.

Another major problem of anaerobic respiration is the low energy yield of the process. Other challenges that flooded plants may have to face are the toxicity of inorganic ions (e.g. iron) in the soil released under the reducing conditions of the flooding and also hormonal imbalance within the root.

A biochemical hypothesis to explain the tolerance of some plants to flooding has been proposed by Crawford (1978), based partly on the responses of animals to anoxia (lack of oxygen). It is suggested that under these conditions, the glycolytic pathway is diversified so that the intermediates of carbohydrate breakdown (e.g. PEP) instead of ending up as toxic ethanol are switched into other products, such as malate, lactate or alanine, which can safely accumulate without harmful consequences. In animals (e.g. diving reptiles, birds and sea molluscs), a variety of metabolites have been detected under conditions of anoxia including lactate, pyruvate and succinate. In the case of plants, the hypothesis is supported by the circumstantial observation of accumulation of certain metabolites during winter flooding of roots: of malate in *Juncus effusus* and of glycerol in the alder, *Alnus incana*.

Unfortunately, this hypothesis did not stand up to detailed biochemical analysis. When three tolerant species were tested under conditions of anoxia, they failed to show any evidence of metabolic diversification and additionally all three increased in alcohol dehydrogenase levels and in the production of ethanol (Smith and Ap Rees, 1979). Again, when the flood-tolerant *Glyceria maxima* was compared with the intolerant *Pisum sativum*, there was no evidence that alcohol dehydrogenase induction was correlated with adaptation (Jenkin and Ap Rees, 1983).

Work has continued on the problem and there are now two further hypotheses proposed to account for the tolerance or lack of tolerance in plants to anaerobic growth. The first theory is 'cytoplasmic acidosis', that in non-tolerant plants, there is a leakage of lactic acid from the vacuole into the cytoplasm which causes the damage (Roberts *et al.*, 1984). It is supported by the observations that during hypoxia, such leakage occurs earlier in pea root tips than in maize root tips and that lactic acid accumulates in maize seedlings that are deficient in alcohol dehydrogenase isozymes.

In the second theory, it is suggested that damage in non-tolerant plants occurs, not during the anoxic conditions, but subsequently when the plants return to aerobic growth. For example, post-anoxic oxidation of anaerobically produced ethanol could cause a surge in acetaldehyde production, formed by catalase oxidation; the acetaldehyde so produced is highly toxic. This idea is supported by experiments which showed an increase in catalase activity in rhizomes of the intolerant *Iris germanica* but no such increase in the anoxia-tolerant *I. pseudacorus* (Monk *et al.*, 1987). Similarly post-anoxic peroxidation of rhizome lipids might produce irreversible cellular damage at the membrane. Here, the enzyme involved might be either catalase or superoxide dismutase. Again, there is experimental evidence of increases in lipid peroxidation in *I. germanica* which do not occur in *I. pseudacorus* (Hunter *et al.*, 1983). Other biochemical work on flood-tolerant plants is admirably summarized in Crawford (1989).

F. Adaptation to Drought

Plants which grow in areas of low rainfall often tolerate drought and are termed xerophytes. Such plants of desert or high plateau areas of the world adapt to drought by morphological or anatomical means. For example, cacti have thick waxy coatings and the production of spines instead of leaves helps to minimize water loss by evaporation.

Plants which resist drought have been broadly divided into two groups: those which conserve water and those which have an increased ability to absorb water (Levitt, 1980). The latter mechanism is largely physiological, while the former has a biochemical element. Thus, one way of conserving water is to reduce the time when the leaf stomata are open or to only open the stomata at night time. The hormone abscisic acid causes stomatal closure and there is circumstantial evidence that drought-resistant plants contain larger amounts of this important hormone. In addition, it has been found that the content of abscisic acid can increase as much as 40-fold within 4 h of wilting in wheat plants; the level continues to rise in such osmotically stressed plants for at least 48 h after wilting induction (Wright and Hiron, 1969; Milborrow and Noddle, 1970).

The effect of abscisic acid is reversible and the hormone drops to normal low levels when the water supply is replenished. There is evidence that the abscisic acid produced during wilting is not subsequently degraded but instead is stored in inactive form in the leaf and presumably becomes available to the plant if another period of water stress is imposed on it. In the control of stomatal closure, abscisic acid can be replaced by at least three related oxygenated sesquiterpenes (Fig. 1.5; in this figure, the structure of (*E*),(*E*)-farnesol is redrawn to indicate its biosynthetic and structural relationship to abscisic acid). In particular, phaseic acid and (*E*),(*E*)-farnesol have been shown to produce stomatal closure in *Vitis vinifera* and *Sorghum* respectively (Loveys and Kriedemann, 1974; Wellburn *et al.*, 1974). The mechanism by which these sesquiterpenoids produce their effect on the stomatal apparatus is not yet entirely clear but there is evidence that abscisic acid is released from the mesophyll chloroplasts and moved to the guard cells when water stress occurs. In this process in *Sorghum*, it is possible that (*E*),(*E*)-farnesol does not substitute for abscisic acid but instead is responsible for altering the permeability of the chloroplast membrane so that abscisic acid is released into the cytoplasm (Mansfield *et al.*, 1978).

The exogenous application of abscisic acid or related sesquiterpenes to plant leaves should theoretically have practical benefits in reducing transpiration and hence the amount of water needed by a given crop. However, such treatments could be very expensive since relatively large amounts of hormone would be needed because of the rapid turnover of abscisic acid when applied externally to crop plants. Furthermore, the reduction in photosynthesis and harvestable yield which accompanies stomatal closure would probably outweigh any savings in water loss.

Another set of biochemical observations on water-stressed plants indicates that proline accumulates during adaptation. Indeed, a comparison of drought-resistant and -susceptible varieties within a species has shown consistently higher levels of proline in the former than in the latter. Production of increased proline levels is so regular in barley that it is possible to use proline concentration for scoring varieties for their ability to resist drought (Singh *et al.*, 1972). The proline increase is usually such that about 30% of the

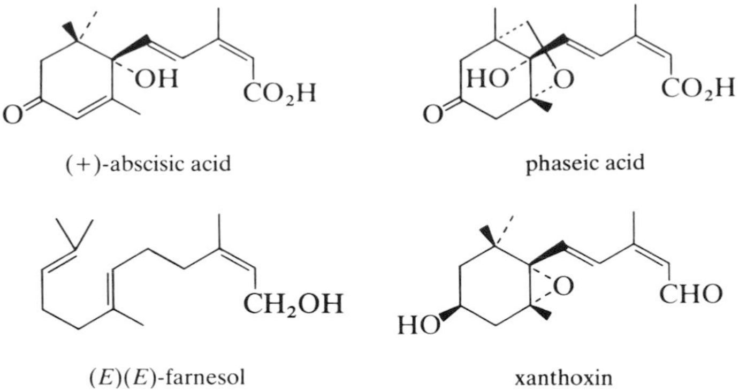

(+)-abscisic acid

phaseic acid

(*E*)(*E*)-farnesol

xanthoxin

Figure 1.5 Sesquiterpenoids capable of initiating stomatal closure in plants

free amino acid pool is made up of this imino acid. In actual figures, an increase to 1.2 mg/g dry wt has been recorded in water-stressed Bermuda grass, *Cynodon dactylon*. In legumes, proline is replaced by pinitol as a drought indicator (Ford, 1984). The increase in proline or pinitol could be simply a symptom of some more fundamental adaptation to water stress. It is also possible that proline itself, because of its special osmotic properties, is able to contribute directly to the retention of water in the plant and hence to drought resistance (see also the role of proline in halophyte adaptation).

Another possibility is that proline may directly contribute to the metabolic adaptation of the plant cell. Thus, plants suffering drought stress are often subject to high temperatures as well (see Section IID) and there is evidence that proline and related stress metabolites can improve the heat stability of certain enzymes. For example, in the sand dune plant, *Ammophila arenaria*, the proline present increases the heat stability of glutamine synthetase and glutamate:oxalacetate aminotransferase, the two key enzymes of ammonium assimilation in plants (Smirnoff and Stewart, 1985).

Biochemical adaptation to produce greater than normal amounts of the imino acid proline requires some changes in enzyme levels in the pathway of proline biosynthesis, utilization or degradation (Fig. 1.6). Points of control along the pathway which may be affected during drought or salt stress include loss of feedback regulation on the first step of the pathway, lowered rate of proline oxidation and slowing of the rate of incorporation of proline into protein. What is true for proline also applies when other primary metabolites accumulate in stressed plants, e.g. pinitol, glycerol and glycinebetaine (see Section IIIC).

One footnote to the effects of biochemical adaptation to drought on plants might be added here. Breeding plants such as barley for drought resistance could mean an increase in proline leaf levels which in turn might make the plant more susceptible to insect feeding. Proline is known to be a phagostimulant to both locusts and grasshoppers. Fortunately, however, tests with a barley mutant containing six times the normal level of proline showed that there was no concomitant increase in pest or disease susceptibility (Bright *et al.*, 1982).

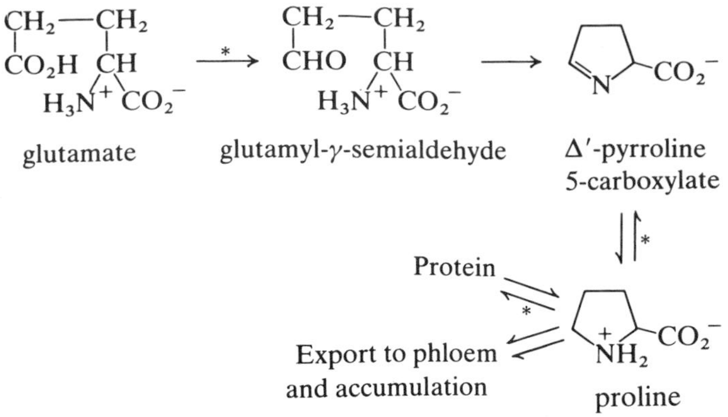

glutamate $\xrightarrow{*}$ glutamyl-γ-semialdehyde $\rightarrow$ Δ'-pyrroline 5-carboxylate $\rightleftharpoons^{*}$ proline

Protein $\rightleftharpoons^{*}$ proline

Export to phloem and accumulation $\leftarrow$ proline

Figure 1.6 Pathway of proline biosynthesis and points of control (*) for proline accumulation

III. Biochemical Adaptation to the Soil

A. Selenium Toxicity

Selenium is an element in the same group 6 of the periodic table as sulphur but unlike sulphur it is not usually essential to plants. Because of its close similarity in properties to sulphur, it can substitute for sulphur in biochemical systems. It is this ability to exchange with sulphur and become incorporated into amino acids and further into protein, as selenoprotein, which is the basis of its toxic properties. In any quantity selenium is thus highly toxic to all living organisms.

Selenium occurs mainly in soils in bound form so that it is not normally a hazard to plant life. There are, however, areas of the world where unusually high levels of soluble selenium are present in the soil and are taken up by plants. These include pasture lands in central Asia, Australia and North America. The effects of these high levels of selenium are revealed in toxic symptoms in grazing animals. The toxicity is expressed in both acute and chronic forms and continued ingestion leads to death. One of the symptoms of toxicity in sheep is the falling out of the woolcoat, with the production of bald patches. Human fatalities have also been recorded. However, it was not until the 1930s that serious attempts were made to investigate this phenomenon. The cause of the toxicity was traced to the ability of certain plants to adapt to selenium by accumulation of the element and it was ingestion of these plants which caused the death of cattle and sheep (Rosenfeld and Beath, 1964).

Many of the plants which have adapted to high levels of available selenium in the soil belong to the legume genus *Astragalus* and most work has been done with these plants. Of the some 500 species of *Astragalus* in the North American flora, some 25 species have adapted to selenium and are called selenium accumulators, as distinct from non-adapted plants which avoid such areas and are called non-accumulators. The ability of accumulators such as *A. bisulcatus* and *A. pectinatus* to sequester selenium is remarkable; such plants have as much as 5000 ppm selenium, compared to non-accumulators which have less than 5 ppm. The dangerous nature of these plants to other life forms can be realized from experiments in which toxic symptoms have been recorded in *Trifolium* plants treated with 5 ppm and in grazing animals fed on a diet containing as little as 1 ppm selenium (Shrift, 1972).

The toxicity of selenium in farm animals is entirely related to its absorption in quantity from a dangerous accumulator plant. Trace amounts of selenium are necessary for animal life, since one of the key enzymes of glutathione metabolism, namely glutathione peroxidase, contains 4 g-atoms of selenium. The borderline between life and death, however, is fairly close. Thus, the minimum desirable pasture content of selenium for livestock is 0.03 ppm, while continual ingestion of fodder containing 1–5 ppm will induce toxicity. A further complication in the intake of selenium by animals is the sulphur/selenium ratio of the fodder plant, since a high sulphur content will limit the availability of the selenium to the animal (Anderson and Scarf, 1983).

The question now arises—how have selenium accumulators been able to absorb so much selenium without any damage to themselves? The answer appears to lie in the ability of accumulators to separate inorganic sulphur (as sulphate) from inorganic selenium (as selenate or selenite) as they enter the plant and to channel the selenium into the synthesis of non-protein amino acid analogues (Fig. 1.7). These amino acids are structurally different from the two standard protein sulphur amino acids, cysteine and methionine, and are not therefore incorporated into protein synthesis. The plant then presumably sequesters them in the vacuole of the leaf and they are perfectly harmless to the plant but, of course, intensely harmful to any unsuspecting grazing animals.

By contrast, the absorption of soluble selenium from the soil into non-adapted plants follows a parallel but fatal pathway, shown in Fig. 1.7, whereby protein amino acids are synthesized in which sulphur is directly replaced by selenium. While selenocysteine and selenomethionine may possibly be harmful as such, the greatest damage must occur after they are mistaken for sulphur amino acids and are incorporated into enzymic protein. Selenium toxicity may then be attributed to the replacement of cysteine by selenocysteine and the production of dysfunctional proteins in which S–S bonds between polypeptide chains are substituted by the more labile Se–Se bonds. Proof that selenocysteine is absorbed into the protein of *Vigna radiata*, which had been fed with selenate, has been recently obtained by Brown and Shrift (1980). Selenomethionine has been successfully fed to the bacterium *Escherichia coli* with the concomitant replacement of as many as 150 methionine residues in the enzyme β-galactosidase. In this case, although the properties of the seleno-β-galactosidase were significantly affected, the enzyme was not completely inactive. Thus selenomethionine is possibly less disruptive of the metabolic activity of the cell than the more deadly selenocysteine.

The non-protein selenium-containing amino acids synthesized by accumulator *Astragalus* include the two shown in Fig. 1.7 but several others are also found, particularly Se-methylcysteine sulphoxide, γ-glutamyl-Se-methyl-cysteine and its sulphoxide. These various acids are found not only in the leaf, but also in the seed, and surveys of *Astragalus* for accumulators and non-accumulators can be carried out on the seed amino acids, as well as on those of the leaf (Dunnill and Fowden, 1967). It is possible that accumulator species

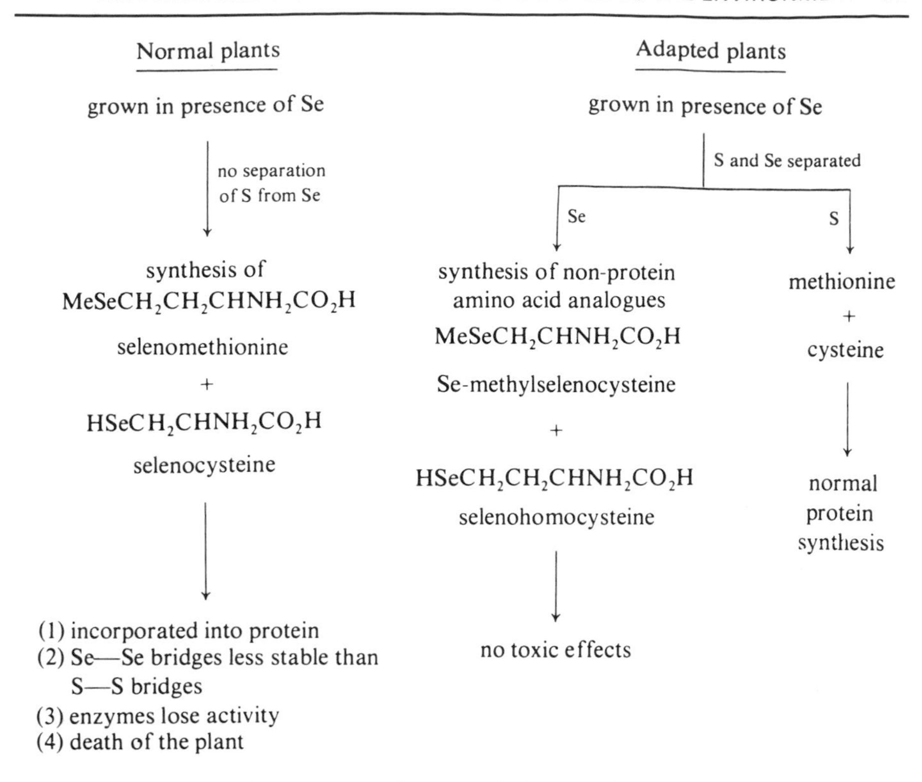

Figure 1.7 Adaptation of plants to selenium

actually thrive on the presence of selenium salts in the soil. In laboratory experiments with accumulators, it is possible to demonstrate stimulation of growth by adding selenate to the soil. This behaviour is similar to that of some halophytes which grow better in the presence of NaCl than in its absence (see p. 22).

B. Heavy Metal Toxicity

One of the most striking examples of the ability of plants to be selected for and adapt to high concentrations of toxic metals in soils is provided by the way in which grasses such as fine bent *Agrostis tenuis* and Sheep's fescue *Festuca ovina* have rapidly colonized waste tips from heavy metal mining. Some strains of *A. tenuis*, for example, will grow successfully on soils containing as much as 1% lead. Genetic aspects of this phenomenon have been studied among the plant communities in mining areas of North Wales by Bradshaw and his associates (e.g. Gregory and Bradshaw, 1965; Smith and Bradshaw, 1970). Populations tolerant to lead, copper or zinc develop rapidly and it is possible to examine these plants in order to determine the biochemical basis of this tolerance. While the precise mechanism of adaptation is not yet known, there are some clues about the processes involved.

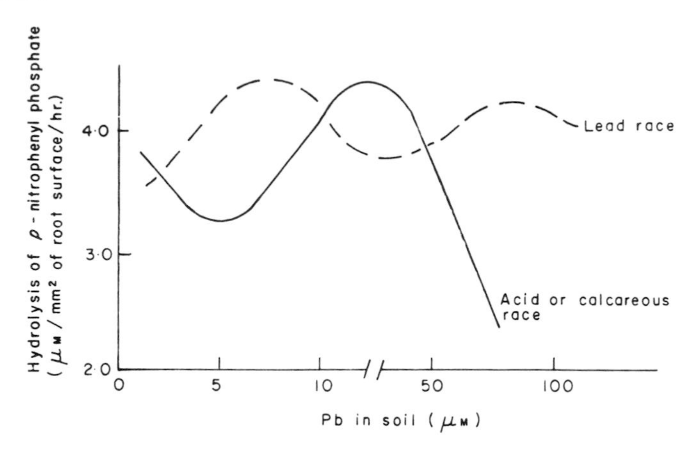

Figure 1.8 Activities of acid phosphatases in roots of *Agrostis tenuis*

The first point where the ability to deal with toxic amounts of heavy metals is important is at the root surface. That biochemical changes occur here has been demonstrated by Woolhouse (1970), who examined the activities of acid phosphatases present on the roots of *Agrostis tenuis*. His experiments (Fig. 1.8) show that the root surface enzymes are adapted to the high metal concentrations and can function in spite of the high toxic metal content of their environment. Presumably several different forms (or isozymes) of acid phosphatase occur on the root surface and particular forms are selected for by the environment. Indeed, it has been found that tolerant clones may have more isozymes present than non-tolerant ecotypes (Cox and Thurman, 1978).

The actual mechanism by which the toxic metal is deactivated or sequestered in a tolerant plant is not yet entirely clear, but there are a series of heavy metal-binding peptides produced in plants which have the ability to chelate these cations. These peptides, called phytochelatins (Grill *et al.*, 1987), are also produced in suspension cultures of plants, where their synthesis can be induced by adding trace amounts of cadmium, zinc, lead, silver or tin. They have the general structure: (glutamylcysteinyl)$_n$ glycine, where n = 2–8. They are formed from glutathione (glutamylcysteinyl glycine) and thus differ biosynthetically from most other peptides. That they may be involved in plant adaptation to heavy metal pollution is indicated by the fact that copper-tolerant strains of *Silene cucubalus* produce higher levels of these phytochelatins than non-tolerant strains.

The discovery of phytochelatins indicates a clear evolutionary divergence in adaptive mechanisms between higher plants and micro-organisms and animals, where heavy metal binding is carried out by the metallothioneins. Metallothioneins are true proteins, rich in cysteine, and they appear to have been strongly conserved over evolutionary time, since the amino acid sequences of microbial metallothioneins do not differ greatly from those recorded from animals (Lurch, 1980). There have been provisional reports of these proteins in plants, but they are probably due to misidentification. Phytochelatins have now been recognized in all plants so far surveyed, from algae to orchids (Grill *et al.*, 1987).

While the precise role of phytochelatins in heavy metal detoxification has yet to be determined, they do clearly provide a means of specifically chelating the metal on its entry into the plant root and of transporting it within the tissues. These peptides are not the only way that metals can be chelated within the plant cell, since organic acid complexes are known. Studies of zinc-tolerant and non-tolerant ecotypes of *Deschampsia caespitosa* (Godbold *et al.*, 1984) show that zinc accumulates in the vacuole as the citrate complex and in the cytoplasm as the malate complex. Similarly, nickel has been identified in nickel-accumulating plants as the citrate complex, while chromium occurs in the chromium accumulator *Leptospermum scoparium* as the trioxalate (Woolhouse, 1983).

In the case of copper tolerance, it appears that several different mechanisms may operate. Copper exclusion is possible and this appears to occur in the alga *Chlorella vulgaris* (Foster, 1977). This does not mean complete exclusion, since minute amounts must be allowed to enter, because copper is an essential micronutrient for life. Some immobilization of copper at the cell wall is part of the tolerance mechanism in some species, while transportation and compartmentation occur in others (Woolhouse, 1983). By contrast, lead is usually bound irreversibly to the cell wall in tolerant ecotypes, e.g. of the grass *Anthoxanthum odoratum*. In lead-sensitive strains, toxicity is probably due to the leakage of some of this lead into the cytoplasm (Qureshi *et al.*, 1986).

In summary, it is patent that biochemical adaptation in plants to heavy metals involves several mechanisms: exclusion, binding at the cell wall, complexing with phytochelatins, complexing with organic acids, transportation and compartmentation. Metal accumulation and hence tolerance is metal specific and each metal cation represents a different problem to the plant. Plants do, however, show co-tolerance, e.g. to zinc and copper, and multiple-metal tolerance has been recorded (Peterson, 1983) so that there may be some mechanisms which are common to more than one metal.

The ability of certain plants such as those mentioned above to adapt to one or other of these heavy metals has been turned to advantage by man in prospecting for new minefields. The use of indicator plants for showing where valuable mineral deposits are present under the soil is now an accepted part of mining techniques. For example, the shrub *Hybanthus floribundus* is an indicator for nickel in Australia; indeed, it can accumulate up to 22% of its ash as this metal. Again, the plant *Eriogonum ovalifolium* is an indicator for silver deposits in Montana, while certain *Astragalus* species are indicators for uranium in Colorado. Finally, several gold-accumulating plants of obvious mining interest have been found in natural vegetation of Southern Africa. They include the plant *Phacelia sericea*, which takes gold in as the cyanide, a mechanism which enables the plant to allow 21 ppb gold to accumulate in the leaves (Peterson, 1983).

The ability to adapt to heavy metal poisoning also extends to marine plants. Thus the brown kelp *Ecklonia radiata* concentrates arsenic in the coastal ecosystem of Australia, storing it within its tissues in organic form as two arsenosugar derivatives. Interestingly, arsenic is also found in bound form as arsenobetaine, $Me_3As^+CH_2CO_2^-$, in associated marine fauna, such as the lobster and whiting. It appears that these sea creatures obtain their arsenic indirectly from the brown kelp via detritovores, the organic form of arsenic being changed during this process (Fig. 1.9) (Edmonds *et al.*, 1982).

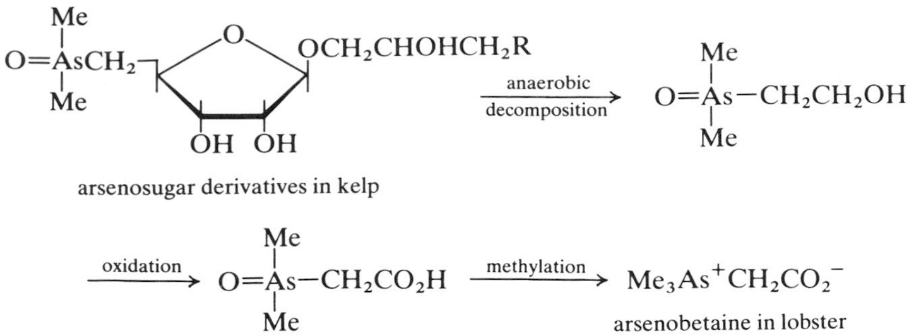

$O{=}As(Me)_2CH_2$–[ribose ring, OH OH]–$OCH_2CHOHCH_2R$ $\xrightarrow[\text{decomposition}]{\text{anaerobic}}$ $O{=}As(Me)_2{-}CH_2CH_2OH$

arsenosugar derivatives in kelp

$\xrightarrow{\text{oxidation}}$ $O{=}As(Me)_2{-}CH_2CO_2H$ $\xrightarrow{\text{methylation}}$ $Me_3As^+CH_2CO_2^-$

arsenobetaine in lobster

Figure 1.9 Interconversion of bound forms of arsenic in the marine food chain

C. Adaptation to Salinity

Plants which grow in saline habitats, in salt marshes, salt deserts or on the sea coast, are termed 'halophytes'. That they have adapted to such conditions of high salinity in the soil is clear when attempts are made to grow nonhalophytes, or glycophytes as they are called, in the presence of increasing amounts of sodium chloride (NaCl). The normal requirement in higher plants for Na in the soil is very low (a few ppm) and toxic symptoms develop quite quickly as soon as this is raised: amounts as low as 0.1% NaCl can be damaging to relatively sensitive glycophytes such as tomatoes, peas and beans. By contrast, true halophytes are resistant to high concentrations of NaCl; indeed presence of 1–2% NaCl may be needed by these plants to stimulate growth. Some halophytes can withstand up to 20% NaCl in the soil, although the levels in most saline soils are lower than this, between 2 and 6% (Levitt, 1980).

Among typical coastal halophytes are such plants as sea thrift, *Armeria maritima*, and sea plantain, *Plantago maritima*, of temperate floras, and the mangroves (e.g. *Rhizophora, Avicennia*) and sea-grasses (e.g. *Thalassia*) of tropical coastlines. Desert halophytes include many *Atriplex* species, e.g. *A. halimus* and *A. spongiosa*, and *Suaeda fruticosa*. The definition of a halophyte is not always clear cut and such plants are not necessarily restricted to growing in saline habitats. With glycophytes, there is a degree of variation in response to NaCl stress, some species (e.g. tomatoes, peas) being particularly sensitive and others (e.g. many grasses) being more resistant. Detailed accounts of the ecology and general biology of halophytes are contained in the books of Reimold and Queen (1974) and of Waisel (1972).

Adaptation is by one of three procedures: accumulation of NaCl within the vacuole (*Salicornia* has a 10% NaCl solution in its tissues); resistance to the entry of NaCl into the cell; and dilution of NaCl after its entry into the plant. Biochemical differences, especially in enzyme levels, have been recorded between halophytes and glycophytes but, in fact, none of these differences seems to be associated with the adaptation process. Thus, enzymes of halophytes are not, *in vitro*, more tolerant of high salt concentrations than those of glycophytes. Furthermore, although it is possible to demonstrate stimulation of activity of certain enzymes, e.g. malate dehydrogenase, by adding NaCl (up to 50

Table 1.3 Glycinebetaine and proline levels (mg/100 g fr wt) in plant shoots grown under low and high salt conditions

Plant type	Plant	GLYCINEBETAINE Low salt	GLYCINEBETAINE High salt	PROLINE Low salt	PROLINE High salt
Salt-sensitive glycophyte	Tomato	2	2	6.9	72
Salt-resistant	Barley 'Arimar'	32	158	0.8	22
glycophyte	*Chloris gayana*	25	106	0.6	48
Halophyte	*Atriplex spongiosa*	177	246	1.3	2.0
	Suaeda monoica	385	462	5.7	3.7

Data from Storey and Wyn Jones (1975). Similar changes were also noted in root levels. Low salt conditions refer to standard Hoagland's solution, high salt conditions to growth in the presence of from 100 to 500 mM NaCl.

mM), such stimulation occurs with enzymes extracted from both halophytes and glycophytes (Flowers *et al.*, 1976). One particular symptom of excess NaCl in the broad bean *Vicia faba* is the accumulation of the diamine putrescine (Strogonov, 1964). This can, however, be explained as an effect of excess Na on limiting the absorption of K by the cell, since K deficiency in a number of plants, including the broad bean, also causes putrescine to accumulate (Smith, 1965).

Recently, two critical biochemical features of halophytes have been discovered: accumulation of two nitrogen compounds, the protein imino acid proline and the quaternary nitrogen compound, glycinebetaine. In the case of proline, concentrations found can be up to ten times the 'normal' level present in the free amino acid pool of glycophytes (Stewart and Lee, 1974). In the extreme case of the halophyte *Triglochin maritima*, free proline accumulates in such amounts that it represents 10–20% of the shoot dry weight. Other halophytes in which high levels have been found are *Aster tripolium* and *Armeria maritima*. It should be noted, however, that a few halophytes, notably *Plantago maritima*, do not show enhanced proline levels.

The importance of the above observations on proline content has been reinforced by the induction of high levels of this acid in glycophytes by subjecting them artificially to high salt levels (see results with tomatoes in Table 1.3). Indeed, increased levels of proline can be induced in both halophytes and glycophytes by growing them under non-saline conditions and then moving them gradually into increasing salinity. That proline is truly an adaptive response to salinity is suggested by the observation of Stewart and Lee (1974) that inland populations of *Armeria* have a relatively normal proline content (1.4 μmol/g fr wt) while nearby coastal populations reach a concentration of 26 μmol/g fr wt.

How such high levels of proline actually provide the basis of resistance to salt accumulation is not known precisely but proline does have valuable osmotic properties so that its presence may enable such plants to withstand the high osmotic forces to which their cells would otherwise be subjected. This accumulation of proline in halophytes seems to be related to the enhanced levels of proline found in drought-resistant plants (see p. 15).

Clearly, there may be a common mechanism in adaptation to saline habitats and to extremes of dry climates.

Evidence that the accumulation of aliphatic quaternary ammonium compounds provides plants with relief from NaCl stress has been obtained by Wyn Jones and his collaborators (see e.g. Storey and Wyn Jones, 1977). Most attention has been given to choline, $Me_3N^+CH_2CH_2OH$, and the related acid betaine, $Me_3N^+CH_2CO_2^-$. Both are widespread in plants but while choline has a recognized function as a constituent of membrane lipid, no similar role has been ascribed to glycinebetaine. Measurements of the concentration of this betaine in plants representing a spectrum of salt sensitivities have shown that increasing the salt stress produces large increases in betaine, without affecting choline levels (Table 1.3). Increases in betaine levels are often accompanied by increases in proline, but there are some exceptions. Thus, while proline levels increase dramatically in the salt-sensitive tomato, glycinebetaine levels are not similarly affected (Table 1.3). Conversely, in the case of the two halophytes chosen for study by Storey and Wyn Jones (1975), betaine levels are significantly raised by high salt concentrations but proline levels do not change.

These data (Table 1.3) indicate that in some ways glycinebetaine may be more important than proline in the adjustment of halophytes to NaCl stress. However, as with proline accumulation, there are some halophytes which do not show glycinebetaine accumulation. One must therefore assume that there is more than one biochemical mechanism for adaptation to salinity. However, the actual role of this betaine in providing osmotic adjustment may well be similar to that of proline. Indeed, a close correlation has been observed between the osmotic pressure of shoots and their betaine content, both when values were compared in a single plant grown at various salinities and when different plant species were compared under similar growth conditions (Wyn Jones *et al.*, 1977). The protective role of both these nitrogen compounds is further emphasized by the fact that neither glycinebetaine nor proline has any inhibitory effect on enzyme activities in halophytes, even when present in high concentration.

While proline and glycinebetaine are the two most widespread compounds involved in adaptation to the halophyte condition, they may be replaced by several related structures in certain plants (Table 1.4). Indeed, there is a significant taxonomic element in the kinds of compound accumulated. Of particular interest are the two sulphonium derivatives, which, like glycinebetaine, are zwitterionic. One of these, S-dimethyl-sulphonium propanoic acid, has also recently been reported as a cytoplasmic osmoticum in the marine alga, *Ulva lactuca* (Dickson *et al.*, 1980). Algae of the family Laminariales have been reported to accumulate lysinebetaine among other osmotica and this was isolated from the seaweeds as the dioxalate (Blunden *et al.*, 1982). The discovery of the aliphatic polyols sorbitol and pinitol in higher plant halophytes (Table 1.4, cf. Ahmad *et al.*, 1979; Gorham *et al.*, 1980) also links up with the report of enormous glycerol accumulation (up to 80% dry weight) in the green alga *Dunaliella parva* in response to a saline environment (Ben-Amotz and Avron, 1973). It is of interest also that polyols have been implicated in frost resistance (see p. 11). The relative cost to the plant in terms of nutrient and energy supply of these different osmotica undoubtedly varies and it is an interesting question whether there is any selective advantage in producing one type over another. Perhaps expectedly, there

Table 1.4 Low molecular weight substances which accumulate in salt-tolerant higher plants

Cytoplasmic osmotica	Formulae	Halophytes in which accumulation occurs
Proline	$N^+H_2—(CH_2)_3—CHCO_2^-$ (ring)	*Triglochin maritima* Juncaginaceae
Glycinebetaine	$Me_3N^+—CH_2CO_2^-$	*Atriplex spongiosa* Chenopodiaceae
β-Trimethylamino-propanoic acid	$Me_3N^+—CH_2CH_2CO_2^-$	*Limonium vulgare* Plumbaginaceae
S-Dimethylsulphonium propanoic acid	$Me_2S^+—CH_2CH_2CO_2^-$	*Spartina anglica* Gramineae
S-Dimethylsulphonium pentanoic acid	$Me_2S^+—(CH_2)_4CO_2^-$	*Diplotaxis tenuifolia* Cruciferae
Sorbitol	$CH_2OH—(CHOH)_4—CH_2OH$	*Plantago maritima* Plantaginaceae
Pinitol	$CHOH—(CHOH)_4—CHOMe$ (ring)	*Spergularia medea* Caryophyllaceae

is some evidence in the case of the quaternary ammonium salts that the nitrogen present is recycled within the plant on a seasonal basis.

By means of NMR measurements directly on plant tissues, Robinson and Jones (1986) have been able to show that in salt-stressed spinach leaves, the concentration of glycinebetaine in the chloroplast is ten times that in the whole leaf. In fact, half the total glycinebetaine is concentrated in the chloroplast and the other half is in the cytoplasm. Localization in these organelles within the cell allows the glycinebetaine to produce an osmotic potential sufficient to balance that due to the storage of sodium cations in the vacuole.

In summary then, recent experiments with halophytes suggest that many, perhaps the majority, are able to adapt biochemically to NaCl stress by accumulating in the cytoplasm particular non-toxic solutes. Such substances exert a protective effect on osmotic regulation by balancing the decreased water potential caused by inorganic ion accumulation in the vacuole. The most common osmotica are proline and glycinebetaine but sulphonium derivatives and polyols may also have this role. Considering the enormous range of plants that are able to establish themselves in saline environments, undoubtedly other mechanisms remain to be discovered.

Plants growing in or near the sea or in salt marshes are subjected to influx of other ions besides sodium and chloride and it is conceivable that adaptation to other inorganic salts present in brackish waters may be necessary for survival of plants in such habitats. One such ion present in seawater in some quantity is inorganic sulphate; one possible route for inactivation or storage is through conjugation with naturally occurring phenolic compounds, and particularly with flavonoids. A large number, over 50, of such conjugates have recently been discovered in plants (Harborne, 1977) and remarkably enough, they occur principally in plants which are subject to water stress, but especially in halophytes.

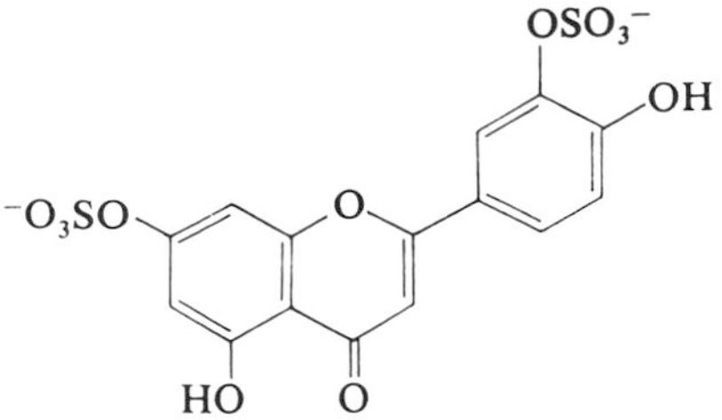

luteolin 7,3′-disulphate
present in *Zostera*

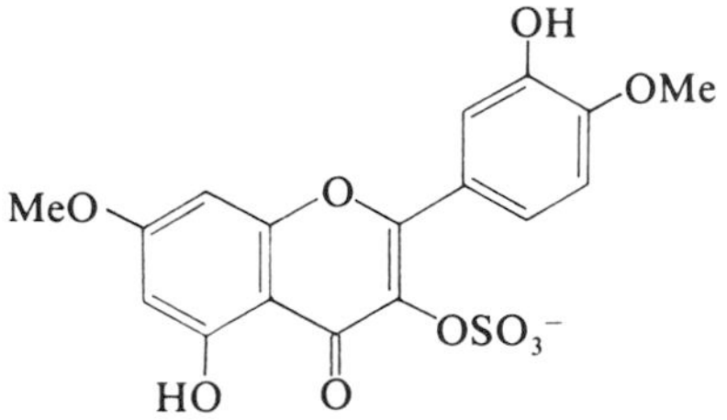

ombuin 3-sulphate
present in Flareria

Figure 1.10 Structures of two flavonoid sulphates found in halophytes

Flavonoid sulphates (for typical structures, see Fig. 1.10) have been identified in such land halophytes as *Suaeda maritima, Armeria maritima, Limonium vulgare, Nypa fruticans* and species of *Atriplex, Frankenia* and *Tamarix*. They also occur abundantly in sea-grasses such as *Thalassia, Zannichellia* and *Zostera*. Just as with proline or betaine accumulation, there are some halophytes (e.g. *Plantago maritima*) in which flavonoid sulphates are apparently absent. This presumably means that more than one system of sulphate conjugation occurs. That these organic conjugates of sulphur have a dynamic role in ion relationships is still largely conjectural, but such a role is under active current study.

IV. Detoxification Mechanisms

A. General

Besides the many stresses which plants are subjected to in the natural environment, they are also today subjected to stresses which are man-made. These are derived from the pollution of the atmosphere with factory fumes and degraded gasoline exhaust vapours, other escapes of organic matter into the environment and the deliberate application of a variety of pesticides to plant crops. It is fortunate that plants are able to cope with these stresses and indeed continue to survive in spite of such bombardment. That they can do so is due at least in part to the fact that they have an efficient system for the detoxification of foreign compounds within their cells. This fact has been established by feeding experiments with toxic compounds and it is also apparent in the way that plants can store toxins within their cells in non-toxic bound form (e.g. HCN bound as cyanogenic glycoside). Even foreign gases entering the plant through the stomata can be resisted. Tolerance to SO_2, which is normally lethal to plant life in any concentration, has been established in perennial ryegrass *Lolium perenne* growing in industrial areas (Horsman *et al.*, 1978). Concentrations of up to 700 $\mu g\ m^{-3}$ of SO_2 in the atmosphere can be dealt with, although the mechanism of detoxification is not yet known.

The key detoxification reaction of organic compounds in plants is glucoside formation, carried out by a glycosyltransferase enzyme in the presence of the energy-rich uridinediphosphateglucose as cofactor. This contrasts with the major detoxification route in animals, which involves conversion to glucuronide, not glucoside, or ethereal sulphate formation (Williams, 1964). The purpose in both plants and animals is the same, to inactivate the toxin and to give it water solubility so that it can be pushed into the cell vacuole in plants or excreted in the urine in animals.

Foreign compounds containing phenolic or nitrogen groups are directly detoxified as glucosides. In other compounds, these functional groups have first to be introduced before conjugation with sugar can occur. Thus, other chemical reactions may take place as part of the detoxification procedure, especially oxidation, decarboxylation, methylation, acylation or esterification. Conjugation with amino acid instead of glucose may also happen occasionally.

B. Detoxification of Phenols

Phenols have been the most widely studied group of compounds from this viewpoint (Towers, 1964). Phenols are highly toxic to both plants and micro-organisms. Feeding experiments have shown that they are rapidly glucosylated, within a few hours of entry into the plant and the products—the glucosides—are either stored in the vacuole or further metabolized within the plant and broken down eventually to CO_2.

Some examples of the products of glucosylation of phenols are shown in Fig. 1.11. The detoxification of hydroquinone has been especially widely studied (Pridham, 1964) and it has been found that the ability to glucosylate it is universally distributed in all higher plants; only fungi, algae and bacteria lack this capacity. Usually, the monoglucoside, arbutin, which occurs naturally in *Pyrus* leaf, is formed. In some plants and in some tissues, the glucosylation reaction is so well developed that higher glucosides are also produced; in wheat germ, for example, both the diglucoside (gentiobioside) and triglucoside have been identified as metabolites. In compounds which have two adjacent phenolic hydroxyls (e.g. aesculetin), isomeric glucosides may be formed; aesculetin is also methylated to scopoletin in some plants and is recovered as the glucoside scopolin.

Compounds containing both phenolic and carboxyl groups, e.g. *p*-coumaric acid (Fig. 1.11), are mainly detoxified via the glucose ester, i.e. the sugar is attached preferentially to the carboxyl and not to the phenolic group. It is interesting, however, that when cinnamic acid is fed, only small amounts of cinnamyl-glucose are recovered. The main reaction is hydroxylation to *p*-coumaric acid, which is subsequently glucosylated at the carboxyl group so that the main product is *p*-coumarylglucose. In this connection, it may be noted that the growth hormone, indoleacetic acid, when fed to plants is not *N*-glucosylated but is detoxified via the glucose ester. In this case, the aspartic acid conjugate, also joined through the carboxyl group, may be formed. Conjugation with this amino acid represents another, but minor, detoxification pathway in plants.

C. Detoxification of Systemic Fungicides

In the last 20 years, systemic fungicides have been developed to control such diseases as powdery mildew on cereals and cucumber. The rationale behind the use of such fungicides is that they should be absorbed by the plant, where they are harmless, and be

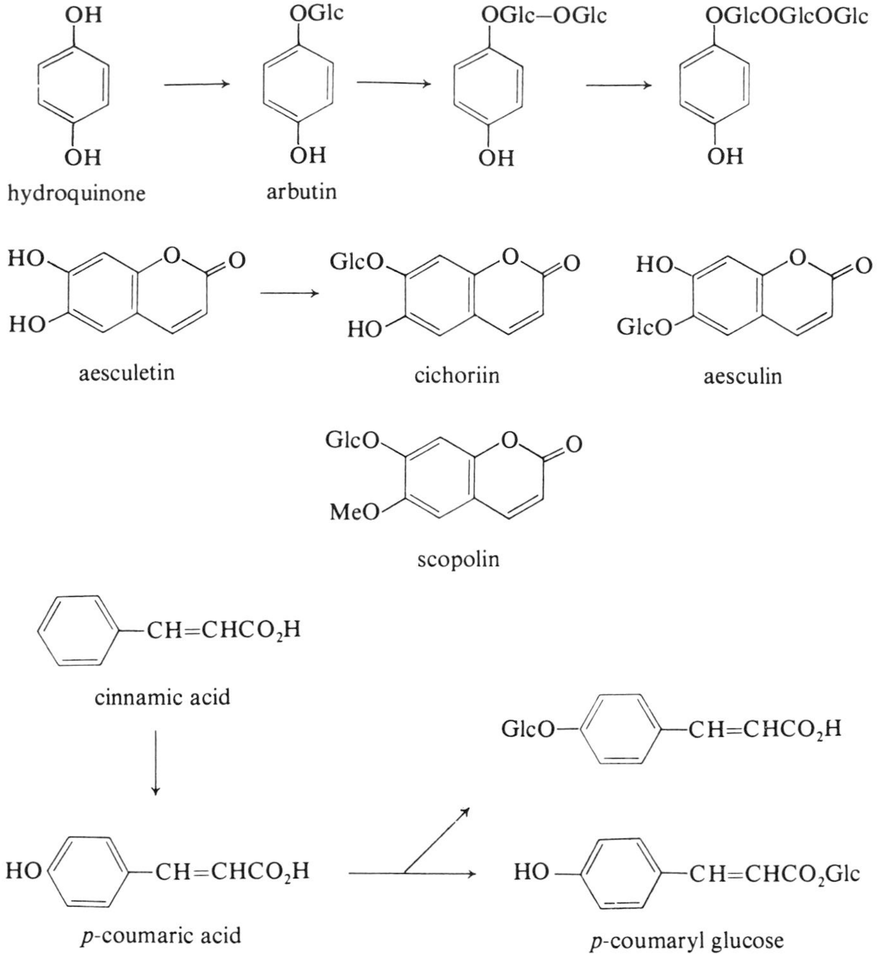

Figure 1.11 Detoxification products of phenols in plants

retained within its tissues to prevent the growth and development of the fungal parasite. While earlier types of fungicide were sprayed onto the mature crop, systemic fungicides can be applied as a seed dressing and persist in the plant for a sufficient time to provide resistance during the whole growing season. Such compounds may of course be modified *in vivo,* the detoxification product itself providing much of the resistance to fungal invasion.

One widely used systemic fungicide on barley is ethirimol, a pyrimidine derivative which presumably acts as an antimetabolite in its effectiveness against the fungus. Its metabolism has been studied in barley leaves and it has been shown by tracer experiments to have a half-life *in vivo* of only 3 days. There is some photochemical reaction, but one of the major metabolites is the expected glucoside. However, it is interesting

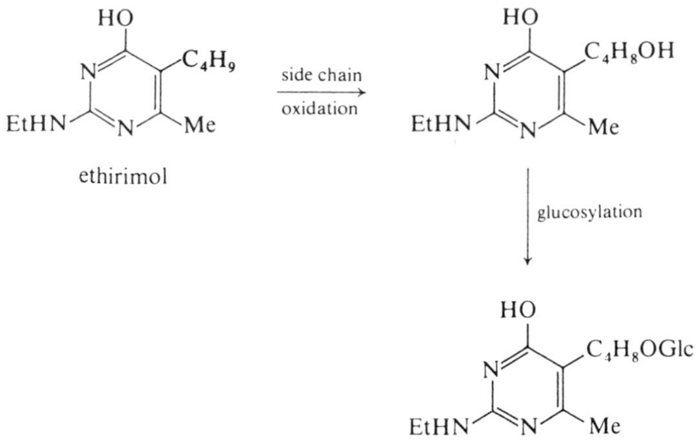

Figure 1.12 Metabolism of ethirimol in barley

that no simple *N*-glucoside is formed. Instead, the aliphatic side chain of ethirimol is first oxidized to the corresponding alcohol and it is the glucoside of this alcohol which is the natural metabolite (Fig. 1.12) (Teal, 1973). For other metabolites formed from this and other pyrimidine derivatives, see the review by Vonk (1983).

D. Detoxification of Herbicides

One final example of the flexibility of plants in handling foreign compounds is taken from the field of practical weed control. Detoxification of herbicides has been widely studied and it appears that the rate and mode of detoxification is often critical to their effectiveness as selective weed killers. Indeed, the selective use of 2,4-dichlorophenoxyacetic acid (2,4-D) to kill weeds among crop plants such as cereals depends on the fact that while the crop can readily metabolize it, the weeds cannot and are killed as a result.

While conjugation at the carboxyl group with glucose or aspartic acid or hydroxylation of the aromatic ring may occur, an important reaction in the detoxification of 2,4-D is oxidation of the side chain (Naylor, 1976). Once this side chain is removed, growth hormone activity is lost, 2,4-dichlorophenol is formed and this is then conjugated with glucose in the normal way (Fig. 1.13). Although the ability to degrade the side chain is widespread in plants, the facility to do it rapidly is limited to a few crop plants. The enzyme machinery is not immediately active in weeds, and it is this variation in *rate* of detoxification which is the key issue. The weed species, unable to detoxify 2,4-D rapidly enough, are killed, not because 2,4-D is toxic as such, but because it is an extremely active auxin. In excess concentration, it causes the plant to grow too fast and it is the upset in growth pattern which causes the death of the plant.

An example of a more unusual pathway of detoxification is the case of the nitrogen-containing herbicide monuron. This compound in the cotton plant is first demethylated, the terminal *N*-methyl group is then oxidized and finally glucosylation occurs (Fig. 1.14). Two complex enzymic changes are needed before the compound can be removed from

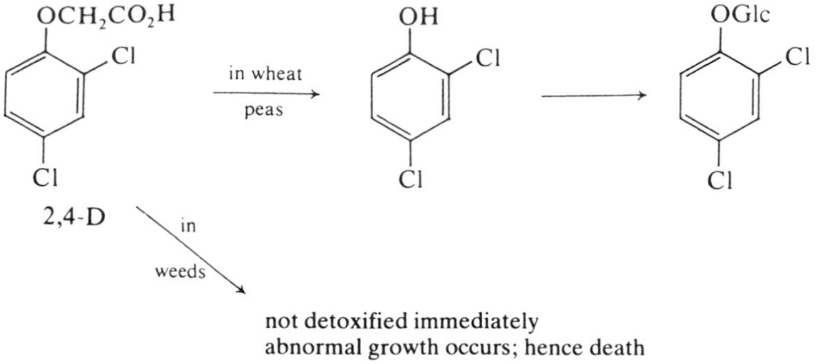

OCH_2CO_2H Cl Cl 2,4-D — in wheat, peas → OH Cl Cl → OGlc Cl Cl

2,4-D — in weeds → not detoxified immediately
abnormal growth occurs; hence death

Figure 1.13 Detoxification and metabolism of 2,4-D in plants

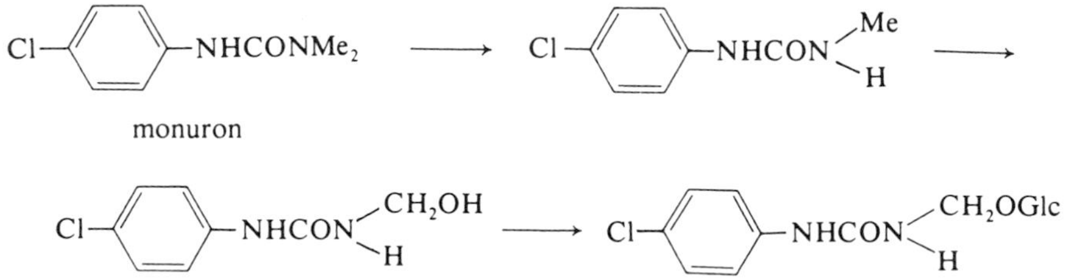

Cl–C$_6$H$_4$–$NHCONMe_2$ (monuron) → Cl–C$_6$H$_4$–NHCON(Me)(H) →

Cl–C$_6$H$_4$–NHCON(CH_2OH)(H) → Cl–C$_6$H$_4$–NHCON(CH_2OGlc)(H)

Figure 1.14 Metabolism of the herbicide monuron in plants

the system via the glucoside (Frear *et al.*, 1972). Further references to the metabolism of fungicides and herbicides may be found in the review series of Hutson and Roberts (1981–85).

V. Conclusion

Biochemical adaptation in plants involves various changes in the basic biochemistry of the cell (Table 1.5). These include:

(1) the development of new metabolic pathways (e.g. photosynthetic adaptation);
(2) the accumulation of low molecular weight metabolites, usually from the carbohydrate or nitrogen pool (such compounds may have protective, e.g. osmotic, properties);
(3) changes in hormone levels (e.g. ABA in droughted plants);
(4) the synthesis of special proteins (e.g. HS or anti-freeze proteins); and
(5) detoxification mechanisms (e.g. the chelating of heavy metals with organic acids or special peptides).

In each case, the biochemical response is probably much more complex than this, involving many more subtle modifications in cellular biochemistry, e.g. at the mem-

Table 1.5 Biochemical responses of plants to different forms of environmental stress

Stress	Biochemical response in plant
High temperature	Long term: C_4 photosynthesis Short term: HS proteins[a]
Drought	CAM photosynthesis Increase in ABA Increase in proline or pinitol
Low temperature	Accumulation of sugars and/or polyols[a] Synthesis of anti-freeze proteins[a]
Salt	Accumulation of cytoplasmic osmotica[a] (polyols, quaternary ammonium compounds, etc.)
Selenium	Accumulation of selenium as non-protein amino acid
Heavy metals	Modification of enzymes in root surface Binding to sites in cell wall of root Accumulation/detoxification as peptide complex[a] Accumulation/detoxification as organic acid complex

[a] Similar responses have been observed in micro-organisms and/or animals.

brane level. Biochemical responses are often correlated with physiological or anatomical adaptations.

The biochemical responses in plants are often similar to those that occur in animals. The response to salt stress is common in microbes, plants and animals. There is a real divergence in heavy metal-binding factors: phytochelatins are formed in plants, while metallothioneins are formed in microbes and animals. Another divergence is in the response to anoxia and the dangers of alcohol poisoning. Plants do not appear to diversify the pathway of glycolysis to accumulate organic or amino acids, as animals do.

There is sometimes a clear distinction between short-term and long-term biochemical adaptation. Thus, in the response to high temperatures, the genes for HS protein synthesis are ready to be 'turned on' and the response is rapid. By contrast, the evolution of the C_4 pathway of photosynthesis in plants of tropical climates occurred over a very long time-scale and is permanently fixed in the genome of C_4 plants. By contrast, the adaptation to heavy metal toxicity and other types of pollution may occur relatively quickly. Plants have a very versatile system of detoxification and can deal with some environmental toxins (e.g. herbicides, fungicides) directly. In the case of lead and zinc in mining waste tips, adaptation requires several mutational events, but tolerant plants emerge within a few years.

It is as well for man and herbivores that plants do have the ability to respond flexibly and rapidly to environmental stresses. Occasionally, this can have toxic consequences (as in the case of livestock suffering selenium poisoning) but generally it is beneficial to man in the long term that plants are able to colonize new habitats or survive successfully in old habitats in spite of the effects of pollution or changing climates.

Bibliography

Books and Review Articles

Anderson, J. W. and Scarf, A. R. (1983). Selenium and plant metabolism. In: Robb, D. A. and Pierpoint, W. S. (eds.), 'Metals and Micronutrients, Uptake and Utilisation by Plants', pp. 241–275. Academic Press, London.

Atsatt, P. R. and O'Dowd, D. J. (1976). Plant defense guilds. *Science*, N.Y. **193**, 24–29.

Baldwin, J. (1937). 'An Introduction to Comparative Biochemistry'. University Press, Cambridge.

Bjorkman, O. and Berry, J. (1973). High efficiency photosynthesis. *Scient. Am.* (October), pp. 80–93.

Crawford R. M. M. (1989). 'Studies in Plant Survival'. Blackwells, Oxford.

Ellis, R. J. and Gray, J. C. (eds.) (1986). 'Ribulose Bisphosphate Carboxylase–Oxygenase', 165 pp. Royal Society, London.

Fitter, A. H. and Hay, R. K. M. (1987). 'Environmental Physiology of Plants', 2nd edn. Academic Press, London.

Florkin, M. and Mason, H. (1960–64). 'Comparative Biochemistry', Vols. 1–8. Academic Press, New York.

Gibbs, M. and Latzko, E. (eds.) (1979). Photosynthesis II. *Encyclopedia of Plant Physiology, new series* **6**, 1–578.

Harborne, J. B. (1977). Flavonoid sulphates: a new class of natural product of ecological significance in plants. *Prog. Phytochem.* **4**, 189–208.

Hatch, M. D. and Slack, C. R. (1970). The C_4-dicarboxylic acid pathway of photosynthesis. *Prog. Phytochem.* **2**, 35–106.

Hutson, D. H. and Roberts, T. R. (1981–85). 'Progress in Pesticide Biochemistry and Toxicology', Vols. 1–4, Wiley, Chichester.

Kluge, M. and Ting, I. P. (1978). 'Crassulacean Acid Metabolism', 209 pp. Springer-Verlag, Berlin.

Levitt, J. (1980). 'Responses of Plants to Environmental Stresses', 2nd edn. in 2 vols. Academic Press, New York.

Luckner, M. (1990). 'Secondary Metabolism in Microorganisms, Plants and Animals', 3rd edn., 576 pp. Springer-Verlag, Berlin.

Naylor, A. W. (1976). Herbicide metabolism in plants. In: Audus, L. J. (ed.), 'Herbicides', Vol. 1, pp. 397–426. Academic Press, London.

Peterson, P. J. (1983). Adaptation to toxic metals. In: Robb, D. A. and Pierpoint, W. S. (eds.), 'Metals and Micronutrients, Uptake and Utilisation by Plants', pp. 51–69. Academic Press, London.

Rathnam, C. K. M. and Chollet, R. (1980). Photosynthetic carbon metabolism in C_4 plants and C_3–C_4 intermediate species. *Prog. Phytochem.* **6**, 1–48.

Reimold, R. J. and Queen, W. H. (1974). 'Ecology of Halophytes', 605 pp. Academic Press, New York.

Robinson, T. (1980). 'The Organic Constituents of Higher Plants', 4th edn., 347 pp. Cordus Press, N. Amherst, Mass.

Rosenfeld, I. and Beath, O. A. (1964). 'Selenium, Geobotany, Biochemistry, Toxicity and Nutrition'. Academic Press, New York.

Schlesinger, M., Ashburner, M. and Tissieres, A. (1982). 'Heat Shock: From Bacteria to Man'. Cold Spring Harbor, New York.

Schoeffl, F., Lin, C. Y. and Key, J. L. (1984). Soybean heat shock proteins: temperature regulated gene expression and the development of thermotolerance. *Ann. Proc. Phytochem. Soc. Eur.* **23**, 129–140.

Shrift, A. (1972). Selenium toxicity. In: Harborne, J. B. (ed.), 'Phytochemical Ecology', pp. 145–162. Academic Press, London.

Smellie, R. M. S. and Pennock, J. F. (eds.) (1976). 'Biochemical Adaptation to Environmental Change', 240 pp. Biochemical Society, London Symposium Vol. No. 41.

Stetter, K. O., Fiala, G., Huber, R., Huber, G. and Segerer, A. (1986). Life above the boiling point of water? *Experientia* **42**, 1198–1191.

Strogonov, B. P. (1964). 'Physiological Basis of Salt Tolerance of Plants'. In: Poljakoff–Mayber, A. and Mayer, A. M. (eds. and trans.). Monson, Jerusalem.

Swain, T. (1974) Biochemical evolution in plants. *Comp. Biochem.* **29A**, 125–302.

Ting, I. P. (1985) Crassulacean acid metabolism. *A. Rev. Plant Physiol.* **36**, 595–622.

Towers, G. H. N. (1964). Metabolism of phenolics in higher plants and microorganisms. In: Harborne, J. B. (ed.), 'Biochemistry of Phenolic Compounds', pp. 249–294. Academic Press, London.

Vonk, J. W. (1983). Metabolism of fungicides in plants. In: Hutson, D. H. and Roberts, T. R. (eds.), 'Progress in Pesticide Biochemistry and Toxicology', Vol. 3, pp. 111–162. Wiley, Chichester.

Waisel, Y. (1972). 'Biology of Halophytes', 395 pp. Academic Press, New York.

Williams, R. T. (1964). Metabolism of phenolics in animals. In: Harborne, J. B. (ed.), 'Biochemistry of Phenolic Compounds', pp. 205–248. Academic Press, London.

Woolhouse, H. (1970). Environment and enzyme evolution in plants. In: Harborne, J. B. (ed.), 'Phytochemical Phylogeny', pp. 207–232. Academic Press, London.

Woolhouse, H. W. (1983). Toxicity and tolerance in the responses of plants to metals. *Encyclopedia of Plant Physiology, New Series* **12c**, 245–300.

Literature References

Ahmad, I., Larher, F. and Stewart, G. R. (1979). *New Phytologist* **82**, 671–678.

Barbehenn, R. V. and Bernays, E. A. (1992). *Oecologia* **92**, 97–103.

Baross, J. A. and Deming, J. W. (1983). *Nature* **302**, 423–426.

Ben-Amotz, A. and Avron, M. (1973). *Plant Physiol.* **51**, 875–878.

Blunden, G., Gordon, S. M. and Keysell, G. R. (1982). *J. Nat. Prod.* **45**, 449–452.

Bright, S. W. J., Lea, P. J., Kueh, J. S. H., Woodcock, C., Hollomon, D. W. and Scott, G. C. (1982). *Nature* **295**, 592–593.

Brown, W.V. (1975). *Am. J. Bot.* **62**, 395-402.

Brown, T. A. and Shrift, A. (1980). *Plant Physiol.* **66**, 758–761.

Caswell, H., Reed, F., Stephenson, S. N. and Werner, P. A. (1973). *Am. Nat.* **107**, 465–480.

Coombs, J. (1971). *Proc. R. Soc. Lond. B.* **179**, 221–235.

Cox, R. M. and Thurman, D. A. (1978). *New Phytologist* **80**, 17–22.

Crawford, R. M. M. (1978). *Naturwissensch.* **65**, 194–201.

Crespo, H. M., Frean, M., Cresswell, C. F. and Tew, J. (1979). *Planta* **147**, 257–263.

Dickson, D. M. J., Wyn Jones, R. G. and Davenport, J. (1980). *Planta* **150**, 158–165.

Dorne, A. J., Cadel, G. and Douce, R. (1986). *Phytochemistry* **25**, 65–68.
Dunnill, P. M. and Fowden, L. (1967). *Phytochemistry* **6**, 1659–1663.
Edmonds, J. S., Francesconi, K. A. and Hansen, J. A. (1982). *Experientia* **38**, 643–644.
Ford, C. W. (1984). *Phytochemistry* **23**, 1007–1015.
Foster, P. L. (1977). *Nature* **269**, 322–323.
Flowers, T. J., Hall, J. L. and Wand, M. E. (1976). *Phytochemistry* **15**, 1231–1234.
Frear, D. S., Swanson, H. R. and Tanaka, F. S. (1972). *Recent Adv. Phytochem.* **5**, 225–246.
Godbold, D. L., Horst, W. J., Collins, J. C., Thurman, D. A. and Marschner, H. (1984). *J. Plant Physiol.* **116**, 59–70.
Gorham, J., Hughes, L. and Wyn Jones, R. G. (1980). *Plant Cell Environ.* **3**, 309–318.
Gregory, R. P. G. and Bradshaw, A. D. (1965). *New Phytol.* **64**, 131–143.
Grill, E., Winnacker, E. L. and Zenk, M. H. (1987) *Proc. natn. Acad. Sci. U.S.A.* **84**, 439–443.
Gutteridge, S., Parry, M. A. J., Burton, S., Keys, A. J., Mudd, A., Feeny, J., Servaites, J. C. and Pierce, J. (1986). *Nature* **324**, 274–276.
Heidorn, T. and Joern, A. (1984). *Oecologia (Berlin)* **65**, 19–25.
Horsman, D. C., Roberts, T. M. and Bradshaw, A. D. (1978). *Nature* **276**, 493–494.
Huner, N. P. A., Williams, J. P., Maissan, E. E., Myseich, E. G., Krol, M., Laroche, A. and Singh, J. (1989). *Plant Physiology* **89**, 144–150.
Hunter, M. I. S., Hetherington, A. M. and Crawford, R. M. M. (1983). *Phytochemistry* **22**, 1145–1147.
Jenkin, L. E. T. and Ap Rees, T. (1983). *Phytochemistry* **22**, 2389–2393.
Knight, C. A., De Vries, A. L. and Oolman, L. D. (1984). *Nature* **308**, 295–296.
Loveys, B. R. and Kriedemann, P. E. (1974). *Aust. J. Plant Physiol.* **1**, 407–415.
Lurch, K. (1980). *Nature* **284**, 368–370.
Mansfield, T. A., Wellburn, A. R. and Moreira, T. J. S. (1978). *Phil. Trans. R. Soc. Lond.* **284B**, 471–482.
Milborrow, B. V. and Noddle, R. C., (1970). *Biochem. J.* **119**, 727–734.
Monk, L. S., Braendle, R. and Crawford, R. M. M. (1987). *J. expl. Bot.* **38**, 233–246.
Murata, N., Nishizawa, O. I., Higashi, S., Hayashi, H., Tasaka, Y. and Nishida, I. (1992). *Nature* **356**, 710–713.
Pridham, J. B. (1964). *Phytochemistry* **3**, 493–497.
Qureshi, J. A. K., Hardwick, K. and Collin, H. (1986). *J. Plant Physiol.* **122**, 357–364.
Raynal, J. (1973). *Adansonia* Ser. 2, **13**, 145–171.
Roberts, J. K. M., Callis, J., Jardetzky, O., Walbot, V. and Freeling, M. (1984). *Proc. natn. Acad. Sci. U.S.A.* **81**, 6029–6033.
Robinson, S. P. and Jones, G. (1986) *Aust. J. Plant. Physiol.* **13**, 659–668.
Rosas, A., Alberdi, M., Delseny, M. and Meza-Basso, L. (1986). *Phytochemistry* **25**, 2497–2500.
Sakai, A. (1960). *Low Temp. Sci. Ser.* **B18**, 15–22.
Sakai, A. (1961). *Nature, Lond.* **189**, 416–417.
Sakai, A. and Yoshida, S. (1968). *Cryobiology* **5**, 160–174.
Singh, T. N., Aspinall, D. and Paleg, L. G. (1972). *Nature New Biol.* **236**, 188–190.
Smirnoff, N. and Stewart, G. R. (1985). *Vegetatio* **62**, 273–278.
Smith, A. M. and Ap Rees, T. (1979). *Planta* **146**, 327–334.
Smith, R. A. H. and Bradshaw, A. D. (1970). *Nature, Lond.* **227**, 376–377.
Smith, T. A. (1965). *Phytochemistry* **4**, 599–607.

Stewart, G. R. and Lee, J. A. (1974). *Planta* **120**, 279–289.
Storey, R. and Wyn Jones, R. G. (1975). *Plant Sci. Lett.* **4**, 161–168.
Storey, R. and Wyn Jones, R. G. (1977). *Phytochemistry* **16**, 447–453.
Teal, G. (1973). Ph. D. Thesis, Univ. Reading.
Warburg, O. (1920). *Biochem. Z.* **103**, 108.
Wellburn, A. R., Ogunkanmi, A. B. and Mansfield, T. A. (1974). *Planta* **120**, 255–263.
Wright, S. T. C. and Hiron, R. W. P. (1969). *Nature, Lond.* **224**, 719–720.
Wyn Jones, R. G., Storey, R., Leigh, R. A., Ahmad, N. and Pollard, A. (1977). In: Marre, E. (ed.), 'Regulation of Cell Membrane Activities in Plants'. North Holland Press, Amsterdam.

2 Biochemistry of Plant Pollination

I. Introduction

When insects, bats and birds visit flowers to feed on (or collect for future consumption) the nectar and pollen, they usually pollinate the flowers in the process, so that both partners clearly benefit from this mutualistic association. There are three biochemical factors in this interrelationship; scent and colour of the flower and the nutritional value of nectar and pollen. As a pollinating animal approaches a flowering plant, one of the signals it receives is an olfactory one, from the flower scent. Animals live in a world of chemical communication, of pheromones, and they are undoubtedly able to detect the terpenes and other volatiles of flower odour at some distance. As the pollinator arrives near the plant, it also receives a visual signal, in the contrasting colour of the flower against the general green leafy background. As it alights on the flower, it may be drawn to the nectar by visual honey guides on the petal, derived from the differential distribution of pigments within the flower tissue. Finally, as it transfers the pollen from anther to stigma, it gains its reward, a nutritional one, based on the sugar and other constituents of nectar and pollen.

In spite of the great amount written on pollination ecology (e.g. Faegri and van der Pijl, 1979; Kevan and Baker, 1983; Proctor and Yeo, 1973; Real, 1983; Richards, 1978), biochemical aspects have rarely been explored in any detail. The present account is an attempt to gather most of the available information on this ecological topic. The subject of pollination biology is vast, largely because this interaction between plant and animal is such a complex and subtle one and also because almost every group of plants has its own method of attracting pollinators and there are an enormous number of morphological adaptations to the various animal pollinators available to plants. Some brief introduction to the subject is needed here, particularly regarding the range of animal pollinators, the varying roles of animal visitors in relationship to flower pollinating processes and the phenomenon of flower constancy.

To the casual observer in a flower garden in temperate latitudes, the pollination of the flowers would largely appear during daylight hours to be the province of the very active bumble and hive bee, with some help being provided by a few smaller insects. This ignores, of course, the much wider range of active pollinators in tropical habitats: the humming birds, an enormous variety of large tropical butterflies, the wasps and the beetles. In addition, some flowers are only pollinated at night by bats or moths. Also, there is occasional pollination by rodents, e.g. by mice and shrews in *Protea* spp. of South Africa and by bushrats in certain *Banksia* spp. in Australia. Finally, there are many smaller fauna, different kinds of flies and fleas, some of which are only apparent as pollinators to the most acute observer. The problem of determining which pollinator or pollinators are active on a particular plant species is difficult, requiring much time-consuming observation by the field naturalist. Some animals may visit flowers for other reasons than pollination; also they may be able to 'steal' the nectar, without carrying out the pollination necessary to the plant. Ants, for example, are well-known nectar thieves and are often so small that they sneak in and out of blossoms without touching the reproductive organs. They do, however, act as genuine pollinators in some cases. Hickman (1974) has shown that the small self-incompatible annual *Polygonum cascadense* is cross-pollinated by the ant *Formica argentea*. Reports of ant-pollinated plant species are, however, still few and far between (see Beattie, 1991).

The need of a plant to attract animals to visit it for purposes of pollination depends quite naturally on its sexual system and floral structure. There are some groups, e.g. the grasses, where pollination is by wind and animal visitations to the inflorescences would be superfluous. However, such angiospermous plant groups are relatively few and the majority of plants clearly require animals to achieve their pollination. This is obvious in plants with single sex flowers, particularly those that are dioecious, i.e. where the male and female flowers are on different plants. It is also obvious in self-incompatible hermaphrodite plants, which account for the majority of angiosperms. Self-incompatibility is essentially a system which ensures out-crossing and hence genetic variability and vigour within a plant population. There are immunological barriers to self-pollination and such plants depend on cross-pollination, i.e. insects travelling from flower to flower and unwittingly transferring pollen from the anther of one plant to the stigma of a second, in order to achieve seed set.

The evolution of the sexual system in the angiosperms has generally progressed from self-incompatibility to self-compatibility (see Crowe, 1964). However, even in those self-compatible species with large, coloured flowers (e.g. the sweet pea) where the floral morphology is such that self-pollination can occur without animal visitors, it is generally agreed that insects are beneficial in increasing seed set. This may be because pollinators increase the amount of self-pollen transferred to the stigma or because, when cross-pollen is available, it grows faster down the style than self-pollen. At least, the theory that many self-pollinated species still gain an advantage from animal pollinators explains why many such plants continue to produce large and brightly coloured petals and fragrant flower scents which attract bees and other visitors.

Finally, there is the phenomenon of flower constancy, a factor of great significance in the co-evolution of angiosperms and their animal partners. It represents the fidelity of a pollinator to regularly visit only a limited number of plant species and in extreme cases, only one. Such fidelity is guided by floral morphology, odour and petal colour. Indeed many plants through evolution of their floral parts have deliberately restricted themselves to pollination by one type of vector so that they have what are called 'bee-flowers' (with short, wide corollas), 'butterfly-flowers' (with medium-length, narrow corollas) or 'humming bird-flowers' (with long, narrow corollas). Animals on their part, within the range of plants they are capable of pollinating, become restrictive and dependent on a small number of species and eventually even a single plant. This may be because of a special blossom fragrance, a richness in nectar or some other lure. This mutual co-evolution has many benefits to both plant and animal. In extreme form, it can be seen in the fig genus, *Ficus*, where almost every species has its own species of chalcid wasp to pollinate it. Similarly, one finds examples in the Orchidaceae, where individual species of *Ophrys* depend on a single pollinator, an *Andrena* bee, to pollinate them. The case of the bee orchids will be considered in more detail in a later section.

II. Role of Flower Colour

A. Colour Preferences of Pollinators

Largely due to the work of von Frisch (1950) and others, much information is available about the colour preferences of bees. They are known to prefer what to us appear as blue and yellow colours. They can also discern differences in absorption in the UV region of the spectrum and are sensitive to the intensely UV-absorbing flavones and flavonols, which are present as such in practically all white flowers and also occur as co-pigments in cyanic flowers. Although bees are insensitive to red colours, they still visit some red-flowered species (e.g. red poppies) guided by the presence of UV-absorbing flavones, which are also present in these blooms.

Hive bees (*Apis mellifera*) are very catholic in the flowers they visit. They do, however, visit some plant families more than others. Families which have many species with typical bee blossoms include the Labiatae, Scrophulariaceae and Leguminosae; blue and yellow flower colour are common in these groups. Honey bees are also regular pollinators of Compositae, a family in which yellow is the dominant flower colour. Other types of bee are more restricted in their choice of flowers, notably those of the genus *Andrena*, which mainly visit orchids such as *Ophrys*.

Hive bees exhibit their colour preferences by visiting blue- and yellow-blossomed flowers if given a choice of other colours as well. Clearly, when nectar is in short

Table 2.1 Colour preferences of different pollinators

Animal	Flower colour preferences	Comments
Bats	White or drab colours, e.g. greens and pale purples	Mostly colour-blind
Bees	Yellow and blue intense colours, also white	Can see in UV, but not sensitive to red
Beetles	Dull, cream or greenish	Poor colour sense
Birds	Vivid scarlets, also bicolours (red-yellow)	Sensitive to red
Butterflies (Rhopalocera)	Vivid colours, including reds and purples	—
Moths (Heterocera)	Reds and purples, white or pale pinks	Mostly pollinate at night
Flies	Dull, brown, purple or green	Chequered pattern may be present
Mice	Whitish interiors, surrounded by dark reddish bracts	Pollination occurs at night time
Wasps	Browns	

Data modified from Faegri and van der Pijl (1979).

supply, bees will visit flowers with other colours (assuming the nectar is available to them) but such plants are at a selective disadvantage, e.g. in a bad summer when bee activity is limited. The operation of natural selection for bee colour can be seen in blue-flowered species (e.g. *Delphinium nelsonii*) which give rise to the white mutant forms in natural populations. Such mutants are unable to maintain themselves, seed set and viability being poor, because they are discriminated against by their pollinators (Waser and Price, 1981). In the case of the larkspur, *D. nelsonii*, discrimination occurs because white flowers have inferior nectar guides and therefore it takes longer for the pollinators to locate the nectar. The pollinators thus experience lower net rates of energy intake than on blue flowers, a sufficient reason for undervisitation by optimally-foraging animals (Waser and Price, 1983).

The colour preferences of other pollinators have been less well studied; present available data are collected in Table 2.1. Humming birds are sensitive to red and their preferences for bright scarlet blooms as in *Hibiscus* is well known. Tropical members of the Bignoniaceae, Gesneriaceae and Labiatae all have characteristic humming bird blossoms with red, orange-red or yellow-red colours. These birds do, on occasion, visit plants with white blooms in special habitats, e.g. in the Hawaiian forests. Some humming birds have brilliant scarlet plumage resembling the colour of the flower they pollinate. This is seen in the flower paraqueet *Loriculus* which feeds from scarlet *Erythrina* blossoms. This is a clear case of protective colouring, since these birds are most vulnerable to predators when hovering by the flower to collect nectar. Other birds which pollinate flowers, e.g. sun-birds and honey-birds, appear to have similar colour preferences as humming birds.

The other classes of pollinator (Table 2.1) show less sensitivity to flower colour. While butterflies are actively attracted to brightly coloured blossoms, moths and wasps prefer dull and drab colours. Finally, there are the beetles and bats, which are visually colour-blind, and which depend mainly on other sorts of signal to draw them to their host plants.

The colour preferences listed in Table 2.1 are only a very general guide to which pollinators are likely to visit a particular plant species. The adaptive significance of certain colours may not be directly attributable to the major pollinator of a species but rather to selection pressures exerted by other floral visitors. For example, it is possible that the scarlet red colour of hummingbird flowers originally evolved as a mechanism to diminish visits by bees, which are insensitive to red colours (Wyatt, 1983).

B. Chemical Basis of Flower Colour

Flower colour is largely due to the presence of pigments present in chromoplasts or cell vacuoles of floral tissues. Colours produced by the reflection and refraction of light from cell surfaces, so important in the animal kingdom, are not apparent in plants. Flower pigments have been widely studied, particularly from the genetical viewpoint and much information is now available about them (see e.g. Goodwin, 1988).

The most important group of flower pigments are the flavonoids, since they contribute cyanic colours (orange, red to blue) as well as yellow and white (Harborne, 1967, 1988). The only other major group are the carotenoids, which provide principally yellow colours, with some oranges and reds. Other classes of much less importance in relation to flower pigmentation are chlorophylls (greens), quinones (occasional reds and yellows), and betalain alkaloids (giving yellow, red and purple colours in Centrospermae). A brief summary of the chemical basis of flower colour is presented in Table 2.2 together with some indication of frequency and importance of the different pigment types.

In the case of cyanic colour, the chemical basis is simple. There are three main pigments, all members of the class of flavonoids known as anthocyanidins: pelargonidin (Pg) (orange-red), cyanidin (Cy) (magenta) and delphinidin (Dp) (mauve). These differ in structure only in the number (one, two or three) of hydroxyl groups in the B-ring (see Fig. 2.1). These three chromophores occur, usually singly or occasionally as mixtures, in angiosperm flowers and provide the whole range of colour from orange, pink, scarlet and red to mauve, purple and blue. Essentially, all pink, scarlet and orange-red flowers contain pelargonidin, all crimson and magenta flowers cyanidin and all mauve and blue flowers delphinidin.

A rare change in hydroxylation pattern is loss of the 3-hydroxyl. This happens infrequently, but when it does, it causes significant shifts to shorter wavelengths. Two such pigments are known, luteolinidin (3-desoxycyanidin) and apigeninidin (3-desoxypelargonidin) which are orange-yellow and yellow respectively (Fig. 2.1). These compounds occur in the New World Gesneriaceae (see p. 47) but hardly anywhere else.

A number of other chemical factors modify the basic anthocyanidin colours (Table 2.3); this is one of the reasons why such a variety of different shades and hues can be found in flowering plants. One of the modifying factors is methylation of one or more of the free hydroxyl groups in the three basic pigments. Only three methylated pigments are at all common: peonidin, petunidin and malvidin (see Fig. 2.1 for structures). While methylation has only a small reddening effect on colour, it is probably important in improving the

Table 2.2 Chemical basis of flower colour in angiosperms

Colour	Pigments responsible[a]	Examples[c]
White, ivory, cream	Flavones (e.g. luteolin) and/or flavonols (e.g. quercetin)	95% of white-flowered spp.
Yellow	(a) Carotenoid alone	Majority of yellows
	(b) Yellow flavonol alone	*Primula, Gossypium*
	(c) Anthochlor alone	*Linaria, Oxalis, Dahlia*
	(d) Carotenoid + yellow flavonoid	*Coreopsis, Rudbeckia*
Orange	(a) Carotenoid alone	*Calendula, Lilium*
	(b) Pelargonidin + aurone	*Antirrhinum*
Scarlet	(a) Pure pelargonidin	Many, inc. *Salvia*
	(b) Cyanidin + carotenoid	*Tulipa*
Brown	Cyanidin on carotenoid background	*Cheiranthus*, many Orchidaceae
Magenta, crimson	Pure cyanidin	Most reds, inc. *Rosa*
Pink	Pure peonidin	Peony, *Rosa rugosa*
Mauve, violet[b]	Pure delphinidin	Many, inc. *Verbena*
Blue	(a) Cyanidin + co-pigment/metal	*Centaurea*
	(b) Delphinidin + co-pigment/metal	Most blues, *Gentiana*
Black (purple black)	Delphinidin at high concn.	Black tulip, pansy
Green	Chlorophylls	*Helleborus*

[a] In the table and elsewhere, anthocyanidins are referred to as pigment chromophores; these pigments actually occur *in vivo* as glycosides (anthocyanins). The nature of the sugar, however, usually has little effect on colour.
[b] One group of ten families in the Centrospermae differ from all other higher plants in having alkaloidal betalains as their yellow and purple pigmentation.
[c] For details of the species concerned, see the text or Goodwin (1988).

stability of the anthocyanidin chromophore; methylated pigments are relatively common in the more highly specialized plant families. All anthocyanidins occur *in vivo* as glycosides (anthocyanins) and have sugars attached to the 3- or 3- and 5-hydroxyl groups. Sugar attachment is probably important (as is methylation) for pigment stability but generally has little effect on flower colour *per se* since glycosylation is the rule rather than the exception. In rare cases, sugar attachment at the B-ring hydroxyls of the anthocyanidin has been observed, giving, for example, cyanidin 3,5,3′-triglucoside which occurs in many bromeliads. This substitution does produce a small colour shift towards shorter wavelengths (Saito and Harborne, 1983).

One of the factors modifying cyanic colour (Table 2.3) needs special mention—presence of flavone and/or flavonol co-pigment. For many years, it was thought that co-pigmentation was a special effect, restricted to plants with blue flowers. The co-pigments were present to form weak complexes with the anthocyanidin, shifting the mauve or purple delphinidin colour to pure shades of blue. However, research (Asen *et al.*, 1972) has now demonstrated that for the full expression of the colour of all three common anthocyanidins—Pg, Cy and Dp—flavones or flavonols are needed to stabilize the pigment chromophore at the pH of the flower cell sap (around 4.5). This explains why, in fact, *all* cyanic flowers, not just those which are blue, contain both anthocyanidin *and*

The three common anthocyanidins

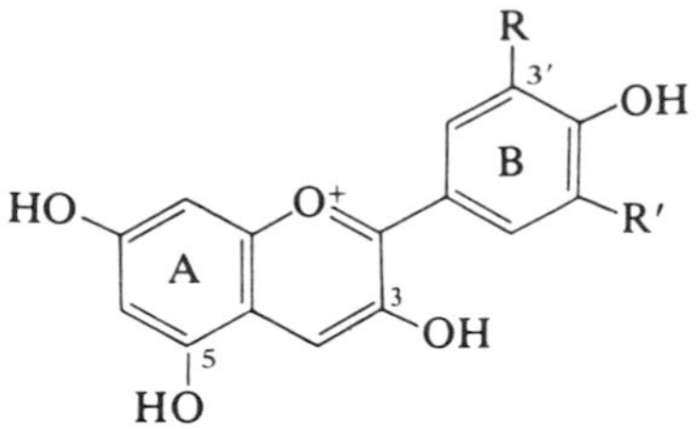

pelargonidin, R = R′ = H (510)
cyanidin, R = OH, R′ = H (525)
delphinidin, R = R′ = OH (535)

Common methylated pigments

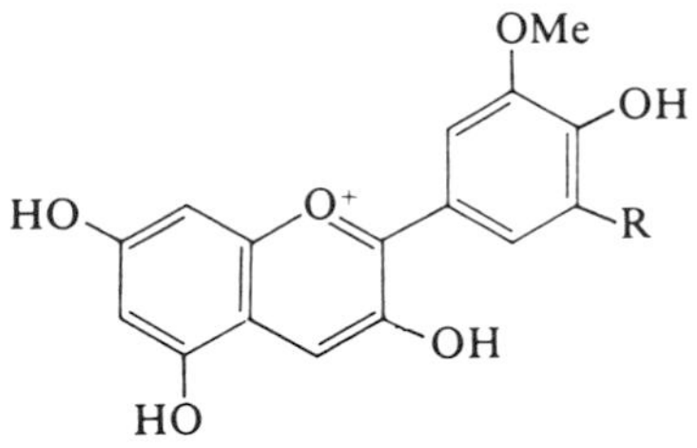

peonidin, R = H (523)
petunidin, R = OH (534)
malvidin, R = OMe (532)

Rare 3-desoxyanthocyanidins

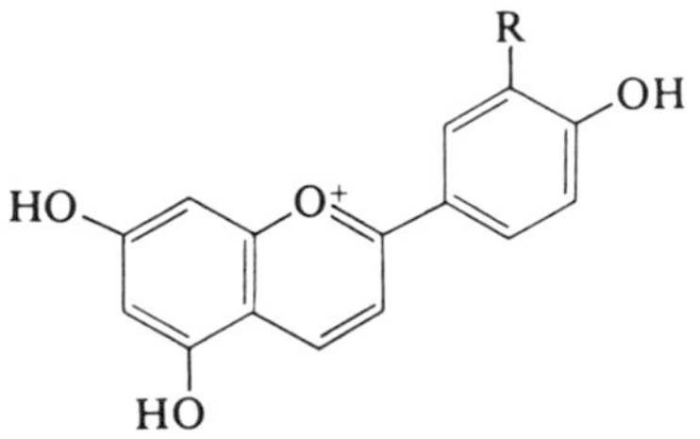

apigeninidin, R = H (477)
luteolinidin, R = OH (495)

Rare methylated pigments

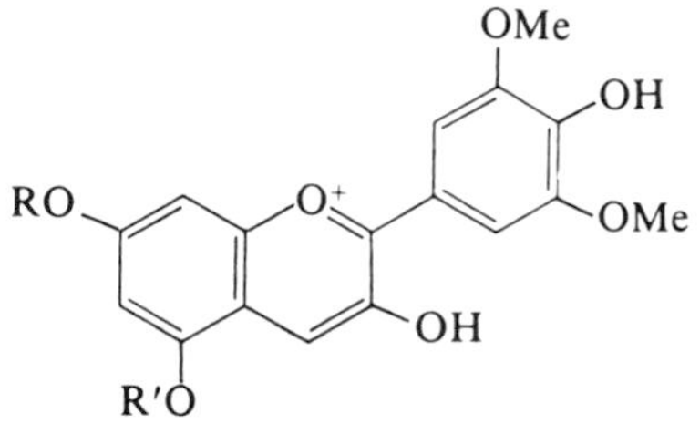

hirsutidin, R = Me, R′ = H (530)
capensinidin, R = H, R′ = Me (529)

Figure 2.1 Anthocyanidin chromophores of the angiosperms

Note: All pigments occur naturally with sugars attached (usually to the 3-OH) as anthocyanins. The values in parentheses refer to the visible wavelength maxima (in nm) of the different pigments in methanolic HCl.

Table 2.3 Factors controlling cyanic colour in flowers

1.[a] Hydroxylation pattern of the anthocyanidins (i.e. based on pelargonidin, cyanidin or delphinidin)
2. Pigment concentration
3. Presence of flavone or flavonol co-pigment (may have blueing effect)
4. Presence of chelating metal (blueing effect)
5. Presence of aromatic acyl substituent (blueing effect)
6. Presence of sugar on B-ring hydroxyl (reddening effect)
7. Methylation of anthocyanidins (small reddening effect)
8. Presence of other types of pigment (carotenoids have browning effect)

[a] In approximate order of importance. There are other minor factors, including pH, physical phenomena, etc.

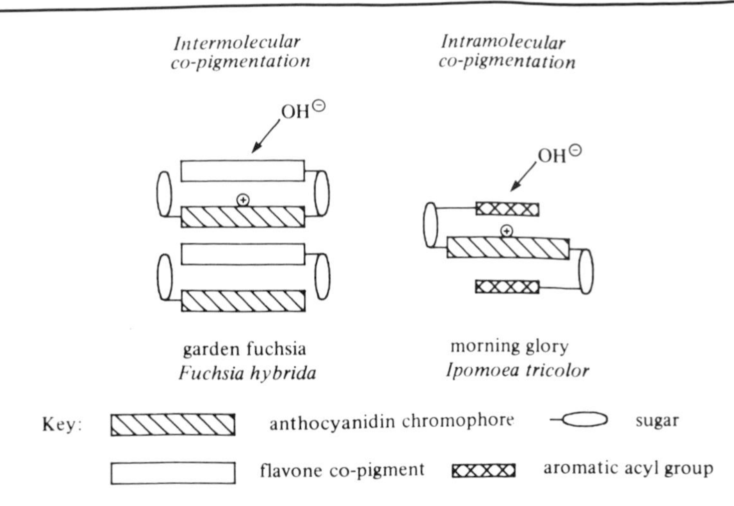

Figure 2.2 Two alternative forms of blue flower pigment where the anthocyanidin chromophore is protected from hydration through co-pigmentation

flavone or flavonol (as glycosides). Then, the blueing effect of flavones in blue flowers is simply due to an increase in the concentration of flavone; i.e. the anthocyanin/flavone ratio is decreased from that in mauve blooms. That this is so has been confirmed by directly comparing the spectra of pigments and co-pigments mixed in the test tube with those given by the pigments in the living flower. Not surprisingly, flavone co-pigments are located with the anthocyanins *in vivo*, usually in the cells of the petal epidermis (Kay *et al.*, 1981).

Other chemical features which are important in providing blue flower colour are the presence of aromatic acylation and the presence of chelating metal. It is now possible to recognize two forms of co-pigmentation, inter- and intramolecular, the first involving a loosely bound flavone, as in *Fuchsia*, and the second the presence of an aromatic hydroxycinnamic acid attached covalently through sugar to the anthocyanidin, as in *Ipomoea*. The acyl group is so arranged that it protects the anthocyanidin chromophore from hydrolytic attack in the same way as the hydrogen-bonded flavone co-pigment (Fig. 2.2). In addition to having co-pigments, some blue pigments occur *in vivo* associated with metal ions; in *Commelina communis*, the metal is magnesium, while in *Hydrangea* flowers it is aluminium.

There are a variety of ways that yellow colour may be produced in flowers (see Table 2.2). Most are due to carotenoids; almost all yellow and lemon-yellow carotenoid-containing flowers have mainly xanthophylls, such as zeaxanthin and its 5,8-epoxides auroxanthin and flavoxanthin. Deep orange flowers may have large amounts of β-carotene (e.g. the orange fringes of *Narcissus majalis*) or alternatively lycopene (*Calendula*) (for carotenoid structures, see Fig. 2.3). The carotenoids in petals are concentrated in the chromoplasts and may be present in bound form linked to protein or esterified with fatty acids.

Flavonoids make minor contributions to yellow colour, through three groups of pigment: yellow flavonols, chalcones and aurones (Fig. 2.4). Yellow flavonols such as gossy-

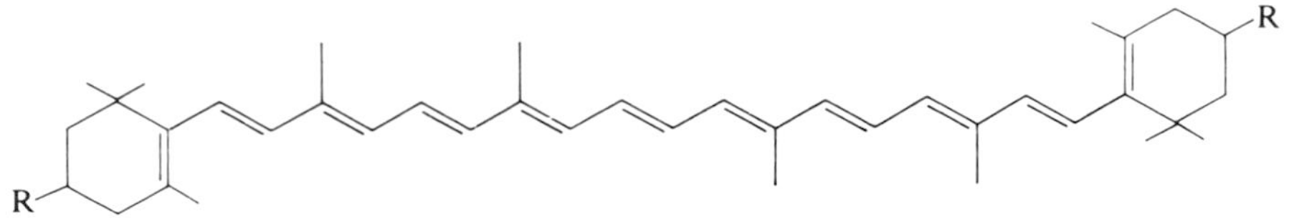

β-carotene, R = H (in daffodils) (451, 482)
zeaxanthin, R = OH (in tulips) (423, 451, 483)

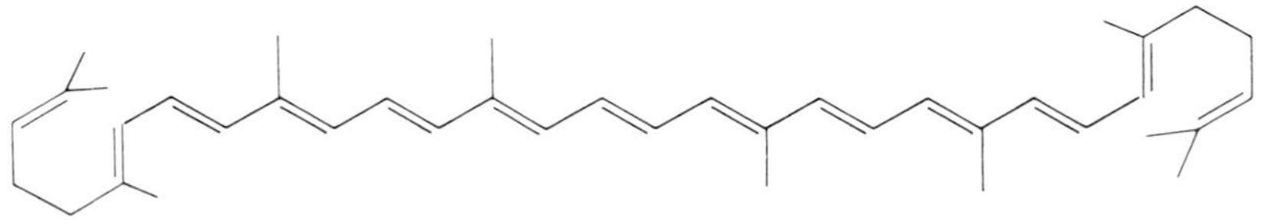

lycopene (in marigolds) (446, 472, 505)

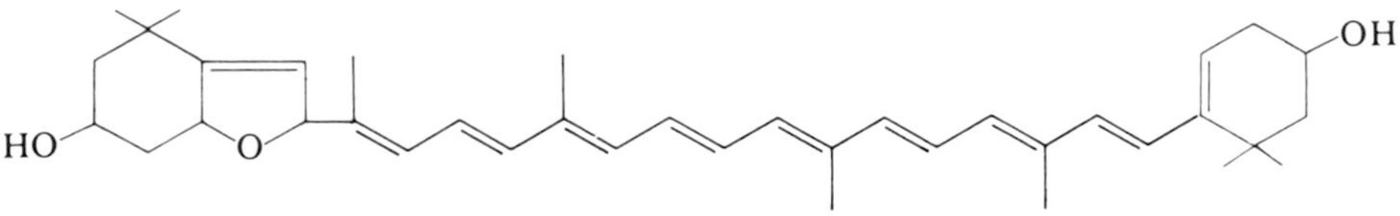

flavoxanthin (in yellow chrysanthemum) (432, 481)

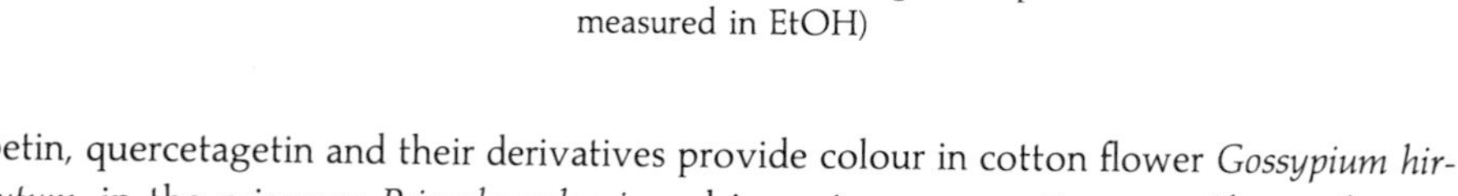

Glc–O–GlcO_2C ... CO_2Glc–O–Glc

crocein (crocus flowers) (411, 437, 458)

Figure 2.3 Some carotenoid pigments of yellow flowers (figures in parentheses = visual maxima measured in EtOH)

petin, quercetagetin and their derivatives provide colour in cotton flower *Gossypium hirsutum*, in the primrose *Primula vulgaris* and in various composites, e.g. *Chrysanthemum segetum*. Yellow flavonols owe their colours to the presence of an extra hydroxyl (or methoxyl) group in the 6- or 8-position of the aromatic A-ring of their structures. The related flavonols without this feature, e.g. quercetin (Fig. 2.5), are more or less colourless. Chalcones and aurones occur especially frequently in another group of composites, including *Coreopsis* and *Dahlia*, but do also occur sporadically in nine other plant families. They are distinguished from other types of yellow pigment in that when petals containing them are fumed with ammonia (or the basic smoke of a cigar) there is a colour change from yellow to red. Chalcones and aurones often occur together in flower petals and are collectively known as anthochlor pigments.

One other class of water-soluble yellow pigment needs to be mentioned: those based

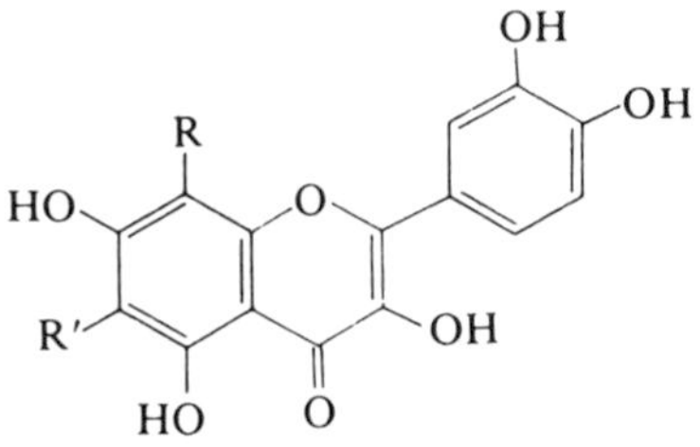

gossypetin, R = OH, R′ = H (388)
quercetagetin, R = H, R′ = OH (367)

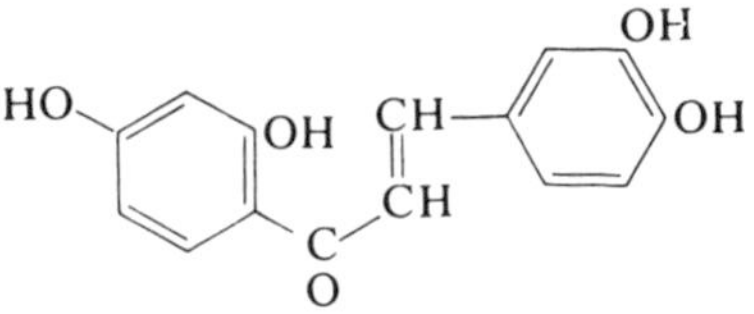

butein, chalcone of *Coreopsis* (382)

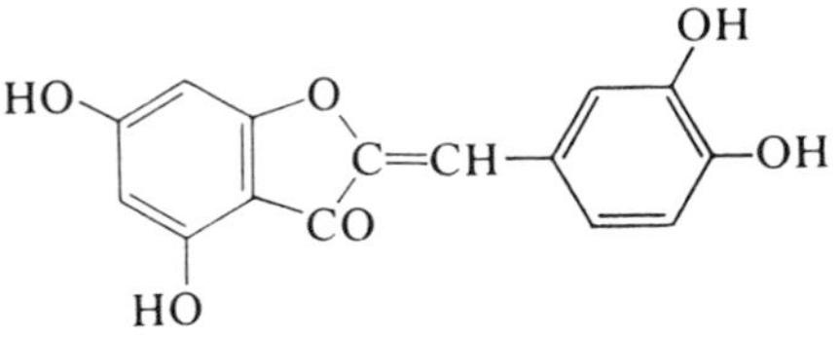

aureusidin, aurone of yellow *Antirrhinum* (399)

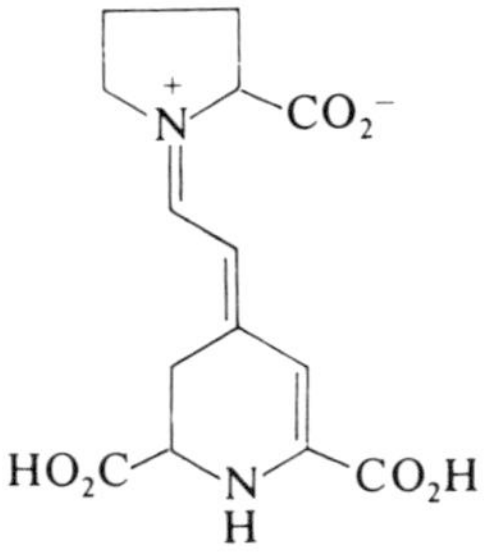

indicaxanthin, betaxanthin of *Opuntia ficus-indica* (480)

Figure 2.4 Yellow flavonoid and alkaloid pigments

on alkaloids. The well-known base berberine, for example, contributes yellow colour in *Berberis* tissues. One important class of yellow alkaloids are the betaxanthins of the Centrospermae. Within this order, all yellows are given by pigments such as indicaxanthin (Fig. 2.4) a betaxanthin based on the amino acid proline linked to a betalamic acid moiety. Eight other betaxanthins are known in the Centrospermae with aliphatic amino acids other than proline as part of their structures (Piattelli, 1976).

One final point may be made about yellow colour. Mixtures of two unrelated classes of

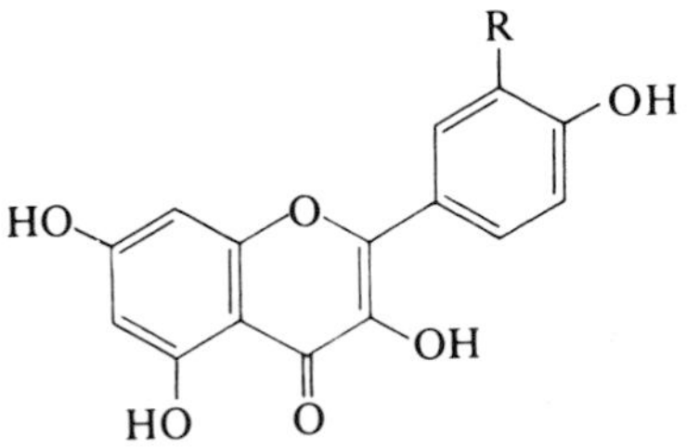

kaempferol, R = H (372)
quercetin, R = OH (374)

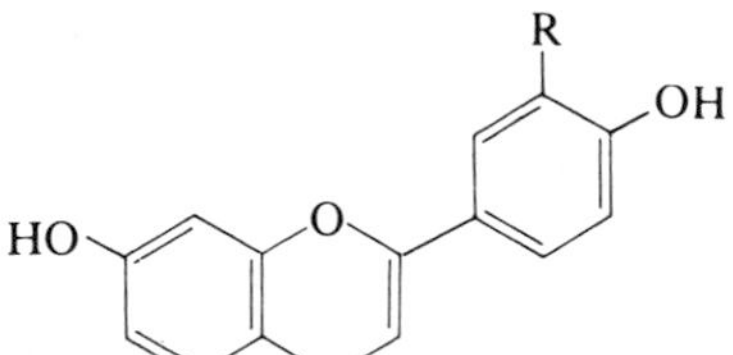

apigenin, R = H (335)
luteolin, R = OH (350)

Figure 2.5 Flavones and flavonols of white petals

yellow pigment are not infrequent in petals, especially of carotenoids and yellow flavonoids in members of the Compositae. It seems peculiarly wasteful in terms of biosynthetic potential for plants to produce two classes of compound to carry out the same function. However, the explanation for this apparent profligacy has been uncovered in relation to guide marks in petals, as will be discussed in a later section.

Finally, there are the compounds which occur in white flowers. They are scarcely colours to human eyes, appearing as pale cream or ivory in the petal. However, as already mentioned, they are clearly discernible by bees and other insects, which can perceive differences in absorption in the UV range of the spectrum. There are two classes: flavones such as luteolin and apigenin, and flavonols such as kaempferol and quercetin (see Fig. 2.5). There seems to be no particular advantage of one over the other, although the flavonols absorb at slightly longer wavelengths (at *c.* 360–380 instead of at 330–350 nm) than the flavones. In fact, flavones are more widely found in the flowers of more advanced plant families than are flavonols.

C. Evolution of Flower Colour

The distribution of cyanic coloration in angiosperms is by no means haphazard. There is a pattern in the relative frequency of delphinidin (Dp), cyanidin (Cy) and pelargonidin (Pg) types. The frequencies vary according to the flora sampled, and there is clear evidence of natural selection for particular colours in different environments, according to the most active pollinators which are present. Analyses of the results of pigment surveys show that selection has worked in two directions, from cyanidin as the basic or more primitive type (see Fig. 2.6). Loss mutations in tropical habitats produce scarlet and orange colours favoured by humming birds; by contrast, gain mutations in temperate climates produce blue colours favoured by bees.

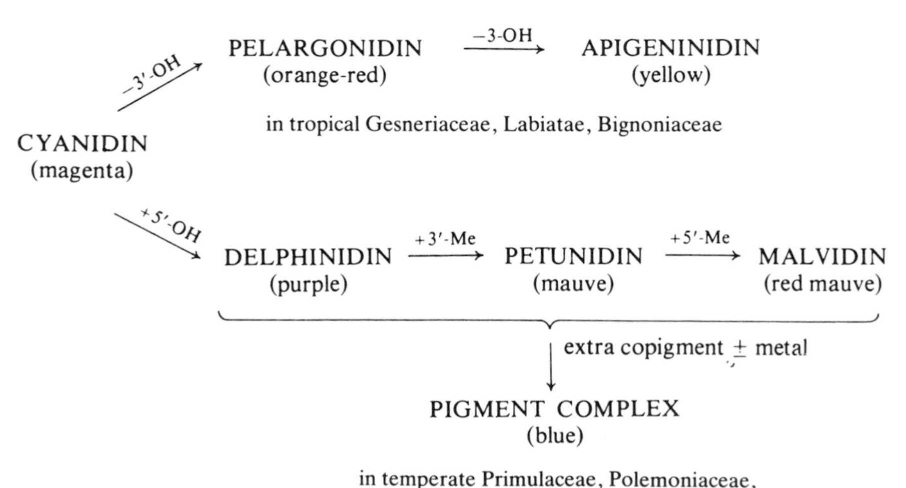

Figure 2.6 Directions of evolution in cyanic colour in angiosperms

Table 2.4 Differences in pigment chemistry of subfamilies of the Gesneriaceae

Subfamily[a]	Desoxyanthocyanin	Presence of yellow pigments Carotenoid	Chalcone	Quinone
New World Gesnerioideae	Present in 29/36 spp.	+	–	–
Old World Cyrtandrioideae	Absent from 0/50 spp.	–	+	+

[a] Generic coverage 74%; data from Harborne (1967).

Evidence that cyanidin is the most primitive pigment type is based on a variety of observations. It is the most common type in the ancestors of angiosperms, i.e. the gymnosperms. It is the major pigment of wind-pollinated groups such as the grasses, where clearly selection for flower colour will not operate. It is also the most common pigment found in tissues more primitive than the flower such as the leaf.

Evidence that pelargonidin is advanced over Cy is based on its regular occurrence in tropical plants and almost complete absence from temperate floras. The further loss mutation to give the 3-desoxyanthocyanidins luteolinidin and apigeninidin (Fig. 2.6) only appears in very advanced angiosperm families, such as the Gesneriaceae and Bignoniaceae. In the former family, 3-desoxyanthocyanidins are clearly restricted to the tropical American New World species and are completely absent from the Old World taxa in the family. This difference in cyanic pigmentation is also correlated with differences in the type of yellow pigmentation in the two geographical groups (Table 2.4).

Finally, evidence that delphinidin and its derivatives are advanced over cyanidin is drawn from distribution patterns in the angiosperms and especially the frequent presence of Dp in advanced bee families, such as the Scrophulariaceae, Boraginaceae, Hydrophyllaceae and Polemoniaceae.

The situation illustrated in Fig. 2.6 is only an evolutionary trend and clearly there will be some exceptions. The position of Cy itself is to some extent ambiguous since with suitable modifying factors it can under different circumstances provide the basis of scarlet colours (e.g. in *Tulipa*, see Table 2.2) or the basis of blue colours, as in the cornflower, *Centaurea cyanus*. However, if one takes the other two pigment types, Pg and Dp, it seems that humming bird flowers almost never have delphinidin, and bee flowers almost never have pelargonidin; such cases, if they exist in nature, are very rare.

The effects of the evolutionary trends portrayed in Fig. 2.6 can be gauged also, by comparing the frequencies of Pg, Cy and Dp types in various floras. In the Australian flora, the relative frequencies based on analyses of wild plant species are Dp 63%, Cy 47% and Pg 2%. The remarkable scarcity of Pg types is presumably due at least in part to the infrequency of species with bird-pollinating mechanisms in the flowers. Also the bird fauna of Australia is different from that in tropical America and there may be different colour preferences. Where bird pollination occurs, the mechanism is often distinctive as in those species of the Myrtaceae which have bright red inflor-

Table 2.5 Correlation between anthocyanidin type, flower colour and pollinator in Polemoniaceae

Plants	Flower colour	Petal anthocyanidin
HUMMING BIRD-POLLINATED SPECIES		
Cantua buxifolia	Scarlet	Cy
Loeselia mexicana	Orange-red	Pg
Ipomopsis aggregata ssp. *aggregata*	Bright red	Pg
I. aggregata ssp. *bridgesii*	Red to magenta	Pg, Cy
I. rubra	Scarlet	Pg
Collomia rawsoniana	Orange-red	Cy
BEE-POLLINATED SPECIES		
Polemonium caeruleum	Blue	Dp
Gilia capitata	Blue-violet	Dp
G. latiflora	Violet	Dp/Cy
Eriastrum densifolium	Blue	Dp
Langloisia matthewsii	Pink	Dp/Cy
Linanthus liniflorus	Lilac	Dp/Cy
LEPIDOPTERA-POLLINATED SPECIES		
Phlox diffusa	Pink to lilac	Dp/Cy
P. drummondii	Pink to violet	Dp/Cy (Pg)
Ipomopsis thurberi	Violet	Dp
Leptodactylon californicum	Bright rose	Dp/Cy
L. pungens	Pink to purple	Dp/Cy
Linanthus dichotomus	Reddish-brown	Cy

Key: Pg = pelargonidin; Cy = cyanidin; Dp = delphinidin.

escences arranged like the bristles of a bottlebrush. On the other hand, the high figure for Dp suggests that pollination by insects attracted to mauve and blue colours must be especially common.

Figures from the contrasting tropical flora of the West Indies are also available. Here a sampling based on both wild and introduced species (and hence not entirely representative of the natural habitat) gave the results of Dp 47%, Cy 70% and Pg 17%. The high figure for Pg and to some extent that of Cy is undoubtedly because bird pollination is a widespread feature in this flora (see van der Pijl, 1961).

The dichotomous nature of evolutionary trends in cyanic colours can also be seen at work in families which have both tropical and temperate members. One of the best examples here is the Polemoniaceae, a family restricted to the New World but present both in northern temperate areas as well as central tropical habitats. The animal pollinators of these plants have been exhaustively studied by Grant and Grant (1965). In their monograph on the family, these authors include two colour plates of typical Polemoniaceae species pollinated by humming birds and by bees respectively. There is a remarkable contrast in colour and flower shape in these plates. The humming bird flowers have long, narrow corollas, mostly scarlet. The bee flowers have wide open short corollas, nearly all blue in colour.

Yet other flowers in the family (not illustrated) have shorter narrow corollas, mauve or pink in colour and are butterfly pollinated. Analyses of the anthocyanidins present in petals of 18 representative members of the Polemoniaceae (Table 2.5) show a clear-cut correlation between flower colour, anthocyanidin type and pollinator (Harborne and Smith, 1978a). Thus humming bird species all have pelargonidin, with but one exception, while bee- or bee/fly-pollinated species all have delphinidin. On the other hand, lepidopteran species are intermediate in containing mainly cyanidin or mixtures of cyanidin and delphinidin.

Similar differences in flower colour types have been observed between tropical and temperate members of the Labiatae. Pelargonidin was found in all six scarlet-flowered species surveyed, cyanidin in 17 species with red-purple flowers and delphinidin in 26 species with violet or blue flowers which are bee pollinated (Saito and Harborne, 1992). The results with the Gesneriaceae, where the main geographical difference is Old World/New World, have already been given in Table 2.4.

Evolutionary changes in flower colours can also be observed at the species level. Plants may have to switch their flower colours within a generation or two in order to adapt to changes in pollinators. Baker and Hurd (1968) have pointed out the considerable differences in dominant flower colour that can exist between two habitats adjacent to each other. In the northern Californian flora, herbaceous species growing in the open prairie are pollinated by bees and have yellow flowers. Close by in the dark Redwood forest, the plants are pollinated by moths and have white or pale pink flowers. Any species migrating across the border from one habitat to another would have to switch flower colour rapidly in order to adapt to the new environment. Species known to be variable in their flower colour (e.g. members of the *Viola* genus) are presumably in a better position than most to achieve emigration from one contrasting habitat to another in this way.

Undoubtedly plant species have considerable flexibility in being able to cease, modify or re-establish anthocyanin synthesis in the flowers depending on what pollinators are present or whether any pollinators are available. In this respect, it is significant that two autogamous species of Polemoniaceae that were analysed, *Allophyllum gilioides* and *Microsteris gracilis* (Harborne and Smith, 1978a), retained the anthocyanin pigments of their out-breeding relatives, in spite of the fact that animal pollination is not needed in these self-compatible species. Thus, the possibility remains in the make-up of these plants of returning to out-breeding in some future generation.

The ability of plants to respond rapidly to changing pollinators is nicely illustrated in another polemoniad, the scarlet gilia *Ipomopsis aggregata*. In populations growing near Flagstaff, Arizona, it has been observed that a minority of plants shift in flower colour during the season from red, through shades of pink to white (Paige and Whitham, 1985). The shift is correlated precisely with the coincident southern emigration of humming birds, which are primary pollinators in mid-July and the need to be attractive to the remaining pollinator, a hawkmoth *Hyles lineata* (Fig. 2.7). Colour shifting, which can occur within the same inflorescence, involves diluting the amount of anthocyanin formed in the petal (see Table 2.5) and eventually turning off anthocyanin synthesis altogether.

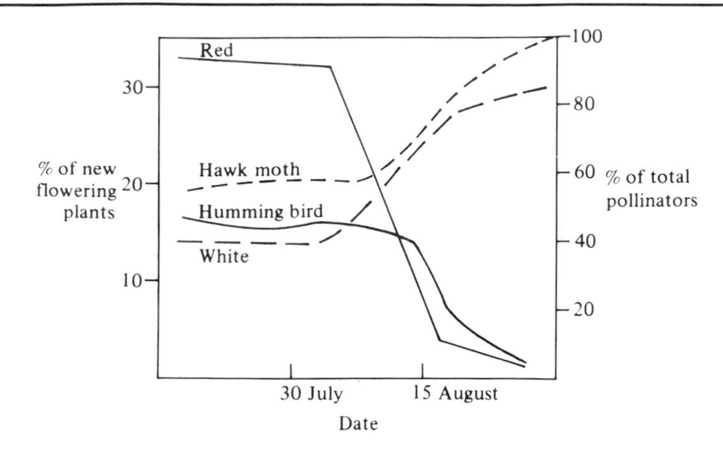

Figure 2.7 Colour shifting in *Ipomopsis aggregata* correlates with change in pollinator

Colour shifting in *Ipomopsis* provides a mechanism for ensuring that the pollinators present at a particular time are most effectively attracted to visit the flower. Another plant which benefits from the visits of two pollinators is *Pedicularis* (Scrophulariaceae) (Macior, 1986). Here the flower has scarlet corollas to attract humming birds and magenta calyces and bracts with high UV-reflective hairs which attract bumble bees. While younger flowers are bee pollinated, the older flowers are bird pollinated. Instead of a shift in colour, there is an increase in nectar sugar (from 15 to 25%) as the flower ages, which correlates with the change in pollinator.

One way that flower colour may be modified is by hybridization, but this may not always be advantageous. In *Penstemon* for example, there is a red humming bird-pollinated species which will hybridize with a blue carpenter bee-pollinated species when growing sympatrically (see Grant, 1971). The hybrid is purple flowered and attracts yet another pollinator, a wasp. The purple intermediate is pigmented by the same delphinidin as the blue form, but with less co-pigment being present; delphinidin would be expected to be dominant to the pelargonidin present in the red-flowering species (Beale, 1941). Such a purple hybrid may not always be fortunate enough as in this case to attract its own pollinator. It clearly could be at a selective disadvantage, since it may not be able to attract either of the pollinators of the parental plants. In other situations, it might then have to revert back quickly to the flower colour of one or other parent. This could clearly be a limiting factor in the success of hybrids in evolving plant populations.

Flowers may change colour after being pollinated, e.g. from yellow (carotenoid) to red (anthocyanin) as occurs in *Lantana camara*. This is triggered by the removal of nectar from the nectary by the pollinator (Eisikowitch and Lazar, 1987). Such colour changes are beneficial to both parties. Thus, they direct the pollinator towards the unvisited (yellow) flowers and improve the efficiency of pollination and nectar gathering. Equally, the retention of the pollinated (red) flowers in the inflorescence increases the attractiveness of the flowers from a distance. These colour changes have now been recorded in at least 74 angiosperm families and the majority of insect pollinators readily discriminate between the different floral colours at close range (Weiss, 1991). Although colour changes may be of several kinds, the biochemistry of the switch in pigment synthesis has yet to be investigated.

D. Honey Guides

Honey guides or guide marks are part of the pigment patterning in flowers, their object being to guide pollinating insects to the centre, where the sex organs and nectar are present. They are particularly prominent in bee flowers and take a variety of forms. Many are visible to the human eye and may be a colour contrast—a yellow spot on the lip of an otherwise blue flower, as in *Cymbalaria muralis*. They also may take the form of coloured dots or lines on the corolla tube. Recent research has shown that some honey guides in yellow flowers are invisible to the human eye, but can be detected by insects due to their intense absorption in the UV; this work has created a new impetus in the study of honey guides in flowering plants.

Visible honey guides are often produced by the local concentration of anthocyanin pigmentation in particular areas of the corolla (see Fig. 2.8). This is true in the foxglove, *Digitalis purpurea*, which has a pink bell-shaped corolla, pigmented by cyanidin, with a series of concentrated areas of the same pigment in the inside of the bell drawing the insect to the stigma and style. In *Streptocarpus* species, a similar situation exists except that the honey guides take the form of lines of pigment inside the tubular corolla.

A slightly different situation holds in the genus *Papaver* where honey guides generally take the form of pigment blotches on the petals (Fig. 2.8). In this case, the pigment in the blotch (cyanidin as the 3-glucoside) is different from that present in the rest of the petal, which is cyanidin as the 3-sophoroside (in *P. rhoeas* or pelargonidin as the 3-sophoroside in *P. orientale*).

The first biochemical evidence that honey guides invisible to the human eye occur in flowers was provided by Thompson *et al.* (1972) in a variety of *Rudbeckia hirta* called Black Eyed Susan. In daylight the petals of this composite are uniformly yellow. However, in UV light, the outer parts of the ray are UV-reflecting and bright, while the inner parts are dark-absorbing (see Fig. 2.9). Chemical analyses reveal that carotenoid pigment

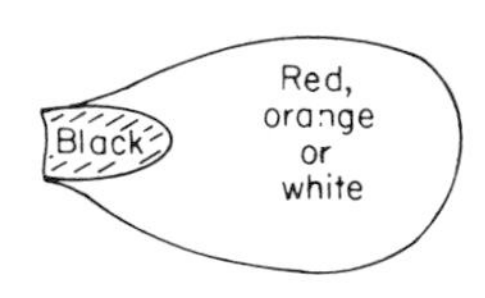

Papaver
superimposition of cyanidin in high concn. on background of cyanidin or pelargonidin

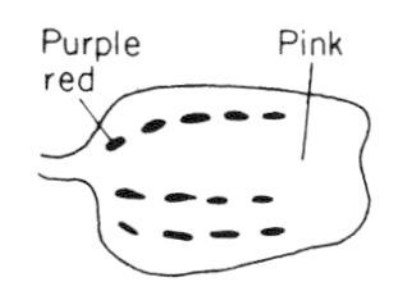

Digitalis
local concentration of corolla pigment (cyanidin) in spots

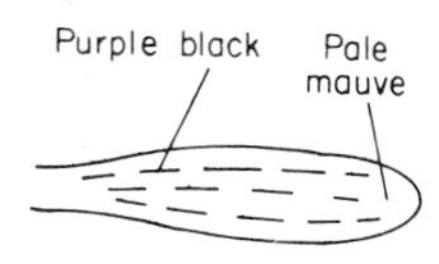

Streptocarpus
local concentration of corolla pigment (malvidin based) in lines

Figure 2.8 Some visible honey guides in cyanic flowers

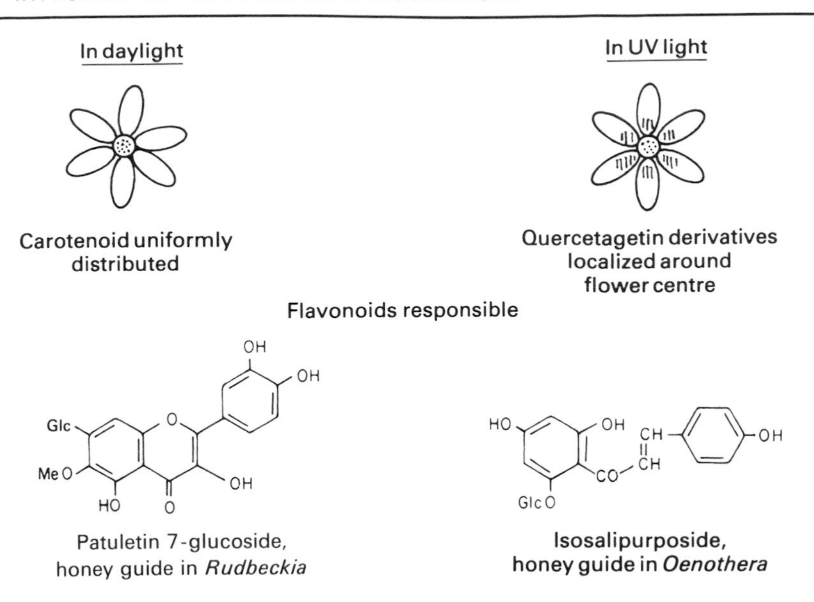

Figure 2.9 Invisible honey guides in yellow flowers

is responsible for the UV reflection of the outer ray and also that this lipid pigment is uniformly distributed throughout the ray. In the inner dark-absorbing zones of the ray, there is a different kind of pigment. In fact, three water-soluble yellow flavonols are present, including especially a derivative of patuletin (see Fig. 2.9).

Thus, in *Rudbeckia*, there is separation of function of the two types of yellow pigment present. The carotenoid provides the general yellow flower colour in the plant, in order to attract the bee from a distance. On the other hand, the water-soluble yellow flavonols, differentially present only in the inner ray, act as a UV honey guide, directing the UV-sensitive bee once it has landed on the flower head to the nectar in the centre of the blossom. This explains why many highly evolved plant species tend to have two types of yellow pigment in the flower; the two types clearly have different functions (see p. 46).

Invisible honey guides can be readily detected in yellow-flowering species by looking at flower heads under UV light and then by confirming the result by photography, using an appropriate filter. Such detection can actually be done on plants from herbarium sheets. However, it is also vital to actually identify that both carotenoids and yellow flavonoids are present in the flower and this can only be done on fresh flowers. Herbarium surveys have shown that UV guides probably occur in a number of other yellow-flowering composites, particularly those in Heliantheae, the same tribe as *Rudbeckia* (Eisner *et al.*, 1973). Honey guides have been confirmed in *Eriophyllum* and *Helianthus* spp. by pigment analyses (Harborne and Smith, 1978b). Detection of both yellow carotenoid and yellow flavonoid in a flower does not *a priori* mean that UV guides are present. We have found that in another tribe of the Compositae, in the Anthemideae, there are many yellow-flowered species with both carotenoid and yellow flavonols (Harborne *et al.*, 1976) but there is no evidence of honey guides in them.

Other types of yellow flavonoid can contribute to honey guides. In *Oenothera* (Onagraceae), Dement and Raven (1974) have shown that the chalcone isosalipurposide is responsible for UV honey guides in these flowers. Similarly, Scogin and Zakar (1976) have found that both chalcones and aurones provide UV absorption patterns in flowers of *Bidens* (Compositae). These patterns are not universally present but occur in five of the seven sections of the genus; there is also variation in how far along the ray the absorbing pigments extend.

There is no necessity for the flavonoids in such flowers to be visibly yellow (although it is obviously a more efficient mechanism), since all flavonoids, irrespective of their visible absorption characteristics, absorb strongly in the UV. Indeed, colourless flavonol glycosides have recently been recognized to provide patterning in some species. Thus in yellow flowers of *Coronilla* (Leguminosae), yellow gossypetin derivatives provide differential UV absorption in the wings of the petal of *C. valentina*, but kaempferol and quercetin glycosides take over this role in the wings of *C. emerus* (Harborne and Boardley, 1983). Similarly, in petals of *Potentilla* (Rosaceae), the yellow chalcone isosalipurposide provides honey guides in six species, while quercetin glycosides are responsible for UV patterning in eight other species (Harborne and Nash, 1983). Honey guides have also been observed in some white flowers (Horovitz and Cohen, 1972) and in such cases there may well be differential distribution of colourless flavonols in these petals too.

III. Role of Flower Scent

A. Types of Scent

The odour or scent of a flower often plays a major role as attractant to pollinating insects in the angiosperms. Bees are especially responsive to flower scents which we would describe as fragrant or 'heady' and many bee flowers are scented, e.g. the garden violet and other *Viola* species. Odour is of special importance in night-flying insects and other animals, where visual stimulus is practically absent; bat-pollinated and moth-pollinated flowers are generally strong smelling.

Because of the sensitivity of insects to small concentrations of volatile chemicals, flower odours are probably effective at relatively low concentrations. Many species which do not appear to be strongly scented to human senses may, in fact, produce sufficient odour to attract bees or butterflies. In many species, maximum scent production is co-ordinated with the time when the pollen is ripe and the flower is ready for pollination. Diurnal variations in production also occur so that scent is produced for day-time pollinators at noon, for night-time pollinators at dusk.

Floral scent may attract a pollinator to the flower for other purposes than food, as happens in the case of the primitive angiosperms *Zygogynum* and *Exospermum*, which are pollinated by a moth of the genus *Sabatinca*. These moths use the flowers, which blossom for only a few days, as a mating site. The fragrance, a mixture of terpenes and aliphatics (see Fig. 2.10), also contains large quantities of ethyl acetate, with its peardrop odour. This latter compound appears to improve the effectiveness of pollination by making the moth drowsy (Thien *et al.*, 1985).

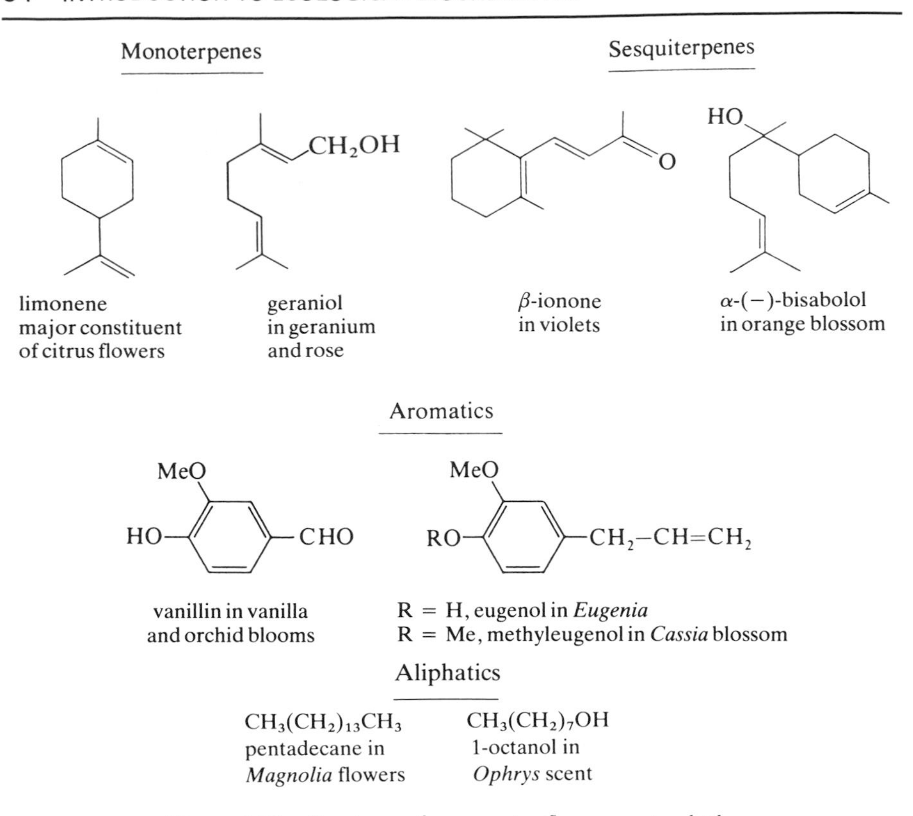

Figure 2.10 Structures of some major flower scent volatiles

Other plant tissues besides the petals give off scents. Indeed, many labiates and other plants have special scent glands on the leaf surfaces which are full of volatile oils. It is not clear in such cases whether leaf odours contribute at all to the attraction of pollinating vectors. Certainly higher animals may be sensitive to leaf odours of the Liabiatae—witness the well-known attraction of the domestic cat for the catmint, *Nepeta cataria*.

From the viewpoint of the human observer, flower scents broadly fall into two classes: those that are pleasant, fragrant or fruity; and those that are distinctly unpleasant or aminoid. While we can make such a classification for our own benefit, the pollinator concerned is clearly attracted to the scent whatever its particular quality to the human nose. Pleasant odours are generally contained in the 'essential oil' fraction of the flower, that part which can be separated by steam distillation or ether extraction and is volatile. Within the essential oils, a range of organic compounds may be present, the majority being mono- or sesquiterpenes. Volatile aromatic substances may be present, as well as simple aliphatic alcohols, ketones and esters. Typical structures of some flower odoriferous principles are illustrated in Fig. 2.10. In some cases, a major constituent may be responsible for a particular flower scent but, more usually, a mixture of components

Monoamines

CH_3NH_2	methylamine
$CH_3CH_2NH_2$	ethylamine
$CH_3(CH_2)_2NH_2$	propylamine
$CH_3(CH_2)_3NH_2$	butylamine
$CH_3(CH_2)_4NH_2$	amylamine
$CH_3(CH_2)_5NH_2$	hexylamine

Diamines

$NH_2{-}(CH_2)_4{-}NH_2$	putrescine
$NH_2{-}(CH_2)_5{-}NH_2$	cadaverine

Indoles

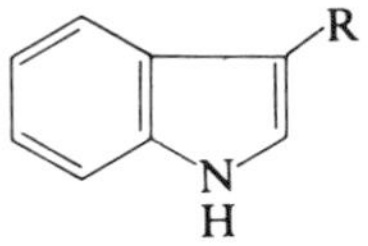

R = H, indole
R = Me, skatole

Figure 2.11 Unpleasant amines in plant odours

provide the scent. An important factor in scent production is that one component may reinforce the effectiveness of a second and third in producing a characteristic odour.

Flower scents have, of course, been utilized for many years in human society as perfume. While most modern perfumes are synthetic in origin, natural flower scent extracts still have an importance for boosting the effectiveness of synthetic mixtures. Roses are still cultivated in Bulgaria for their scent. Modern research by perfumers has shown that even the simplest flower scent may have many, indeed a hundred or more, constituents. In view of this complexity, the way in which many plant species have recognizably different flower scents may readily be appreciated. Some idea of the range of substances detected in flower scents can be gathered from the recent monograph by Kaiser (1993) on orchid volatiles.

Unpleasant aminoid odours in plants have, perhaps not unnaturally, been poorly studied. Our knowledge of the chemistry of highly repulsive and distasteful plant odours is therefore limited. Three typical plants with obnoxious odours are the hogweed, *Heracleum sphondylium*, stinking hellebore, *Helleborus foetidus*, and cuckoo pint, *Arum maculatum*. Other examples occur in the families to which these species belong, especially the Umbelliferae and Araceae. Unpleasant smells, in fact, represent a chemical mimicry by which the plant produces the smell of decaying protein or faeces to persuade carrion and dung insects to transfer their attention to the flower heads. Indeed, the chemicals produced are very similar to those given off by carrion or dung.

Major constituents of aminoid plant odours are monoamines, which have unpleasant fishy smells. They are fairly volatile and range from methylamine to hexylamine (Fig. 2.11). Some free ammonia may also be present. Even more offensive to some are the two diamines, putrescine and cadaverine, which as their names imply are characteristic

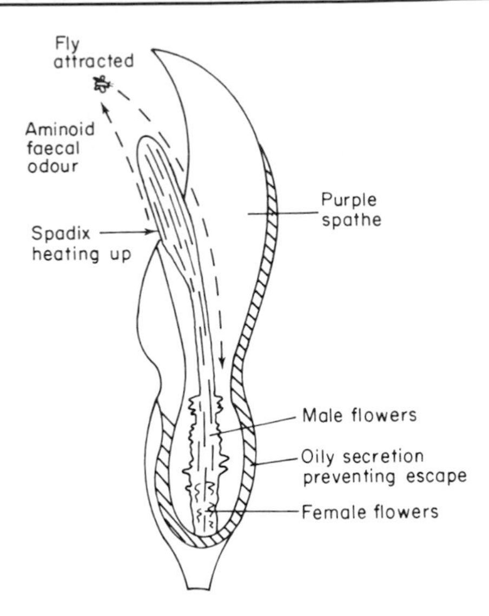

Figure 2.12 Biochemistry of pollination in *Arum*

breakdown products of decaying protein. Putrescine and the monoamine isobutylamine have, for example, been reported in *Arum* floral odours. Other unpleasant compounds that may be present are skatole and indole (which have faecal odours) and odd-chain aliphatic organic acids, such as isobutyric acid with its rancid smell.

The way by which plants use such smells as a trap for insects has been investigated in some detail (see Faegri and van der Pijl, 1979). In *Arum nigrum* and *A. maculatum*, for example, the bright purple spathe opens overnight to reveal the spadix in which respiration is unusually rapid and temperatures of 30°C have been measured (Fig. 2.12). The heat thus generated in the spadix aids the volatilization of the amine into the objectionable odours which are then released. Dung beetles and flies attracted by the amines alight on the spadix, and fall into the bottom of the flower, where they are trapped. The insects cannot escape because of the slippery surface of the inner spathe and are kept prisoner for 24 h, during which time they transfer pollen to the receptive styles; rapid anatomical changes then occur (including wrinkling of the spathe surface) and the insect is eventually released. The generation of heat in the spadix is a truly remarkable feature of these plants and it must increase the effectiveness of the faecal odours. The rapid respiration that occurs uses up a quantity of starch in the process but presumably this is offset in terms of metabolic efficiency by the fact that only small amounts of nitrogen compounds are needed in the scent.

In the related voodoo lily, *Sauromatum guttatum*, the trigger for heat production and volatilization of the aminoids has been identified as salicylic acid (2-hydroxybenzoic acid), the concentration of which rises 100-fold in the upper spadix some 12 h beforehand (Raskin *et al.*, 1987). Production of heat and faecal stench reaches a maximum between 3 and 5 h after dawn on the day of pollination and drops back to normal by late afternoon.

A second heating phase in the following night, also triggered by salicylic acid, stimulates the flies trapped in the floral chamber to carry out the pollination (Diamond, 1989).

B. Insect Pheromones and Flower Scents

Insect behaviour is known to be controlled by chemical signals, which take the form of volatile organic constituents released by one insect to affect another. These substances are active in very small amounts and have been termed 'pheromones', to indicate their relationship to hormones. Pheromones are involved in almost every aspect of insect life: feeding, sex, aggregation, oviposition, defence and laying trails. Chemically, many of the pheromones are simple aliphatic alcohols, acids or esters; others are closely related to the plant scents in being terpenoid in nature. A more detailed account of insect pheromones is given in Chapter 8, Section II.

Pheromones will be discussed *inter alia* in several later chapters of this book. Mention of them here is pertinent, since the action of a flower in producing a fragrant scent to attract a pollinator can be a similar signal to that of a pheromone released by one insect to attract another. Indeed the signals may sometimes get crossed, with interesting results, as will be mentioned below.

Because insects depend on volatile compounds for social communication, they can clearly become sensitive to similar molecules which may be present in flower scents. Plants may occasionally deceive insects by producing attractive odours to trap them (see above) or to draw them away from more rewarding pursuits (e.g. feeding). Insects 'learn' to recognize the smells of individual flowers and it is this factor, perhaps more than any other, which is responsible for the phenomenon of flower constancy, where insects limit their attention to a few or only one plant species. It has even been suggested that some plants produce hallucinogenic or narcotic substances in their scents, so that insects become 'hooked' on them and a close symbiotic relationship may develop. Such a floral reward appears to operate in *Datura inoxia*, where the nectar presumably contains the typical hallucinogenic tropane alkaloids of this genus. Grant and Grant (1983) have observed that the hawkmoth pollinators are erratic in flight after visiting the flowers at dusk and show all the signs of being hooked on this 'fix'.

Three examples will now be given where pheromones and flower scents have become interwoven in insect behaviour. The first refers to the oriental fruit fly, *Dacus dorsalis*, which has the phenylpropanoid eugenol methyl ether (see Fig. 2.10) as sex pheromone. Pheromonal activity is exhibited both in feeding and in male aggregation. This same compound is produced in the blossoms of several plants, especially in the golden shower tree *Cassia fistulosa* (Leguminosae), where its prime purpose is to attract pollinators. Methyleugenol is also incidentally produced in the volatile leaf oils of other species, and in *Zieria smithii* (Rutaceae) it constitutes as much as 85% of the total essential oil. Because of the striking attraction of this chemical to the fruit fly, it has been proposed that *Cassia* blossoms or *Zieria* leaves be employed in insect traps, e.g. when *Dacus* develops to pest proportions in fruit orchards.

This fruit fly is perhaps unusually sensitive to this flower or leaf scent. As little as 0.01 μg methyleugenol will produce a response from a single fly in a cage. The structure is highly specific; synthesis of 34 analogues failed to produce another compound as active and most analogues were essentially inactive. Its pheromonal effect in the male fly is in part to stimulate feeding and this can have drastic effects if wrongly applied. Thus in laboratory experiments, adult fruit flies continuously exposed to traces of methyleugenol will engorge so much food that they will actually die from overfeeding!

The second example of pheromone–flower scent interaction is taken from studies on bee orchids by Kullenberg and Bergström (1975). The specific way that certain wild solitary bees of the genus *Andrena* are attracted to orchid flowers of the genus *Ophrys* for purposes of pollination has been appreciated for some time. The shape and colour of the orchid flower closely resemble that of the female bee of the species and the male descends on the plant, performing what is termed 'pseudocopulation', and pollinating the flower in the process. What has not been realized before (Kullenberg, 1952) is that the visual lure of the orchid flower shape is closely associated with an olfactory attraction and that the orchid scent, in fact, mimics the sexual odours of the female bee, thus ensuring the presence of the male bee to trigger orchid pollination. Significantly, the orchid flower does not offer any nectar for the insect and neither does the female bee visit it.

Following from these field observations, the scent of *Ophrys* has been analysed. The major constituents are short chain aliphatic compounds, monoterpenes and some bicyclic sesquiterpenes of the cadinene series. Both (+)- and (−)-γ-cadinene have been identified and it is interesting that it can be shown in laboratory experiments that the male bee is excited by the (−)-isomer, but not by the (+)-form. Thus only compounds with the right stereochemistry produce maximal behavioural response in these insects.

The odour glands of the female *Andrena* bees have also been extracted and examined. Their 'Dufour's gland' secretions contain open chain mono- and sesquiterpene esters. (*E*)-Farnesyl and geranyl hexanoates are the major substances in five *Andrena* species; the corresponding octanoates occur in a sixth species. The mandibular gland secretions are species-specific and are composed of fatty acid derivatives, especially short to medium chain alcohols and esters, and monoterpenes. Some of these, like octanol, decyl acetate and linalool, are present also in the *Ophrys* perfumes and help to 'fool' the male bee into thinking he is approaching a real female.

The co-evolution between these orchids and bees is fairly complex in terms of the species involved. Pseudocopulation takes place in at least 15 *Ophrys* species or forms and several genera of bees, including *Andrena* and *Colletes*, and sphecid wasps are concerned. Pollination is decidedly assortative in that different orchid species or species groups are preferentially visited by different aculeate hymenopteran species or species groups. However, a considerable number of different volatiles have been variously detected in the orchid flower scents and in the cephalic glands of the female bees (Bergstrom, 1978), sufficient at least to account for the specific interactions that have been observed in nature. *Ophrys lutea* appears to differ from other related species in producing a much wider range of scent compounds and then benefits by receiving visits from males of more than one *Andrena* species. Among the 150 scent constituents of *O. lutea* are several which are present in the gland secretions of the corresponding female bees, so that the plant is directly mimicking the female pheromones (Borg-Karlson *et al.*, 1985).

A special feature of the odour compounds in the *Andrena* female bee secretions is that

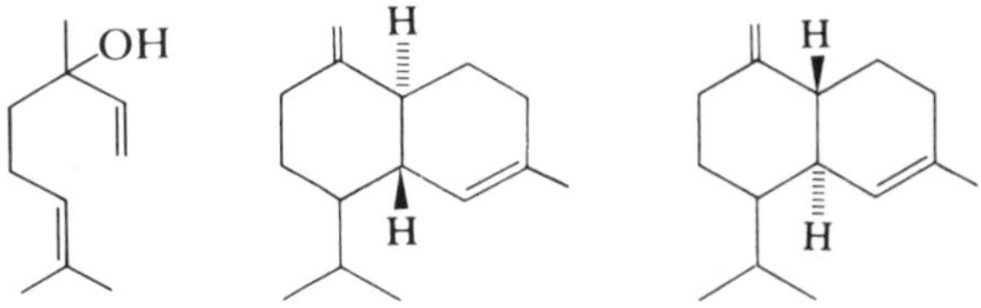

linalool (+)-γ-cadinene (−)-γ-cadinene

Ophrys scent constituents

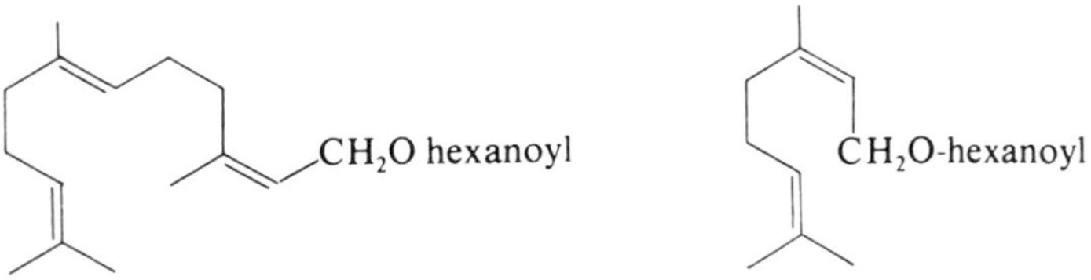

(*E*)-farnesyl hexanoate geranyl hexanoate

Female bee scents, genus *Andrena*

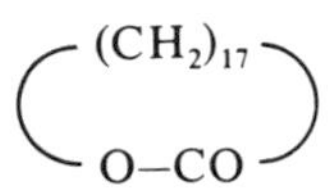

18-octadecanolide
Female bee scent, genus *Colletes*

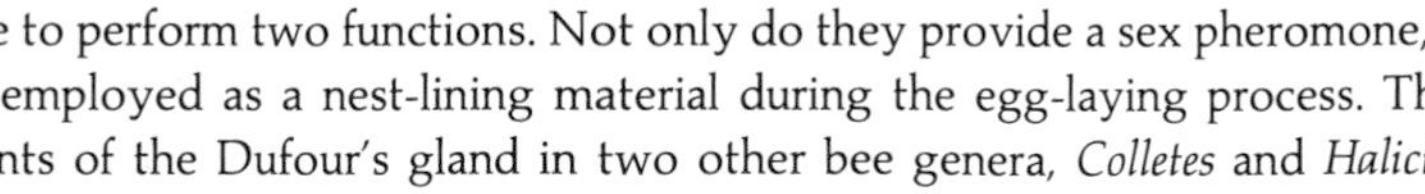

civetone muscone

Figure 2.13 Terpenoids and macrocyclic lactones involved in bee–orchid interactions

they have to perform two functions. Not only do they provide a sex pheromone, but also they are employed as a nest-lining material during the egg-laying process. The active constituents of the Dufour's gland in two other bee genera, *Colletes* and *Halictus*, have also been studied and characterized as macrocyclic lactones, which have distinctive musky odours. One such compound is 18-octadecanolide. Again, these substances are dual-purpose. It is appropriate here to mention that very similar lactones have sex attractant properties in mammals, for example civetone and muscone, the active odour compounds of the civet cat, *Viverra civetta*, and of the musk deer, *Moschus moschiferus*, respectively (see also Chapter 8). Chemical structures of most of the compounds mentioned in the above paragraphs are shown in Fig. 2.13.

A further twist in the story of the *Andrena* female bee secretions has been discovered by Tengo and Bergstrom (1977). It is that the Dufour's gland compounds are utilized by yet other bees (of the genus *Nomada*), which parasitize *Andrena* nests, as chemical cues to specifically locate the nests. The *Nomada* female lays her egg in the next cell, and the larva which hatches then kills the host egg and consumes its food supply. In spite of the harm caused by this parasite in the nest of the *Andrena* bee, an encounter between females of the two very different species never apparently provokes any aggressive behaviour. This lack of aggression, however, is readily understandable when it is realized that the *Nomada* females smell the same as their *Andrena* hosts. Very remarkably, the odour compounds of the *Andrena* female—farnesyl hexanoate or geranyl octanoate—are manufactured by the male *Nomada* bee, who sprays his mate with them during copulation. The chemical links in this co-evolutionary adaptation are almost as striking as those of the original orchid–*Andrena* interaction.

A final, third example, also taken from the bee orchid literature, records a case where male bees make use of flower scents as their sex pheromones. This happens with male euglossine bees which live in the tropical forests of Central and South America. These bees are highly unusual in their mating behaviour; the males are brilliantly coloured and during mating congregate together into small swarms or 'leks' in order to attract the females to them. The orchids which are pollinated by them have evolved a wide range of different floral scents, as part of their extensive speciation in the neotropics. At least 60 chemically distinct fragrances have been recognized in these plants. While pollinating the orchids, several species of *Eulaema* bees collect the odour fragrances in their hind legs and use them to attract other males of the same species. They then form into swarms or leks, and when this is done, the females are attracted by visual means and mating takes place. Orchid compounds used in this way include eugenol, vanillin, cineole, benzyl acetate and methyl cinnamate. Different bee species are differentially attracted by only some of these odours. Thus, isolation mechanisms preventing different bee species from mating may be due to varying preferences for orchid scent compounds (Dodson, 1975).

A close co-evolutionary relationship between bees and orchids has been observed in central Panama, where 11 orchid species are assortatively pollinated by five species of *Eulaema* and three species of *Euglossa* male bees. The chemical link is the major floral volatile carvone epoxide, which is collected by the males and used in lek formation. What is remarkable here is that there is a single species of Euphorbiaceae, *Dalechampia spathulata*, in the same habitat which produces the same floral reward as the orchids and is pollinated by the same bees. Thus, through biochemical convergence in floral odours, an unrelated euphorb has managed to break into the orchid–bee pollination syndrome (Whitten *et al.*, 1986). In a similar way, through visual mimicry, *Campanula rubra* growing in southern Sweden resembles a red helleborine orchid in order to persuade male solitary bees away from the orchid to pollinate it (Nilsson, 1983).

IV. Role of Nectar and Pollen

A. Sugars of Nectar

One of the main reasons why animals visit flowers is to obtain the nectar and its

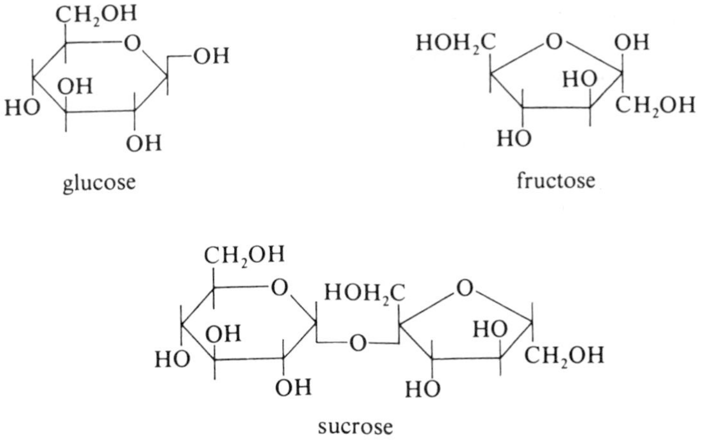

glucose

fructose

sucrose

Figure 2.14 Sugars of nectars

nutritional properties are important to most pollinators, especially those who do not obtain nourishment in any other form (e.g. butterflies). Nectar clearly has no other function in the angiosperm flower other than to attract pollinating animals.

The majority of nectars that have been examined consist simply of a solution of sugars. Most are very sweet to taste, varying in sugar content from 15 to 75% by weight. The compounds present are the three common sugars of plant metabolism: glucose, fructose and sucrose (Fig. 2.14). Oligosaccharides also occur, usually in traces, in a number of plant nectars. Of these, the trisaccharide raffinose (6^G-α-galactosylsucrose) is the most frequent, occurring in nectars of Ranunculaceae, Berberidaceae and related families. Other sugars reported on occasion are the disaccharides maltose (glucosyl-$\alpha 1 \rightarrow 4$-glucose), trehalose (α-glucosyl-α-glucose) and melibiose (galactosyl-$\alpha 1 \rightarrow 6$-glucose) and the trisaccharide melezitose (2^F-α-glucosylsucrose).

The distribution of the three common sugars in nectars has been surveyed in over 900 species (Percival, 1961) and it has been found that there are distinct quantitative differences between species. Indeed, angiosperm nectars can be divided into three broad groups: those in which sucrose is dominant (e.g. *Berberis, Helleborus*); those in which all three sugars occur in about equal amounts (*Abutilon*); and those in which glucose and fructose are dominant (crucifers, umbellifers, some composites). From these results, it could be concluded that there is an evolutionary trend within the angiosperms from nectars with mainly sucrose to those with mainly glucose and fructose. The advantage of this to the pollinator would be the more readily assimilable sugar mixture, i.e. sucrose has to be broken down to glucose and fructose since it cannot be absorbed directly into the blood.

An analysis of nectar types by Baker and Baker (1990) suggests that there is a relationship between the ratio of sugars present and the type of pollinator that visits the flower (Table 2.6). This is particularly striking in the genus *Erythrina*, where flowers pollinated by passerine (perching) birds are uniformly high in glucose and fructose whereas flowers

Table 2.6 Relationship between nectar classes and pollinator types

Sugar ratio[a]	Pollinators
High sucrose (≥ 0.5)[a], e.g. average for 27 species of humming bird-pollinated *Erythrina* = 1.3	Big bees Humming birds Lepidoptera
Low sucrose (< 0.5), e.g. average for 23 species of passerine bird-pollinated *Erythrina* = 0.04	Small bees Passerine birds Neotropical bats

[a] Ratio by weight of sucrose to hexose sugars, glucose and fructose.

pollinated by humming birds are high in sucrose. In favourable cases, such correlations can be used to predict the likely pollinator of a plant. Thus *Luehea speciosa*, which grows in Costa Rica, has a low sucrose value (Table 2.6) and predictably could be bat pollinated, although no records were available in that country. Subsequent observations of the same plant in Brazil showed indeed that it was visited by bats.

A remarkable feature of these quantitative variations in nectar sugars is that the groupings remain consistent within the species and are not subject to diurnal or seasonal variations. Since sucrose is readily converted to glucose and fructose via enzymic reaction with invertase, one might expect considerable changes with age. Although invertase has been detected in nectars, it is presumably not present in sufficient quantity or sufficiently frequently or not present at the right time to change the patterns seriously. A survey of nectars for proteins (Baker and Baker, 1975) showed them to be generally absent, detectable amounts only being present in 14% of the sample.

B. Amino Acids of Nectar

It is curious that until recently, the presence of amino acids in plant nectars lay largely undetected. Apart from a few earlier isolated reports (e.g. Ziegler, 1956), clear-cut proof that amino acids are regular constituents of nectars did not appear until 1973 (Baker and Baker, 1973a, b). These two authors found them in minor but significant amount in 260 of 266 plant nectars surveyed.

The Bakers first looked for amino acids following the logical argument that certain pollinators, especially higher butterflies, were almost completely dependent on nectar for their nutrition; since they survived as adults for several months, they must clearly need nitrogen as well as sugar. These authors were also guided in their search by a number of naturalist observations, all indicating that these same butterflies took advantage of any nitrogen that might be available to them. Thus, tropical forest butterflies have been known to feed both on decaying crocodiles on the banks of the Amazon and on rotting, putrescent fruit of tropical legume trees. They have even been observed to absorb human

Table 2.7 Amino acid concentration of nectars according to plant family

Relative advancement	Plant family	Total amino acid on histidine scale[a]
MORE	Asclepiadaceae	8.4
	Liliaceae	7.4
	Campanulaceae	7.0
	Leguminosae	6.9
	Amaryllidaceae	6.9
	Compositae	6.3
LESS	Rosaceae	3.9
	Myrtaceae	3.1
	Saxifragaceae	2.7
	Caprifoliaceae	2.2

[a] Ninhydrin colour on paper of single drops of nectar compared with same colours of histidine solutions. A score of 2 corresponds to a 98 μM solution, of 4 to 391 μM, of 6 to 1.56 mM and of 8 to 6.25 mM (= *c.* 1 mg/ml). Data from Baker and Baker (1973b).

sweat for its nitrogen content. There is an authenticated story of a hiker in Arctic Canada who took off his boots, while resting at midday, only to be invaded by a swarm of butterflies collecting around his sweaty feet and socks.

The amounts of amino acid present in most nectars, although small, are sufficient to provide insects with a useful nitrogen supply. Thus 0.4 ml of a butterfly flower nectar contains about 840 nmol of amino acid, a daily intake of which would probably be sufficient to meet the nitrogen requirements. Baker and Baker (1973b), in their quantitative analyses of nectar nitrogen, noted significant variations in different angiosperms. Indeed, increase in amino acid content was correlated with increasing evolutionary advancement, woody primitive families tending to have lower amino acid scores than advanced herbaceous groups (Table 2.7). This is also correlated with the fact that the lower scoring families tend to be pollinated by bees, insects which can obtain nitrogen from other sources (e.g. pollen). By contrast, the higher scoring families have significantly more taxa which are pollinated by butterflies and, to a lesser extent, by moths. It thus appears almost as if plants have evolved to produce larger amounts of nitrogen in the nectar in response to the nutritional needs of their chosen pollinating vectors (Table 2.8).

All the common protein amino acids are present in nectars (Baker and Baker, 1975). There is considerable qualitative variation and the number in easily detectable amounts may vary from one to twelve. The ten amino acids essential for insect nutrition (arginine, histidine, lysine, tryptophan, phenylalanine, methionine, threonine, leucine, isoleucine and valine) are better represented than others; glutamic and aspartic acids are also frequent. The variation in detectable amino acids between species is consistent and may be useful as a chemotaxonomic character (Baker and Baker, 1976). Nectars of hybrids between species with different amino acid profiles contain all the amino acids of the two parents; inheritance of nectar nitrogen is thus additive.

Table 2.8 Amount of amino acids in nectars of plants with different animal visitors

Animal group	Amount of amino acid on histidine scale	Other sources of nitrogen
Carrion and dung flies	9.0	None, flowers mimic carrion or dung
Butterflies	5.4	Pollen not eaten
Settling moths	5.4	
Wasps	5.2	
Bees	4.6	Pollen eaten
Hawkmoths	4.4	Ingest large amounts of nectar
Birds	3.9	Insects eaten
Bats	3.6	

Data from Baker and Baker (1986).

C. Lipids in Nectar

On a weight basis, lipids and their component fatty acids are more energy rich than sugars so that there may be some advantage to a plant in using this nutritional attractant rather than sugar since less need be made. Such a substitution has occurred during angiosperm evolution but it appears to have been a rather infrequent event. Indeed, lipids were only recognized recently to be nectar components, after their discovery by Vogel (1969) in a few bee-pollinated members of the Scrophulariaceae. Subsequently, lipids have been recognized in species of 49 genera from five families. Besides the Scrophulariaceae, these are the Iridaceae, Krameriaceae, Malpighiaceae and Orchidaceae (Vogel, 1974).

Lipid production appears to be closely related to certain species of solitary bee of the Anthophoridae which act as pollinators to the above plants. The oil is mainly used by the bees for feeding their young, although its consumption by adult males has been recorded. In *Centris* spp., it is collected exclusively by the female, who carries it back to the nest, mixes it with pollen and places an egg on the mixture. The young hatch out and have a special lipid-rich diet to develop on. These lipids thus play a role in the co-evolutionary relationship between plant and pollinator. The bees benefit from an energy-rich diet, the plants through fidelity of pollination by these bees.

Chemical studies of lipid-containing nectars have been few and it is not yet clear whether they are generally of the common triglyceride type. In the few cases where analyses have been carried out, the oils have been found to be unusual. The oil is usually located within the flower in trichomes and the exudates of these trichomes in *Calceolaria pavonii* were examined by Vogel (1976), who identified the major component as a diglyceride of acetic acid and β-acetoxystearic acid. The oils of *Krameria* spp. have proved to be even more unusual in consisting of free fatty acids (Simpson *et al.*, 1977). The fatty acids were all saturated with chain lengths between C_{16} and C_{22} and all contained an acetate substitution at the β-position. One is β-acetoxystearic acid, $CH_3(CH_2)_{14}CHOAcCH_2CO_2H$, as in *Calceolaria*. Not only are free fatty acids rare in plants, but these particular β-acetoxy acids seem to be unique to floral nectars. It will be interesting to see if subsequent studies confirm the special nature of the oils present in other plants which have lipid-based insect lures.

D. Nectar Toxins

Occasionally, nectars contain toxins, presumably derived initially from other plant parts. Honey produced by bees foraging on unusual plant sources is sometimes tainted by such compounds. The toxic diterpene acetylandromedol has actually been characterized in the nectars of *Rhododendron*. Alkaloids also occasionally appear and in the case of the nectar of *Sophora microphylla*, have been present in sufficient concentration to cause toxicity to the honey bees (Clinch *et al.*, 1972). Alkaloid toxicity may be more serious as a possible hazard to humans consuming the honey. Bees feeding on the ragwort *Senecio jacobaea* are known to produce honey contaminated with the pyrrolizidine alkaloids present in the plant nectar (Deinzer *et al.*, 1977). The alkaloid concentration in the honey may vary from 0.3 to 3.9 ppm. Fortunately such honey has a bitter taste and is off-colour so that it is normally avoided.

There is also the other side of the coin: substances which are harmless to man may be toxic to bees. The glucoside arbutin from *Arbutus unedo* honey is apparently harmful to bees (Pryce-Jones, 1944). Also, the simple sugar galactose has been detected in the stigmatal exudates of tulip flowers and is toxic to bees (Barker and Lehner, 1976). Again, the toxicity of *Tilia* nectar and pollen to bees is attributable to the presence of the sugar mannose. In this case, it is known that the insects cannot fully metabolize it. They lack the enzyme mannose phosphate isomerase so that mannose 6-phosphate accumulates, causing paralysis (cf. Vogel, 1978).

The presence of nectar toxins, of course, may occasionally be advantageous to the pollinator. Adult Ithomiine butterflies in South America have a requirement for pyrrolizidine alkaloids, which are used for pheromone production. This can be met by sucking withered borage leaves but also in part by collecting nectar from the flowers of those *Eupatorium* species which secrete these alkaloids in their nectar (see Chapter 3, Section VI).

E. Extrafloral Nectaries

Extrafloral nectaries are sugar-producing glands which occur on the bracts, leaves, petioles or stems of a significant number of angiosperm plants. In families like the Leguminosae, Orchidaceae and Passifloraceae, they are found with some frequency. Although these nectaries are very similar to floral nectaries and attract many insect visitors, they are not normally concerned at all in pollination. Indeed their function has for a long time remained unclear and controversial. However, recent biochemical analyses have indicated that they contain very similar nutrients to those of the flowers, with the same sugars and amino acids being present. Thus, they represent an important source of food to insects. Furthermore, they do seem to have a purpose. There is increasing ecological evidence that they have value to the plant in attracting particular ant species, the colonies of which then provide the plant with antiherbivore defence (Bentley, 1977).

The mutual interaction that occurs is therefore not all that different from the more widespread plant–pollinator system.

In ant-*Acacia* species, the beneficial interactions between plant and ant are highly developed (see also Chapter 3, Section IIIC), the ants being particularly vicious in their attack on any animal visitors to the plant. The ecological importance of nectar-feeding ant colonies may, however, vary from plant to plant. In *Ipomoea leptophylla*, for example, there are both foliar and sepal nectaries fed on by ants. Here, however, the ants' protective role is to save the plant from seed loss due to bruchid beetle activity and from flower damage by grasshoppers (Keeler, 1980). The extrafloral nectaries thus make an indirect positive contribution to pollination biology by keeping away harmful visitors.

In general, plants with extrafloral nectaries continue to secrete sugars and amino acids irrespective of any mutual association with a colony of ants. This apparently wasteful process was one of the reasons why the ecological significance of these organs was not appreciated until recently. Physiologists argued that the flow of sugar from such nectaries simply represented a method of shedding excess sugar during periods of growth when sugar synthesis was at its maximum. In fact, there is now evidence that certain plants can actually restrict the secretion of food in such organs to times that active insect association is present.

It has been observed, for example, in the plant *Piper cenocladum*, which has a beneficial relationship with the ant *Phoidole bicornis*, that the production of food bodies is directly linked with the presence of ants. If ants are removed, the food, which contains protein and lipid as well as sugar, is no longer produced (Risch and Rickson, 1981). Food bodies in this plant are exactly equivalent to extrafloral nectaries in other species, so that it is likely that a range of these plants may exercise a similar economy in producing nutrients only when their associated ants are in residence.

F. Nutritive Value of Pollen

Any account of the nutritional benefit gained by animal pollinators from plants would be incomplete without some mention of pollen. The pollen is usually more accessible than the nectar and is collected and used by many flower visitors. Pollen is particularly fed on by beetles, who have to chew it in order to break open the tough pollen walls. Bees benefit enormously from pollen, which they are able to digest. Pollen occasionally becomes mixed with the nectar, and in such conditions, the nutritional benefits may be available to animals (e.g. *Heliconius* butterflies) which feed solely on the nectar.

The chemistry of pollens has been exhaustively investigated (Barbier, 1970; Stanley and Linskens, 1974). Nutritionally, pollen is a rich source of food with 16–30% protein, 1–7% starch, 0–15% free sugar and 3–10% fat. Trace constituents present include various vitamins and inorganic salts. There are also varying amounts of secondary substances. Pollen is often coloured, especially by carotenoid but also by flavonoids, and this is probably a signal to indicate its availability to insect feeders. The carotenoids of pollen are usually α- and β-carotene, lutein, zeaxanthin and their various epoxides. Deep red and purple pollens often have anthocyanidin for pigmentation (e.g. *Anemone*). Other flavonoids, especially the flavonol isorhamnetin, are frequently present in pollens and contribute to pale yellow colours.

Pollen has primary importance as the carrier of the male gametophytes, so that all use of pollen by animals for feeding is secondary and represents 'pollen theft' as far as the plant is concerned. Competition between the two contrasting uses of pollen is rarely a problem, since the majority of angiosperms are overabundant in pollen production. If insects did not capitalize on the excess pollen available to them, it would go to waste in other ways.

There is plenty of evidence that pollen is deliberately provided by some plants to their pollinators as a reward. That animals can 'home in' on plant pollen is apparent from the fact that pollens produce characteristic odours which are often different from the scents of the flowers. Furthermore, honey and solitary bees are able to discriminate between plant species on the basis of pollen odours. That pollens differ in their volatile constituents has been confirmed in the case of *Rosa rugosa* and *R. canina* where 31 terpenoids, aliphatics and aromatics were variously detected (Dobson *et al.*, 1987).

V. Summary

In this chapter, the different biochemical aspects of plant pollination have been considered in turn: the role of flower colour, flower scent, nectar and pollen. In the field, all these

Table 2.9 Plant–pollinator interactions and their biochemical features

Plant and family	Major pollinator(s)	Biochemical factors
Delphinium nelsonii Ranunculaceae	Bumble bee and humming bird	Blue flower colour essential; mutation to white leads to pollinator neglect
Rechsteineria macrorhiza Gesneriaceae	Humming bird	Desoxyanthocyanin provides scarlet-orange flower colour
Ipomopsis aggregata Polemoniaceae	Humming bird and hawkmoth	Colour shift from red to white in response to availability of pollinator
Rudbeckia hirta Compositae	Bee	UV patterning by patuletin increases efficiency of pollination
Arum maculatum Araceae	Dungfly	Skatole in scent attracts fly away from dungheap
Datura inoxia Solanaceae	Hawkmoth	Alkaloids in nectar act as addictive drug to pollinator
Ophrys spp. Orchidaceae	*Andrena* male bee	Flowers resemble female bees in shape, colour and odour
Catasetum spp. Orchidaceae	Euglossine male bee	Flower scent collected by male to use as pheromone
Calceolaria pavonii Scrophulariaceae	Solitary bee	Energy-rich lipid nectar used by female to feed the young
Passiflora spp. Passifloraceae	Heliconiad butterfly	Pollen mixed with nectar increases N content of floral reward

factors may come together in particular plant–pollinator interactions (Table 2.9). One or other biochemical factor may dominate in a particular interaction, whereas in others, flower colour, odour and nectar may all be required to attract a particular pollinator.

While the main purpose of flower colour, scent and floral reward is to attract an available pollinator, it is worth remembering that plants have also to protect themselves from animal visitors which may steal the floral reward without pollinating the flower. Thus, biochemical (and structural) features may be important in repelling visitors as well as for attracting them. For example, bees are not generally attracted to red flowers, because they are insensitive to this colour, and likewise they will avoid plants which have sugars (e.g. mannose) in the nectar which they cannot metabolize.

One final point, illustrated here by several examples, is the dynamic nature of the association between flower and pollinator. The situation is continually changing with evolutionary time and some associations which seem to be tightly knit (e.g. bee–orchid and wasp–orchid associations) may be open to variation. Witness the ability of members of the Euphorbiaceae and Campanulaceae (Section IIIB) to mimic orchids biochemically in order to be visited by their highly specific male bee pollinators. Members of the Orchidaceae provide a rich source of bizarre variations in plant–pollinator interactions and biochemical studies of further orchids are bound to be rewarding.

Bibliography

Books and Review Articles

Baker, H. G. and Baker, I. (1973b). Amino acid production in nectar. In: Heywood, V. H. (ed.), "Taxonomy and Ecology', pp. 243–264. Academic Press, London.

Baker, H. G. and Baker, I. (1975). Nectar constitution and pollinator–plant co-evolution. In: Gilbert, L. E. and Raven, P. H. (eds.), 'Coevolution of Animals and Plants', pp. 100–140. Texas Univ. Press, Austin.

Baker, H. G. and Hurd, P. D. (1968). Intrafloral ecology. *A. Rev. Entomol.* **13**, 385–414.

Barbier, M. (1970). Chemistry and biochemistry of pollens. In: Reinhold, L. and Lipschitz, Y. (eds.), 'Progress in Phytochemistry', Vol. 2, pp. 1–34. Wiley, London.

Beattie, A. J. (1991) Problems outstanding in ant–plant interaction research. In: Huxley, C. R. and Cutler, D. F. (eds.) 'Ant–Plant Interactions', pp. 559–576. Oxford Science Publications, Oxford.

Bergström, G. (1978). Role of volatile chemicals in *Ophrys*–pollinator interactions. In: Harborne, J. B. (ed.), 'Biochemical Aspects of Plant and Animal Co-evolution', pp. 207–232. Academic Press, London.

Crowe, L. (1964). Evolution of outbreeding in plants. I. The angiosperms. *Heredity* **19**, 435–457.

Dodson, C. H. (1975). Coevolution of orchids and bees. In: Gilbert, L. E. and Raven, P. H. (eds.), 'Coevolution of Animals and Plants', pp. 91–99. Texas Univ. Press, Austin.

Faegri, K. and van der Pijl, L. (1979). 'Principles of Pollination Ecology', 3rd edn. Pergamon Press, Oxford.

von Frisch, K. (1950). 'Bees, Their Vision, Chemical Senses and Language'. Cornell, Ithaca, New York.

Goodwin, T. W. (ed.) (1988). 'Plant Pigments'. Academic Press, London.
Grant, V. (1971). 'Plant Speciation', 435 pp. Columbia Univ. Press, New York.
Grant, V. and Grant, K. A. (1965). 'Flower Pollination in the Phlox Family'. Columbia Univ. Press, New York.
Harborne, J. B. (1967). 'Comparative Biochemistry of the Flavonoids', 383 pp. Academic Press, London.
Harborne, J. B. (1988). The flavonoids: recent advances. In: Goodwin, T. W. (ed.), 'Plant Pigments', pp. 299–344. Academic Press, London.
Kaiser, R. (1993). 'The Scent of Orchids: Olfactory and Chemical Investigations'. Elsevier, Amsterdam.
Kevan, P. G. and Baker, H. G. (1983). Insects as flower visitors and pollinators. *A. Rev. Entomol.* **28**, 407–453.
Kullenberg, B. and Bergström, G. (1975). Chemical communication between living organisms. *Endeavour* **34**, 59–66.
Percival, M. S. (1961). Types of nectar in angiosperms. *New Phytol.* **60**, 235–281.
Piattelli, M. (1976). Betalains. In: Goodwin, T. W. (ed.), 'Chemistry and Biochemistry of Plant Pigments', 2nd edn., pp. 560–596. Academic Press, London.
Proctor, M. and Yeo, P. (1973). 'The Pollination of Flowers'. Collins. London.
Real, L. (ed.) (1983). 'Pollination Biology', 338 pp. Academic Press, Orlando.
Richards, A. J. (ed.) (1978). 'The Pollination of Flowers by Insects', 213 pp. Academic Press, London.
Stanley, G. and Linskens, H. F. (1974). 'Pollen: Biology Biochemistry and Management', 307 pp. Springer-Verlag, Berlin.
Vogel, S. (1978). Floral ecology. *Prog. in Bot.* **40**, 453–481.
Wyatt, R. (1983). Pollinator–plant interactions and the evolution of breeding systems. In: Real, L. (ed.), 'Pollination Biology', pp. 51–96. Academic Press, Orlando.

Literature References

Asen, S., Stewart, R. N. and Norris, K. H. (1972). *Phytochemistry* **11**, 1139–1144.
Baker, H. G. and Baker, I. (1973a). *Nature Lond.* **241**, 543–545.
Baker, H. G. and Baker, I. (1976). *New Phytol.* **76**, 87–98.
Baker, H. G. and Baker, I. (1986). *Plant Syst. Evol.* **151**, 175–186.
Baker, H. G. and Baker, I. (1990). *Israel J. Botany* **39**, 157–166.
Barker, R. J. and Lehner, Y. (1976). *Apidologie* **7**, 109–111.
Beale, G. H. (1941). *J. Genet.* **42**, 197–213.
Bentley, B. L. (1977). *A. Rev. Ecol. Syst.* **8**, 407–427.
Borg-Karlson, A. K., Bergström, G. and Groth, I. (1985). *Chemica Scripta* **25**, 283–294.
Clinch, P. G., Palmer-Jones, T. and Forster, I. W. (1972). *N.Z. J. Agric. Res.* **15**, 194–201.
Deinzer, M. L., Thomson, P. A., Burgett, D. M. and Isaacson, D. L. (1977). *Science, N. Y.* **195**, 497–499.
Dement, W. A. and Raven, P. H. (1974). *Nature, Lond.* **252**, 705–706.
Diamond, J. M. (1989) *Nature* **339**, 258–259.
Dobson, H. E. M., Bergström, J., Bergström, G. and Groth, I. (1987). *Phytochemistry* **26**, 3171–3174.
Eisikowitch, D. and Lazar, Z. (1987). *Botan. J. Linn. Soc.* **95**, 101–111.

Eisner, T., Eisner, M., Hyypio, P. A., Aneshansley, D. and Silbersgleid, R. E. (1973). *Science, N.Y.* **179**, 486.

Grant, V. and Grant, K. A. (1983). *Bot. Gaz.* **144**, 280-284.

Harborne, J. B. and Boardley, M. (1983). *Z. Naturforsch.* **38c**, 148–150.

Harborne, J. B. and Nash, R. J. (1983). *Biochem. Syst. Ecol.* **12**, 315–319.

Harborne, J. B. and Smith, D. M. (1978a). *Biochem. Syst. Ecol.* **6**, 127–130.

Harborne, J. B. and Smith, D. M. (1978b). *Biochem. Syst. Ecol.* **6**, 287–291.

Harborne, J. B., Heywood, V. H. and King, L. (1976). *Biochem. Syst. Ecol.* **4**, 1–4.

Hickman, J. C. (1974). *Science, N.Y.* **184**, 1290.

Horovitz, A. and Cohen, Y. (1972). *Am. J. Bot.* **59**, 706–713.

Kay, Q. O. N., Daoud, H. S. and Stirton, C. H. (1981). *Bot. J. Linn. Soc.* **83**, 57.

Keeler, K. H. (1980). *Am. J. Bot.* **67**, 216–222.

Kullenberg, B. (1952). *Bull. Soc. Hist. Nat. Afr. du Nord* **43**, 53.

Macior, L. W. (1986). *Bull. Torrey Bot. Club* **113**, 101–109.

Nilsson, L. A. (1983). *Nature* **305**, 799–800.

Paige, K. N. and Whitham, T. G. (1985). *Science, N.Y.* **227**, 315–317.

Pryce-Jones, J. (1944). *Proc. Linn. Soc. Lond.* 129–174.

Pijl, L. van der (1961). *Evolution* **15**, 44–59.

Raskin, J., Ehman, A., Melander, W. R. and Meeuse, B. J. D. (1987). *Science* **237**, 1601–1602.

Risch, S. J. and Rickson, F. R. (1981). *Nature* **291**, 149–150.

Saito, N. and Harborne, J. B. (1983). *Phytochemistry* **22**, 1735–1740.

Saito, N. and Harborne, J. B. (1992). *Phytochemistry* **31**, 3009–3015.

Scogin, R. and Zakar, K. (1976). *Biochem. Syst. Ecol.* **4**, 165–168.

Simpson, B. B., Neff, J. L. and Seigler, D. (1977). *Nature, Lond.* **267**, 150–151.

Tengo, J. and Bergström, G. (1977). *Science, N.Y.* **196**, 1117–1119.

Thien, L., Bernhardt, P., Gibbs, G. W., Pellmyr, O., Bergstrom, G., Groth, I. and McPherson, G. (1985). *Science* **227**, 540.

Thompson, W. R., Meinwald, J., Aneshansley, D. and Eisner, T. (1972). *Science, N.Y.* **177**, 528–530.

Vogel, S. (1969). *Abstracts XI Int. Bot. Congr.* Seattle, p. 229.

Vogel, S. (1974). *Trop. subtrop. Pflanzenw.* (7), 283–547.

Vogel, S. (1976). *Naturwissenschaften* **63**, 44.

Waser, N. M. and Price, M. V. (1981). *Evolution* **35**, 376–390.

Waser, N. M. and Price, M. V. (1983). *Nature* **302**, 422–424.

Weiss, M. R. (1991). *Nature* **354**, 227–229.

Whitten, W. M., Williams, N. H., Armbruster, W. S., Battiste, M. A., Strekowski, L. and Lindquist, N. (1986). *Systematic Bot.* **11**, 222–228.

Zeigler, H. (1956). *Planta* **47**, 447–500.

3 Plant Toxins and Their Effects on Animals

I. Introduction

As Feeny (1975) has put it, the most conspicuous non-event in the history of the angiosperms is the failure of insects and other herbivores to attack plants on a wide scale. Green plants still dominate the landscape, in spite of the proven ability of insects—witness the devastation of locusts—to overeat and destroy them. It follows that all plants must be broadly repellent to animals as food and toxic in the widest sense. It is the selective ability of insects and grazing animals to overcome these defence mechanisms that allows them the limited feeding that we witness today.

The plant's defence is based on both physical and chemical factors. Physical defence mechanisms against herbivores are readily appreciated; tough epidermises, cuticular deposits, spines, thorns, prickles and stinging hairs. Defence may be purely strategic, as in the case of grasses which adapt to grazing by clinging close to the soil and by vegetative reproduction under the soil surface. Nevertheless, chemical defences are also very important, provided by toxins and repellent substances of one type or another within the plant itself. These toxins have had and continue to have a key role in protecting plants from overgrazing.

Our views on plant toxins, the subject of the present chapter, have long been blinkered by the limited viewpoint that the only plants that are toxic are those that are dangerous to ourselves or to farm and domestic animals. On this view, relatively few plants are really poisonous and the toxin present is usually alkaloidal. This, however, completely neglects the fact that plants which are relatively harmless to us may be highly toxic to other groups of animals—birds, fish and especially insects. Plant insecticides such as nicotine, the pyrethrins and rotenoids are well known but less is known of the many other plant toxins inimical to insect life. McIndoo in 1945 drew up a list of 1180 plant species containing insect poisons, the majority of which remain uninvestigated to this day. We are thus concerned here with toxicity in the very widest sense—to all animals which eat plants, from man to the humblest insect. Toxicity is present in most plants, not just those listed in the Flora with a P for poison against them.

The toxicity of a chemical is always relative, dependent on the dose taken in a given time period, the age and state of health of the animal, the mechanism of absorption and mode of excretion. The steroidal alkaloid solanine, for example, is present in all domestic potatoes but the amount present is so infinitesimal that it is rarely a dietary hazard. It is only when enormously large amounts of solanine accumulate in tubers that have been exposed above the soil surface and become 'greened' that death from solanine poisoning is a reality. In such cases, victims have no time to adapt to dealing with the toxin and, unless they are sick, they die from respiratory failure. Whether death occurs on intake of a toxin depends therefore on whether the animal has time to become accustomed to small amounts of the poison in the diet, i.e. whether it has been able to develop a detoxification mechanism.

Toxins often have the role of feeding repellents, since plants usually advertise their presence by a warning signal of a visual or olfactory nature. Thus animals may be made aware of the presence of the toxins even before they start feeding. Mustard oils, for example, which occur in crucifers in bound form and are toxic to most insects, have a pungent acrid smell and are probably emitted continuously in trace amounts from the living plant. Other immediate warnings of danger may be provided through visual means by toxins deposited on the surface of leaves and other organs. Potentially toxic secondary compounds may occur in the surface waxes. Alternatively, glandular hairs on the leaf may secrete a toxic quinone, as in *Primula obconica*, or there may be a deposit of quinone on the leaf undersurface, as in a number of labiates. Again, chemical defence is often 'advertised' in woody plants when they exude resins from bark and fruit.

In the case of HCN, intact cyanophoric plants release no prussic acid, since the substrates and enzymes for HCN production are located in different organelles. It is only when the leaf is damaged by herbivores that the substrate and enzyme come together to produce the poison, which has a clear warning in its 'odour of bitter almonds'. In the case of alkaloids and saponins, the warning signal is only received after the animal has started feeding, in the form of a bitter taste. Most alkaloids and saponins are known to be bitter. Indeed, the standard for bitterness is quinine, the alkaloid of cinchona bark, which is still bitter to humans at a concentration of 1×10^{-6} M. Many other plant compounds are bitter, especially the triterpenoid cucurbitacins of the cucumber family, which clearly provide the basis of repellency to herbivores in these plants. The latexes which flow within plants such as chicory, dandelion, and other composites also have an obvious role in herbivore deterrence, since they often contain bitter toxins among their constituents.

Finally an association of toxicity with a warning coloration can be seen in the brightly coloured, evil-looking purple-black berries of the deadly nightshade, *Atropa belladonna*. The colour signal here is dual-purpose. It is a warning to predators (e.g. grazing mammals) which may be killed by the tropane alkaloids richly present in the berry. It is also a feeding signal to those animals which can safely tolerate the toxin, e.g. birds, which can then distribute the seed for the benefit of the plant.

The objectives in this chapter, then, are to present a brief account of plant toxins, with emphasis on more recently discovered classes, and to consider the ecological role of these toxins in plant–animal interactions.

II. Different Classes of Plant Toxins

A. Nitrogen-based Toxins

Of the various nitrogen-based plant toxins (Table 3.1), the simplest in structure are the non-protein amino acids. These are widely present in plants and may be directly toxic inasmuch as they are anti-metabolites of one or other of the 20 protein amino acids (Fig. 3.1). In the simplest case, e.g. of azetidine 2-carboxylic acid, they may be mistakenly incorporated into protein synthesis, the organism produces unnatural enzymic protein which cannot function properly and death of the organism ensues. The toxic effect of other non-protein amino acids may sometimes be more complex. 3,4-Dihydroxyphenylalanine or L-DOPA, which is harmful to insects, interferes with the activity of tyrosinase, an enzyme essential to the hardening and darkening of the insect cuticle.

There are about 300 known structures of these plant amino acids (Bell, 1980; Rosenthal, 1982). While they are found in a number of unrelated families, they are particularly characteristic of legumes and occur mainly in the seeds. One which has been widely studied is azetidine 2-carboxylic acid, first isolated in quantity from *Convallaria majalis* (Liliaceae). It has a relatively wide distribution and occurs in several legumes. It is toxic because it interferes with proline synthesis or utilization. Plants which manufacture it in quantity like *Convallaria* are protected from its harmful properties because their protein-synthesizing machinery and more especially the proline *t*RNA-synthetase enzymes can recognize it and do not incorporate it into protein. Unadapted plants mistake it for proline and incorporate it into protein, with fatal consequences.

The ecological function of toxic amino acids in legume seeds is fairly clear in that these seeds are particularly large and provide an especially rich source of nutrient to any herbivore. If there was no means of protection, the seeds would undoubtedly be overeaten. Toxicity in these plants is provided by a variety of chemical structures (Bell, 1972). In *Vicia* species, β-cyanoalanine (or its γ-glutamyl derivative) is present; it is toxic to mammals, causing convulsions and death when injected into rats at a concentration of 200 mg/kg body wt. A more widespread legume toxin is canavanine, which occurs in seeds of the jackbean *Canavalia ensiformis* to the extent of 4–6% fr. wt and of *Dioclea megacarpa* to the extent of 7–10% fr. wt. Canavanine is as toxic to mice as β-cyanoalanine is to rats, so presumably it provides protection in the seeds to a range of mammalian herbivores.

The major protective role of legume non-protein amino acids is, however, probably

Table 3.1 Some nitrogen-based toxins in plants

Class of compound	Example(s)	Toxicity
Non-protein amino acids	L-DOPA in *Mucuna* seed	Insects, espec. bruchid beetles
	β-Cyanoalanine in *Vicia* seed	Fatal dose in rats 200 mg/kg body wt
Cyanogenic glycosides	Linamarin and lotaustralin in *Lotus corniculatus*	Universal; fatal dose of HCN in man *ca* 50 mg
Glucosinolates	Sinigrin in *Brassica*	Cattle and insects
Alkaloids	Senecionine in ragwort *Senecio jacobaea* leaves	Espec. cattle
	Atropine in *Atropa bella-donna* berries	Mammals, but not birds; LD_{50} in rats 750 mg/kg
Peptides	Amanitine in *Amanita phalloides*	Mammals
	Viscotoxin in *Viscum album* berries	Animals, but not birds
Proteins	Abrin in *Abrus precatorius*	Lethal dose in man 0.5 mg
	Phytohaemagglutinin in *Phaseolus vulgaris*	Bruchid beetles

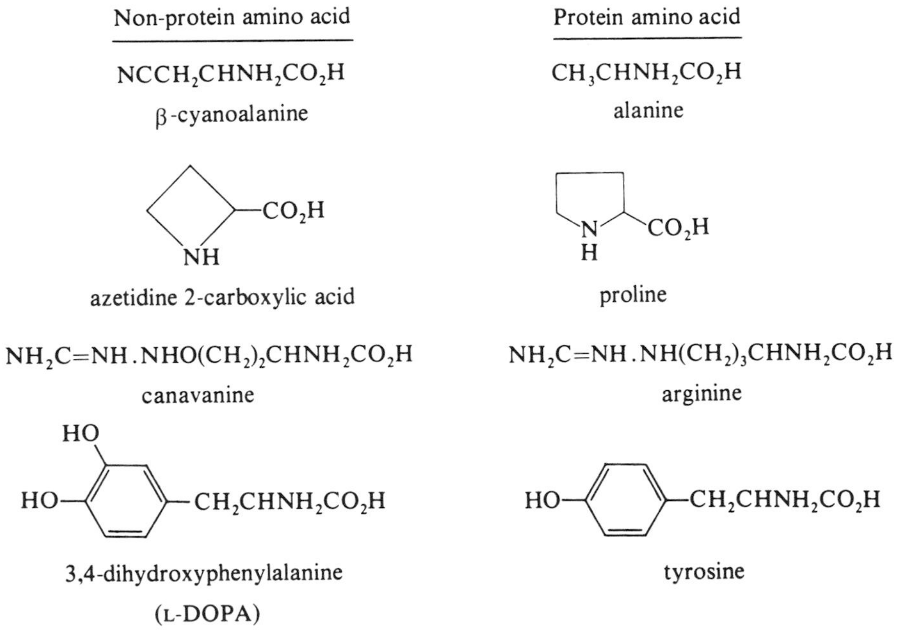

Non-protein amino acid	Protein amino acid
$NCCH_2CHNH_2CO_2H$ β-cyanoalanine	$CH_3CHNH_2CO_2H$ alanine
CO_2H, NH azetidine 2-carboxylic acid	N, H, CO_2H proline
$NH_2C{=}NH.NHO(CH_2)_2CHNH_2CO_2H$ canavanine	$NH_2C{=}NH.NH(CH_2)_3CHNH_2CO_2H$ arginine
HO, HO, $CH_2CHNH_2CO_2H$ 3,4-dihydroxyphenylalanine (L-DOPA)	HO, $CH_2CHNH_2CO_2H$ tyrosine

Figure 3.1 Toxic non-protein amino acids and their protein amino acid analogues

against insect feeding. One example is provided by the occurrence of L-DOPA in *Mucuna* seeds to the extent of 6–9%. While the substance is relatively non-toxic to mammals, being used medicinally in man for treating Parkinson's disease, it is dangerous to insects. On feeding, it causes mortality in the southern army-worm larvae, *Prodenia eridania*. Its ecological role has been discussed by Janzen (1969), in relation to the feeding of bruchid beetles, which attack *Mucuna* and other legume seeds in the natural habitat of the Brazilian forests. Two legume tree species may be growing adjacent to each other; the seeds of one, protected by L-DOPA, are essentially free from bruchid infestation, while the seeds of a second lacking a protective chemical are riddled with bruchid borings. Measurements of seed size and number in these contrasting situations indicate that the trees modify their seed production in relation to their content of deterrent chemical and its effectiveness in preventing beetle predation. Proof that many legume seeds contain insecticidal components has been further obtained in feeding experiments of seed meal to army-worms (Rehr *et al.*, 1973).

While the presence of non-protein amino acids in a seed may provide a general defence against insect attack, it is always possible that an individual insect species may overcome the barrier by detoxifying or otherwise dealing with the toxin. This is true of the larvae of the bruchid beetle, *Caryedes brasiliensis*, which in Costa Rica feeds exclusively on seeds of *Dioclea megacarpa*, which, as mentioned above, contain huge amounts of canavanine. Biochemical studies have shown that this beetle has two-fold protection against this potential toxin. First, the protein-synthesizing machinery of the beetle larvae is able to avoid incorporating canavanine in spite of its similarity in structure to the protein amino acid arginine (see Fig. 3.1), i.e. the arginyl *t*RNA synthetase of the beetle discriminates between the two amino acids and canavanyl-protein is not made (Rosenthal *et al.*, 1976). Second, the larvae possess an extraordinarily high level of urease activity. This enzyme facilitates the conversion of canavanine to ammonia via urea, following the attack of arginase on the canavanine substrate. Thus in one and the same process, the canavanine is both detoxified and turned to good purpose by the larvae as a source of nitrogen (Rosenthal *et al.*, 1977).

Another structurally simple class of nitrogenous toxins are the cyanogenic glycosides. They are toxic not as such but only when broken down enzymically with release of HCN or prussic acid. The primary site of action of HCN is on the cytochrome system; terminal respiration is inhibited, oxygen starvation occurs at the cellular level and rapid death ensues. The distribution and ecological role of the bound forms of HCN will be discussed in a later section of this chapter. Like HCN, nitrite is toxic to a wide range of organisms and some plants, notably species of *Astragalus*, accumulate glucosides of organic nitro compounds, which are toxic due to the release of nitrite according to the scheme:

$$\text{Glc–O–}(CH_2)_3NO_2 \rightarrow HO(CH_2)_3NO_2 \rightarrow NO_2^-$$

Curiously, the toxicity of the above glucoside, miserotoxin, mainly affects cattle, although the human nervous system is not completely immune to nitrite poisoning. Incidentally, the legume genus *Astragalus*, which contains many species with miserotoxin or related nitro compounds (Stermitz *et al.*, 1972), is remarkably heterogeneous in its toxic components. Other species accumulate selenium amino acids (e.g. *A. racemosus*) (see Chapter 1) and yet others (e.g. the loco weed *A. mollisimus*) so far unidentified animal poisons.

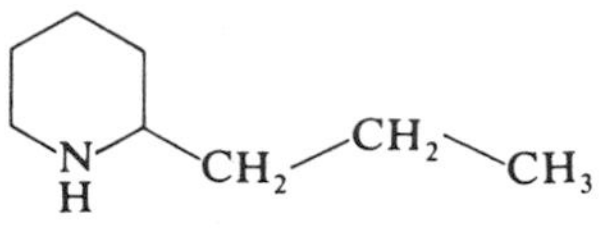

coniine, of *Conium maculatum* (Umbelliferae)

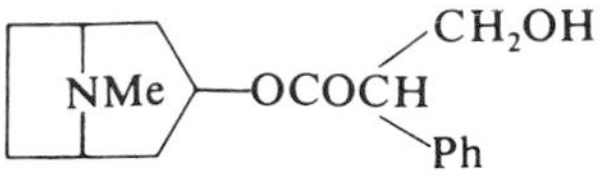

atropine, of *Atropa belladonna* (Solanaceae)

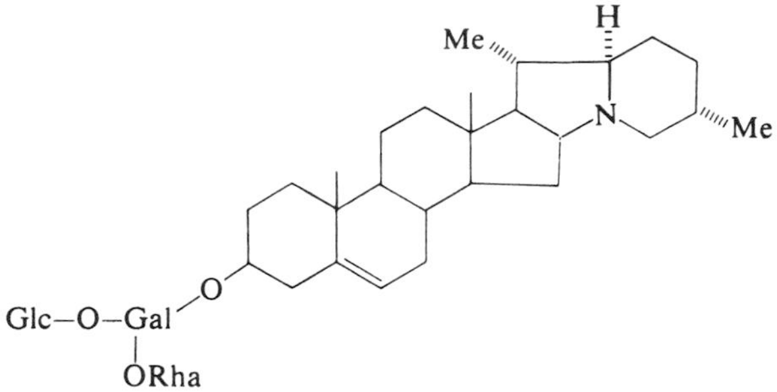

solanine, of *Solanum tuberosum* (Solanaceae)

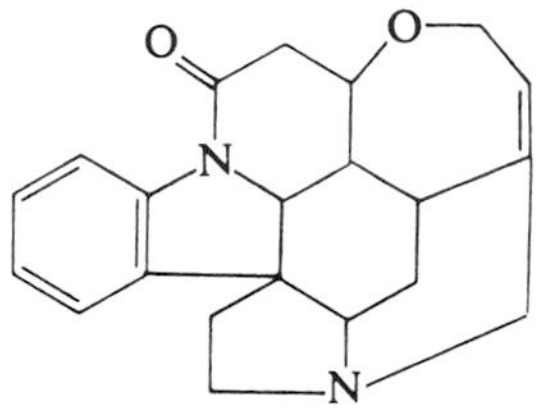

strychnine, of *Strychnos nux-vomica* (Loganaceae)

Figure 3.2 Some characteristic alkaloids of plants

Glucosinolates (mustard oil glycosides) are closely related biosynthetically to cyanogenic glycosides and they can also be toxic to animals, when they occur in sufficient amount in plants, as in wild species of *Brassica*. Toxic symptoms include severe gastro-enteritis, salivation, diarrhoea and irritation of the mouth. Toxicity is actually due to the release of isothiocyanates (mustard oils), which are highly vesicant in their action. A further hazard of these substances is due to the fact that during their release from bound forms, the isothiocyanates produced can undergo rearrangement in part of the corresponding thiocyanates:

$$R{-}N{=}C{=}S \rightleftharpoons R{-}S{-}C{\equiv}N \quad (R = \text{alkyl or benzyl})$$

The latter substances are harmful because they are goitrogenic, and produce hyperthyroidism in mammals. That glucosinolates are toxic to insects has been demonstrated by Erickson and Feeny (1974) who found that caterpillars of the black swallowtail butterfly (*Papilio polyxenes*) were killed by being fed on celery leaves which had been infiltrated with sinigrin (at a concn. of 0.1%/fresh wt leaf).

The most familiar class of plant toxins are the alkaloids (Fig. 3.2). These substances have been used since time immemorial for poisoning purposes, an extract of hemlock leaves being used by the ancient Greeks to put the philosopher Socrates to death. The physiological effects of alkaloids on the central nervous system in man have been widely studied and alkaloids are utilized in modern medicine for a variety of purposes. There are at least 10,000 alkaloids of known structure and many more await structural elucidation. These bases occur widely, albeit sporadically, in the angiosperms, being present in about 20% of higher plant families. The term alkaloid covers an enormous range of chemical structures, from the simple monocyclic piperidine, coniine of hemlock *Conium maculatum*, to the hexa- and heptacyclic alkaloids like solanine of *Solanum tuberosum* and strychnine of *Strychnos nux-vomica*. Not all alkaloids are highly toxic; few are as dangerous as, say, atropine, the principal toxin of deadly nightshade, *Atropa belladonna*. Nevertheless, most alkaloids are liable to show some toxic effects if ingested in any quantity over an extended period of time.

One highly toxic group are the pyrrolizidine alkaloids. They are well known as the toxic principles of plants poisonous to cattle, e.g. species of the genus *Senecio*. They are also dangerous to man. Their identification (Culvenor *et al.*, 1980) in leaves of a reputedly harmless plant, comfrey (*Symphytum officinale*), should be noted since comfrey has not only been used medicinally but also has been recommended by 'natural food' promoters as an item of human diet. The alkaloid content in the leaves of both *S. officinale* and the hybrid *S.* × *uplandicum* (known as Russian comfrey) can be as much as 0.15% dry weight; furthermore the crude leaf extracts have been shown to produce chronic hepatotoxicity in rats. Clearly, to avoid liver damage, one should avoid eating these plants in any quantity on a regular basis. The ecological function and toxicity of these alkaloids will be mentioned in more detail later in this chapter (Section V).

While the general toxicity of plant alkaloids in mammals, and especially in man and farm animals, is widely recognized, their teratogenic effects have only recently been recorded. Adult female cattle and sheep may imbibe alkaloids in their diet in insufficient amount to cause their death, but as a result of feeding on alkaloid-containing plants, congenital defects may occur in their offspring. Among alkaloids implicated in this way are the pyrrolizidine group, the nicotine group, those of *Lupinus* and also the simple piperidine derivative, coniine of hemlock (Keeler, 1975). The malformed offspring usually suffer various skeletal damage and defects of the digits or of the palate. Such livestock have very limited survival rates. In humans, teratogenic effects including skeletal damage as exhibited in the condition known as 'spina bifida' have been attributed to overconsumption by pregnant women of potato tubers, which contain the alkaloid solanine. The relationship between this human congenital defect and diet is, however, a complex one and the role of potato constituents as the causative agents is still far from proven (see Kuc, 1975).

A new and ecologically interesting series of polyhydroxy alkaloids have been encountered in plants which have a structural resemblance to sugars (Fellows *et al.*, 1986). They are simple molecules in which the oxygen atom of a monosaccharide is replaced by nitrogen (Fig. 3.3). Two typical structures are deoxymannojirimycin, which was found in the seeds of the legume *Lonchocarpus sericeus*, and castanospermine, which occurs in the seed of *Castanospermum australe* to the extent of 0.06%. These alkaloids have been termed 'sugar-shaped weapons of plants' because as sugar analogues they are able to inhibit the enzymes of animal carbohydrate metabolism. Deoxymannojirimycin thus inhibits α-mannosidase and castanospermine α-glucosidase. Toxic effects in animals ingesting them may be due to their ability to arrest carbohydrate breakdown. Indeed, swainsonine, a third member of the group which occurs in leaves of *Swainsona*, a legume pasture plant, is toxic to grazing cattle, causing neurological symptoms. This is due to the accumulation of mannose-based oligosaccharides, which the animal's α-mannosidase is prevented from breaking down.

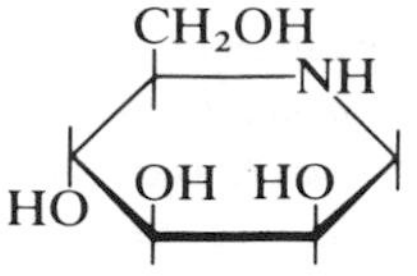

deoxymannojirimycin

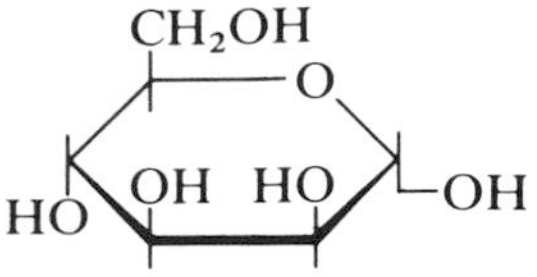

α-D-mannose

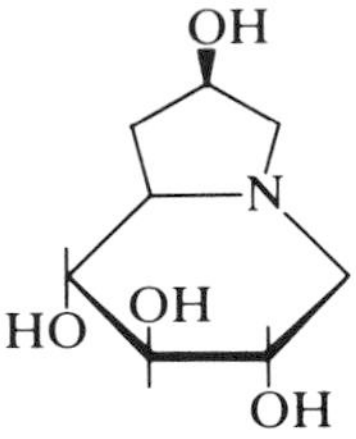

castanospermine

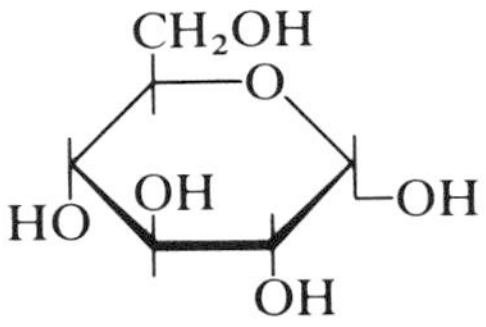

α-D-glucose

Figure 3.3 Alkaloids which resemble simple sugars

While plant proteins are not usually thought of as being toxic, there are a few which are highly dangerous to animals. One is abrin, the main protein of the seed of *Abrus precatorious* (Leguminosae), the lethal dose of which in man is as little as half a milligram. Since the seeds, which are attractively coloured red and black, are employed by African natives for making necklaces, fatalities due to abrin poisoning do occasionally occur. Like most proteins, abrin can be denatured by heating and the toxic effects disappear when the temperature is raised above 65°C. A second well-known protein toxin is ricin, the protein of the castor bean, *Ricinus communis*. It is a protoplasmic poison, the lethal dose, as measured in mice, being 0.001 μg ricin nitrogen/g body wt.

A number of legume seeds, e.g. soybean *Glycine max*, contain proteins which are trypsin inhibitors. While these are not toxic as such, they presumably have a protective role against animal feeding, since they reduce the nutritional value of the protein in seeds containing them. Another class of proteins present in legume seeds are the phytohaemagglutinins, so called because of their ability to coagulate the erythrocytes of human blood. These glycoproteins, which are used routinely in the identification of certain human blood groups, are widely present in plant seeds, both in the Leguminosae and in the angiosperms generally (present in 79 of 147 families tested) (Toms and Western, 1971). The ecological significance of these proteins has largely been neglected, but a study by Janzen *et al.* (1976) suggests that they, too, may provide protection in seeds to insect attack. These authors were able to show that the reason why bruchid beetles (*Caliosobruchus maculatus*), can eat cowpeas (*Vigna unguiculata*) and not black beans (*Phaseolus vulgaris*) is because while the former are phytohaemagglutinin-free, the latter are rich in these proteins. Indeed, 'artificial' seeds made from cowpea flour containing from 1 to 5% phytohaemagglutinin of black beans were found to be lethal when eaten by the beetles. In this case, trypsin inhibitors were present in seeds of both legumes and thus were not implicated in insect repellency.

How far protein toxins generally protect angiosperm seeds from overpredation by animals has not really been studied, but the above examples illustrate the potentialities of plant proteins in this ecological role. Peptides may also be utilized in a similar way and toxic peptides are known both from higher plants (viscotoxin from mistletoe, *Viscum album*) and from fungi (the cyclic heptapeptide amanitine of *Amanita phalloides*).

B. Non-nitrogenous Toxins

It is not always appreciated that a plant substance does not have to be an alkaloid or even to contain a nitrogen atom in the structure to be toxic to animals. There are many poisonous compounds which are terpenoids or even fairly simply hydrocarbons. For example, many of the plant extracts used as arrow poisons by natives in Africa contain cardiac glycosides, such as ouabain, as active ingredients. These steroidal substances are heart poisons. Again, the so-called 'five-finger death' caused by the consumption by humans or cattle of the oddly shaped roots of water dropwort, *Oenanthe crocata*, is caused by the presence of polyacetylene hydrocarbons, such as oenanthetoxin, not of alkaloids.

One of the simplest of all non-nitrogenous toxins is monofluoroacetic acid, CH_2FCO_2H, which occurs in the South African plant, *Dichapetalum cymosum*. It is poisonous because it stops respiration, through inhibition of the Krebs tricarboxylic acid cycle; the fatal dose in man is 2–5 mg/kg body wt. Fluoracetic acid is taken into the cycle instead of acetic, and is metabolized to fluorocitric acid, and it is the enzyme aconitase which refuses this substrate as a substitute for citric that causes the respiratory cycle to stop. Another toxic organic acid in plants such as rhubarb (in the leaves) is oxalic acid $(CO_2H)_2$. Oxalate is only really toxic when associated with the sodium or potassium ion to give a soluble salt; the calcium salt, by contrast, is insoluble and may pass through the animal body without being absorbed. In spite of its simple structure, its mode of action is poorly understood, although it is conceivable that it interferes with terminal respiration by inhibiting the key enzyme succinic dehydrogenase. The fatal dose of oxalic acid is quite high; only plants which contain 10% or more as dry weight are likely to be harmful to mammals (cf. Keeler *et al.*, 1978).

A selection of the known non-nitrogenous toxins are listed in Table 3.2 and some of their structures are given in Fig. 3.4. Among the terpenoids, two particularly toxic groups are mentioned—the cardiac glycosides (or cardenolides) and the saponins. There are other structures in addition; for example, the toxic principles of *Rhododendron* leaves and flowers are diterpenes. Also, the sesquiterpene lactones, compounds widely distributed in the Compositae, include some substances which are either toxic in insects (Burnett *et al.*, 1974) or repellent in having allergenic skin effects in animals (Mitchell and Rook, 1979). A few sesquiterpene lactones are poisonous to farm animals (e.g. hymenovin in *Hymenoxys odorata*), while quite a number are cytotoxic, having anti-tumour activity (Rodriguez *et al.*, 1976). The sesquiterpene lactones are often bitter-tasting as are another group of terpenoid toxins, the monoterpene lactones or iridoids. Iridoids occur in plants both in the free state (e.g. nepetalactone in the volatile oil of catmint) and in glycosidic form (e.g. aucubin) from which the free toxin is liberated after enzymic hydrolysis.

Some non-nitrogenous toxins in plants are also notable for causing photosensitization in farm animals. The quinone hypericin (see Fig. 3.4) of *Hypericum perforatum*, for example,

Table 3.2 Some non-nitrogenous toxins in plants

Class of compound	Example	Toxicity
Iridoids	Aucubin in *Aucuba japonica* leaves	Insects, birds
Sesquiterpene lactones	Hymenovin in *Hymenoxys odorata*	Livestock and insects
Cardiac glycosides	Ouabain in *Acokanthera ouabaio*	Heart poison, LD_{50} in rats 17.2 mg/kg
Saponins	Medicagenic acid in *Medicago sativa* leaves	Fish, insects
Furanocoumarins	Xanthotoxin from *Pastinaca sativa*	Insects
Isoflavonoids	Rotenone in derris root	Mainly insects and fish
Quinones	Hypericin in *Hypericum perforatum* leaves	Mammals, especially sheep
Polyacetylenes	Oenanthetoxin in *Oenanthe crocata* roots	Mammals
Aflatoxins	Aflatoxin B_1 from *Aspergillus flavus* infection on peanut	Birds and mammals

is a photodynamic compound which is absorbed by the animal and enters peripheral circulation. When exposed to sunlight, the animal as a result becomes susceptible to sunburn and other damage; serious necrosis of the skin can occur, with subsequent infection and starvation. Among other photodynamic compounds present in plants are furanocoumarins such as psoralen, which are responsible for photosensitization in sheep which have fed on spring parsley *Cymopterus watsonii* (see Keeler, 1975). Through their photodynamic properties, furanocoumarins are also toxic to most insects. For example, xanthotoxin, a linear furanocoumarin in leaves of the parsnip, *Pastinaca sativa,* when fed to army-worm larvae, *Spodoptera uridania,* at 0.1% level in the daylight, causes 100% mortality; when the same compound is fed in the dark, 40% of larvae survive. However, some species have apparently circumvented this barrier to feeding by rolling up the leaves (e.g. of umbellifer plants rich in these coumarins) before eating so that the effects of sunlight on the ingested toxins are avoided (Berenbaum, 1978). Other plant molecules recently shown to cause toxicity in animals through photosensitization are the polyacetylenes and thiophenes (Towers, 1980).

One further group of plant toxins—the aflatoxins—are exceptional in that they are not higher plant products at all but of microbial origin. The first aflatoxins were discovered in peanuts, which were incorporated into the diet of turkeys and other poultry and caused what at first seemed to be a mysterious 'turkey-X' disease. The cause of death was traced to the contamination of the peanuts, after harvesting, by a fungus *Aspergillus flavus*. Indeed, the fungus when grown on peanuts produces a series of oxygen heterocyclic compounds, and it was these substances that killed the poultry. There are at least four major aflatoxins produced by *A. flavus* of which aflatoxin B_1 is representative (see Fig. 3.4). The recognition of the cause of poisoning as being due to mould products

medicagenic acid, from *Medicago sativa* (lucerne)

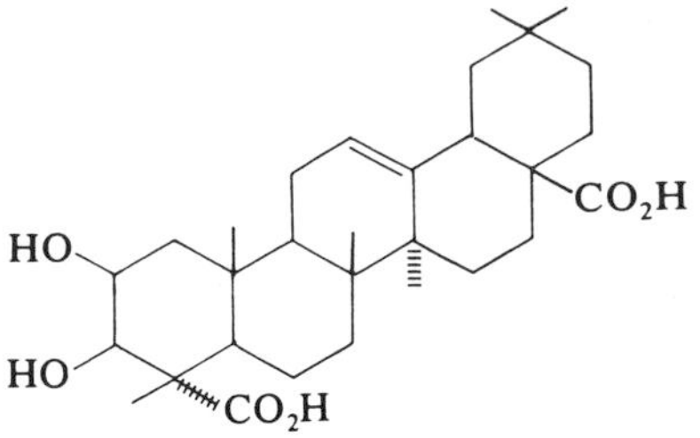

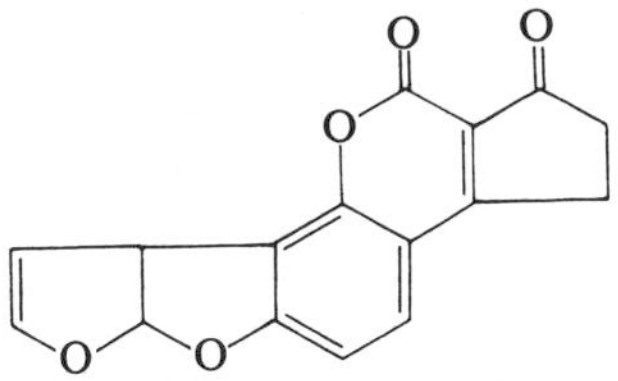

aflatoxin B_1, from *Aspergillus flavus* growing on peanuts *Arachis hypogea*

$$HOCH_2CH{=}CH(C{\equiv}C)_2(CH{=}CH)_2(CH_2)_2CHOH(CH_2)_2CH_3$$

oenanthetoxin, from *Oenanthe crocata*

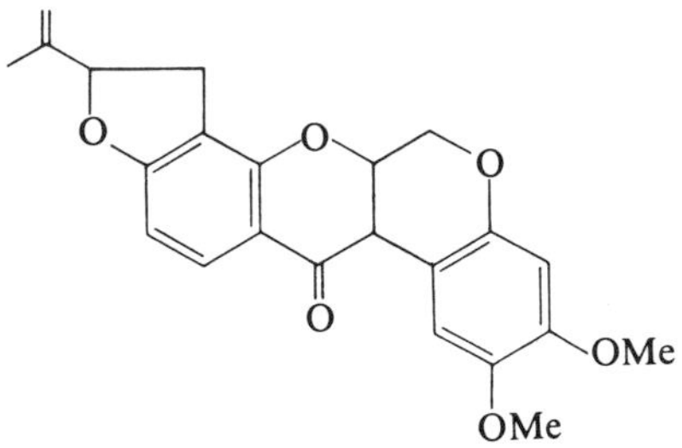

rotenone, from *Derris* root

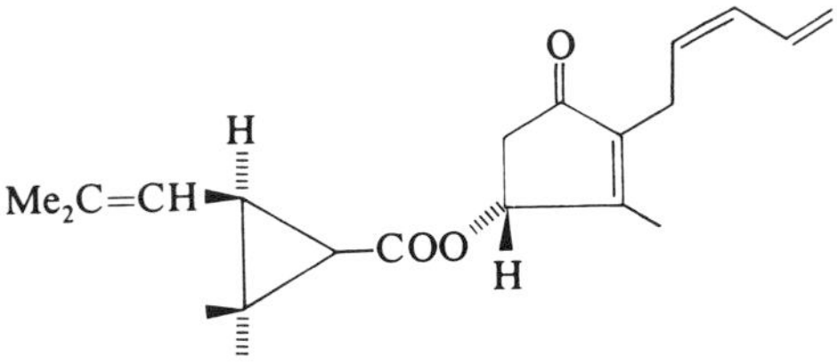

pyrethrin I, from *Chrysanthemum cinearifolium*

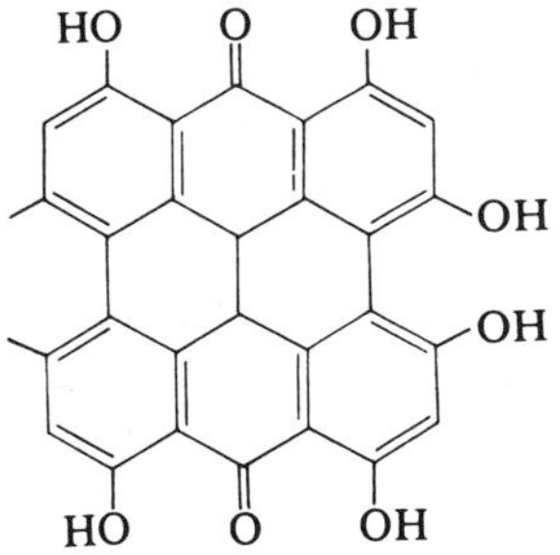

hypericin, from *Hypericum perforatum* (St. Johns wort)

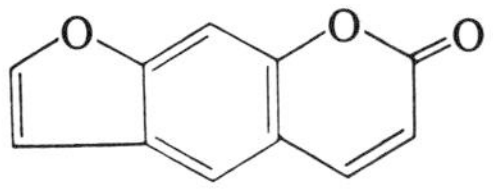

psoralen from umbellifer leaves and stems

Figure 3.4 Some non-nitrogenous plant toxins

was made easy in this case by the intense UV fluorescence of the toxins. The pure compounds are carcinogenic in higher animals, with death being due to liver damage. The lethal dose in ducklings is 20 μg, with death occurring after 24 h. The LD_{50} in mg/kg body wt ranges from 0.35 in ducks and 0.5 in dogs to 9.0 in mice. While both pigs and cattle suffer from the toxin, sheep are relatively impervious.

Following the discovery of aflatoxins in peanuts infected with *A. flavus*, it has been realized that a number of other fungi which are capable of infecting plant foodstuffs produce similar toxins and the general term mycotoxins is now in regular use (Smith and Moss, 1985). These substances represent a significant threat as dangerous contaminants in foodstuffs of plant origin. Whether such toxins have any importance in natural ecological systems is not yet clear but the fact that higher plants live in a symbiotic relationship with many kinds of lower plant suggests that it is at least conceivable for a higher plant to protect itself from animal predators by harbouring a fungus or bacterium within its tissues which is able to manufacture a lethal toxin of this type.

All the toxins so far mentioned have their effects on the higher forms of animal life; further details of the effects of plant toxins on livestock are reported in Keeler *et al.* (1978) and in Keeler and Tu (1983). There are also a range of non-nitrogenous substances which appear to be synthesized by plants specifically to ward off insect attack. Two of the best known groups of insecticide of plant origin are the rotenoids, which occur in legume roots, and the pyrethrins, from *Chrysanthemum cinearifolium* flower heads. Other lesser known groups of compounds, not obviously poisonous to higher animals, have also been shown to have toxic consequences in insects. Several common flavonol glycosides including rutin, quercitrin and isoquercitrin, are toxic to a number of insects, including *Heliothis zea*, *H. virescens* and *Pectinophora gossypiella* (Shaver and Lukefahr, 1969). The role of these substances in insect feeding and deterrence to feeding will be discussed later in Chapter 5.

C. Fate in Animals

As mentioned in the introduction, the harmful effects of plant toxins in animals are closely related to their fate *in vivo*. Within animal tissues, most plant constituents are likely to undergo metabolism. Such metabolism is part of the detoxification process and represents the efforts of the animal system to eliminate a potentially harmful material. The metabolite may then be conjugated, particularly if it is lipid soluble, in order to make it water soluble and non-toxic. Detoxification systems are either very well developed in animals or can be induced by the presence of elevated levels of a toxin in the diet. Ultimately the toxin is likely to be excreted in harmless conjugated form through the urine or faeces. Toxicity of a plant substance may simply reflect the failure of the detoxification processes at some stage to cope with a particular organic structure.

In order to appreciate the ecological significance of plant toxins, some knowledge of the detoxification processes of animals is essential. The subject has been extremely widely studied, particularly in relation to drug metabolism and pesticide residues. Both natural toxins and foreign compounds (synthetic drugs, insecticides, etc.) have been investigated and the all-embracing term 'xenobiotic' is used to refer to an organic compound whose *in vivo* fate is under investigation. A general account of the biotransformation of xenobiotics has been provided by Millburn (1978). Detailed reviews of detoxification processes in mammals (Scheline, 1978) and in insects (Dauterman and Hodgson, 1978) are also available. Here, only a brief outline is necessary; appropriate examples will appear in later sections of this chapter and in later chapters.

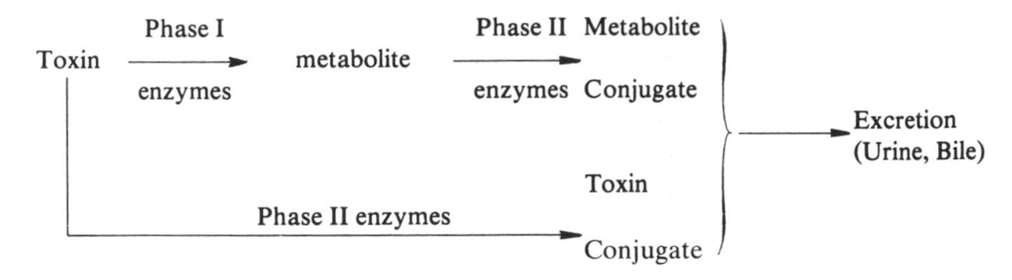

Figure 3.5 Outline of detoxification processes in mammals

As in plant detoxification (see p. 26), a major objective in animal tissues is the solubilization of what is usually a lipophilic compound so that it can be excreted in water-soluble form (Fig. 3.5). For this purpose, the so-called Phase I enzymes are available to convert the original toxin to a metabolite which is capable of being conjugated with sugar or sulphate anion. The most important Phase I enzymes are mono-oxygenases, which are present in liver microsomes and involve cytochrome P-450 as electron carrier. These enzymes, which are capable of catalysing a wide variety of oxidations, are referred to as polysubstrate mixed-function oxidases (PSMOs). In the case of aromatic substances, PSMOs will catalyse the introduction of a phenolic hydroxyl group. Thus benzene when fed to animals is oxidized to phenol which can then be removed from the system as the water-soluble *O*-glucuronide. Any aliphatic compound present in the diet is similarly susceptible to oxidation to give a corresponding alcohol. In phytophagous insects, it is possible to demonstrate the rapid induction of PSMOs by introducing various secondary compounds (e.g. α-pinene, sinigrin) into an otherwise neutral artificial diet (Brattsten, 1992). Although PSMOs are the best studied type of Phase I enzyme, other reactions besides oxidation may occur, such as reduction and/or hydrolysis (see Section VB).

In the second stage of detoxification, conjugation occurs with either the fed compound, if it contains a suitable site for conjugation, or else with the metabolite from a Phase I reaction (Fig. 3.5). Stage II enzymes are generally transferases and require a supply of energy in the form of an activated nucleotide intermediate. In mammals and other vertebrates, the most common conjugates are the glucuronides or ethereal sulphates, but conjugation with amino acids, especially with glycine or ornithine, is regularly observed. In insects (and other invertebrates) glucoside formation via a glucosyltransferase and UDP glucose is a major process; sulphate and phosphate conjugates have also been observed (Dauterman and Hodgson, 1978).

The final stage is excretion through the urine or bile, the molecular size of the excreted compound being the major factor in determining which of these routes is the main pathway. Compounds with two or more aromatic or saturated rings tend to be found in the bile and such compounds undergo an enterohepatic circulation prior to elimination in the faeces and are also subject to metabolism by intestinal microflora. Flavonoids, including isoflavones, for example, are generally broken down by gut bacteria and are recovered as small aromatic fragments. The common plant flavonol quercetin is thus converted to 3-hydroxyphenylacetic acid. This further metabolism becomes important when considering the oestrogenic effects of ingested isoflavonoids in farm animals (see Chapter 4, Section II). Sequestration of a metabolite in the liver may occur and in the case of pyrrolizidine alkaloids in mammals, this can have fatal consequences (see Section VB). In phytophagous insects, sequestration of natural products is a fairly regular event (Duffey, 1980). It may occur with harmless substances, such as the flavone and carotenoid pigments found in butterflies, as well as with plant toxins. In the latter case, such sequestration may be of survival value when a poisonous alkaloid is stored or sequested by an insect in order to avoid predation by other animals (see Section IV).

One final point may be made about detoxification processes in animals: there can be considerable intra- and interspecific variation. The different fates of ingested compounds, for example, represent an interesting area of comparative biochemistry in the animal kingdom. One illustrative example is the fate of benzoic acid. It is excreted as the glycine conjugate by mammals, amphibia, fish and insects, as the ornithine conjugate by birds (Galliformes and Ansiformes) and reptiles and as the arginine conjugate by arachnids and myriapods (Millburn, 1978). Variation in detoxification also occurs at the populational level within a given species. A familiar example in man is the phenomenon of beeturia, whereby 14% of the adult population in the British Isles cannot metabolize betanin, the red betacyanin pigment of the beetroot, and excrete it unchanged in the urine (Watson, 1964). Betanin is actually an alkaloid, as distinct from the flavonoid-based anthocyanin pigments present in most plants. It is an interesting question whether or not people suffering from beeturia may be more susceptible to certain types of plant alkaloid poisoning than the rest of the population.

III. Cyanogenic Glycosides, Trefoils and Snails

A. Occurrence of Cyanogenic Glycosides in Plants

One of the most intriguing examples of plant toxins affecting plant–animal interactions is that involving cyanogenic glycosides, their variable occurrence in clover and birdsfoot trefoil and the differential eating of these plants by slugs and snails. We owe the major development of this work to D. A. Jones, who has described his results in several major reviews (see e.g. Jones, 1972, 1988). Only a brief outline can be given here, the reader being referred to these reviews for further details.

Cyanogenesis is the ability of plants to synthesize compounds (cyanogenic glycosides) which liberate prussic acid or hydrogen cyanide (HCN) upon hydrolysis. One of the classic sources of HCN is the seed of the bitter almond *Prunus amygdalus*, which contains the glycoside amygdalin. The characteristic odour of HCN, which not everyone can smell, is thus 'of bitter almonds'. The toxicity of HCN is such that many cases of livestock poisoning and occasional human deaths are recorded each year. HCN is toxic to a wide spectrum of organisms, since its site of action is inhibition of the cytochromes of the electron transport system.

The fact that HCN is easily detected by a spot test using picrate paper (it turns from yellow to red or brown in the presence of the gas) has meant that the distribution of

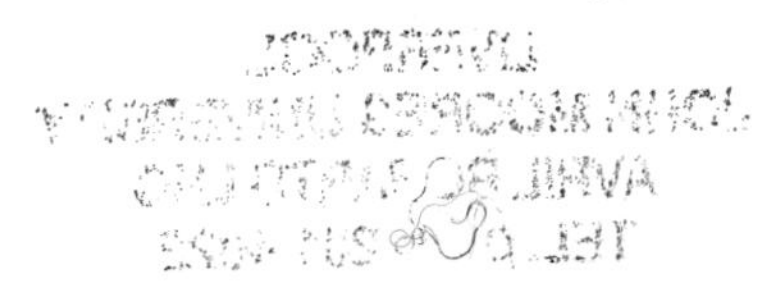

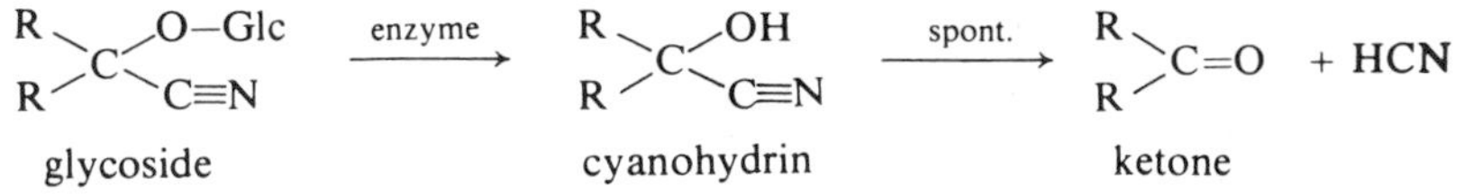

$$\underset{\text{glycoside}}{\mathrm{R_2C(O{-}Glc)(C{\equiv}N)}} \xrightarrow{\text{enzyme}} \underset{\text{cyanohydrin}}{\mathrm{R_2C(OH)(C{\equiv}N)}} \xrightarrow{\text{spont.}} \underset{\text{ketone}}{\mathrm{R_2C{=}O}} + \mathrm{HCN}$$

Figure 3.6 Pathway of release of HCN from cyanogenic glycosides

cyanogenesis has been widely studied. Besides their presence in clover and birdsfoot trefoil, at least 2000 species representing 100 families contain cyanogens. The actual structures of the substances releasing HCN have been studied in a smaller sample and some 50 compounds have been fully characterized. Apart from some cyanogens which are part lipid in structure from the Sapindaceae (Siegler, 1975), the known cyanogens have the same general structure, shown in Fig. 3.6.

The enzymic release of HCN within the plant is strictly controlled and all plants which make a glycoside contain a specific glycosidase which will hydrolyse it. These glycosidases differ in their substrate specificities from the common β-glucosidase and are usually named according to their substrate, e.g. linamarase for linamarin and so on. Hydrolysis of the glycoside releases the sugar (usually glucose) and an intermediate cyanohydrin, which then spontaneously decomposes, producing a ketone or aldehyde and HCN (Fig. 3.6). This second step can also be catalysed by the enzyme α-hydroxynitrile lyase, which is widespread in cyanogenic plants. Its presence presumably speeds up the otherwise spontaneous breakdown of the cyanohydrin.

Probably linamarin (dimethyl substituted) and lotaustralin (methyl, ethyl substituted), the two glycosides of clover and birdsfoot trefoil, are the two commonest cyanogens in nature. These occur in other legume fodder plants, in flax (*Linum*) and in several Euphorbiaceae and Compositae. Aromatic substituted cyanogenic glycosides are also known, such as amygdalin, present in seed of bitter almonds to the extent of 1.8%, and dhurrin, occurring in the cereal *Sorghum vulgare*.

One final point about the cyanogenic glycosides needs stressing—their biosynthetic origin from protein amino acids. It is difficult to believe that plants directly utilize significant amounts of essential amino acids in this way unless some functional benefit is attached to these products. The frequent earlier suggestions that cyanogens are waste products of primary metabolism seems particularly untenable in the case of substances so immediately formed from protein amino acids and containing their amino acid nitrogen locked up in this way.

The pathway to linamarin synthesis begins with the amino acid valine and takes place in five steps as outlined in Fig. 3.7. Lotaustralin is similarly derived from isoleucine and the aromatic cyanogenic glycosides (Fig. 3.8) come from phenylalanine or tyrosine.

B. Polymorphism of Cyanogenesis

The two plants used by Jones (1972) in the study of the ecological role of cyanogenesis are the clover *Trifolium repens* and birdsfoot trefoil *Lotus corniculatus*, two common pasture plants of temperate grassland. The enormous advantage of using these two plants is that cyanogenesis is a genetically variable character in populations of both species. The

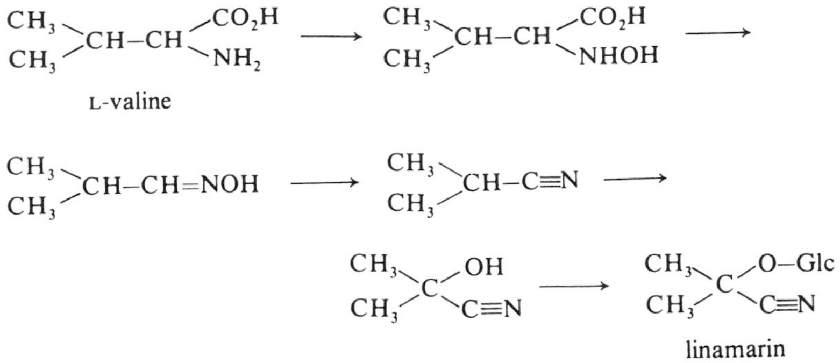

$(CH_3)_2CH{-}CH(CO_2H)NH_2$ (L-valine) $\longrightarrow$ $(CH_3)_2CH{-}CH(CO_2H)NHOH$ $\longrightarrow$

$(CH_3)_2CH{-}CH{=}NOH$ $\longrightarrow$ $(CH_3)_2CH{-}C{\equiv}N$ $\longrightarrow$

$(CH_3)_2C(OH)C{\equiv}N$ $\longrightarrow$ $(CH_3)_2C(O{-}Glc)C{\equiv}N$ (linamarin)

Figure 3.7 Pathway of biosynthesis of linamarin from valine

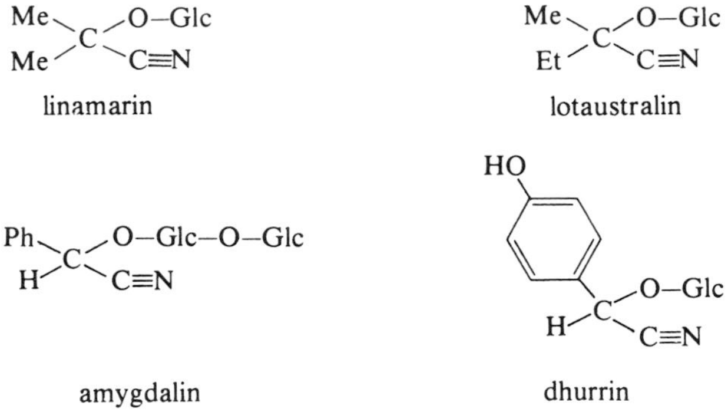

linamarin: $Me_2C(O{-}Glc)C{\equiv}N$

lotaustralin: $Me(Et)C(O{-}Glc)C{\equiv}N$

amygdalin: $Ph(H)C(O{-}Glc{-}O{-}Glc)C{\equiv}N$

dhurrin: $HO{-}C_6H_4{-}C(H)(O{-}Glc)C{\equiv}N$

Figure 3.8 Structures of representative cyanogenic glycosides

function of cyanogenesis should thus be revealed by comparing plants with and without the character.

This variability or chemical polymorphism in clover was early recognized by geneticists and breeding experiments between cyanogenic and acyanogenic forms soon showed that two genes Ac and Li controlled its production. Ac controls the synthesis of cyanogenic glycoside (e.g. linamarin) and Li the enzyme (e.g. linamarase) needed to break it down to give HCN. Natural populations fall into four genotypes (AcLi, Acli, acLi and acli) which can be identified phenotypically by suitable chemical tests.

Fresh leaves are placed in a test tube, crushed briefly with a glass rod in the presence of a drop of chloroform and the tube is stoppered, with a piece of filter paper soaked in picric acid solution hanging down from the stopper. Coloration (to red brown) within an hour indicates cyanogenesis and the dominant type AcLi. Coloration after 24 h standing indicates the genotype Acli, since some HCN is liberated non-enzymically from the glycoside during this period. The process can be speeded up by adding some linamarase after the first 1-h period.

If no colour is given after the 24-h period, the leaf sample must be acLi or acli. In order to finally distinguish between these two possibilities, it is necessary to add a small amount

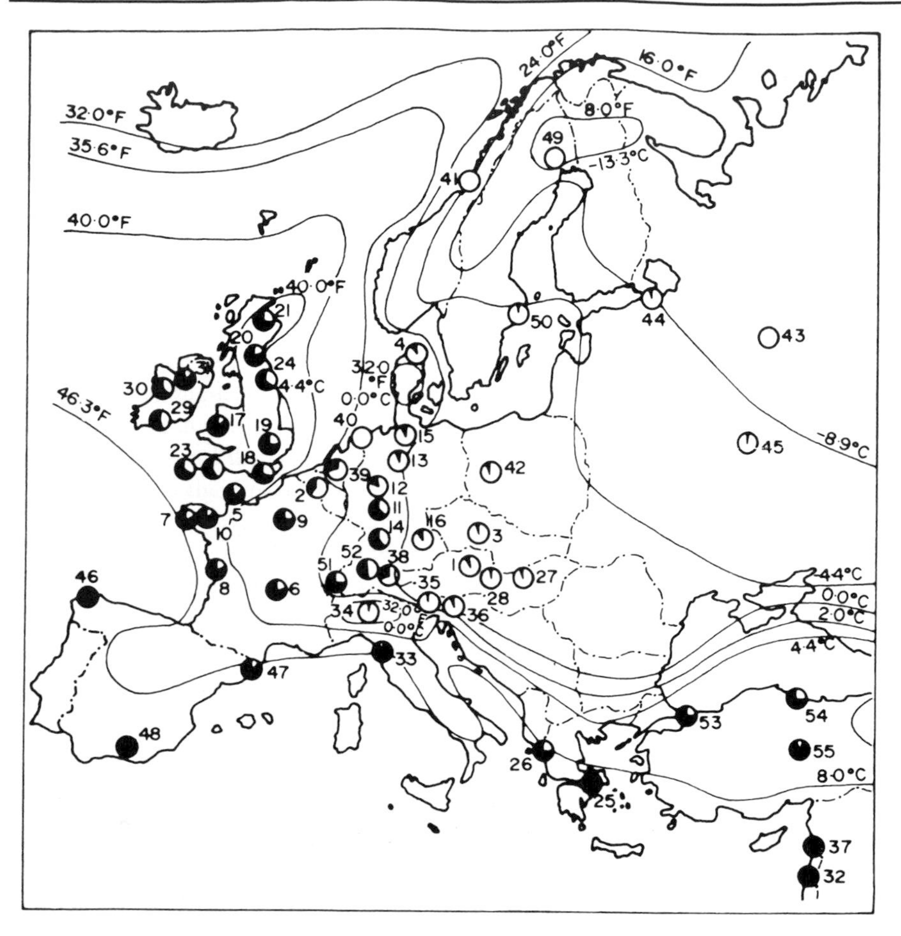

Figure 3.9 Correlations between cyanogenic frequency and January isotherms of clover populations in Europe (from Jones, 1972)

of linamarin to the tube. A colour change now indicates acLi; samples still giving no colour can be registered acli, since they lack both glycoside and enzyme. Only the doubly dominant type AcLi is registered as cyanogenic, all the other three types being acyanogenic in the field.

The fact that clover populations are so readily scored for this character must have been at least partly responsible for Daday (1954) choosing to study the frequency of cyanogenesis in different populations. He soon found remarkable differences in frequency between different European populations. The only factor linking frequency in any given population seemed to be the January mean temperature. A remarkably close correlation was indeed apparent in his results between frequency of cyanogenesis in a given clover population and the winter isotherm at that geographical site. Thus in British populations when the January isotherm is higher than 5°C cyanogenesis runs from 70 to 95% frequency, while in central Russia where the winter temperature is very low, clover populations are acyanogenic. Mid-European populations are intermediate in both senses, with cyanogenic frequencies of 20–50% (Fig. 3.9).

No one was successful in explaining this peculiar correlation between two such apparently unrelated phenomena until Jones (1966) showed conclusively that slugs and snails, which are significant predators on both plants, show preferential eating for acyanogenic forms of *L. corniculatus*. Taking this new factor into account, one can now interpret Daday's data coherently. The reason why cyanogenesis is high in British populations is probably because the high winter isotherm means that predators like slugs and snails are active all the year round. When clover and the trefoil start germinating in the spring, the young seedlings are very vulnerable to herbivores at the critical stage in their development and HCN is of considerable selective advantage to deter feeding. By contrast, cyanogenesis is low in frequency or absent from Russian populations of these plants because the icy cold winter forces most animal predators to hibernate. By the time the herbivores are active in the late spring, the clover seedlings will have developed sufficient leaf material not to require protection from feeding, as in England.

An alternative explanation for the infrequency of cyanogenesis in eastern European populations is the possibility that when low temperatures and frost are frequent, cyanogenesis confers a reduced physiological fitness to the plant. The system of storing cyanogen and hydrolytic enzyme becomes unstable, HCN is released by frost action and the cyanogen becomes an autotoxin (Brighton and Horne, 1977). However, this effect of low temperatures could be overcome quite simply by phenotypic variation, i.e. the avoidance of cyanogenesis production at unfavourable seasons of the year. Indeed, this probably happens in *Lotus corniculatus*, since both stable and unstable phenotypes exist among natural populations (Jones *et al.*, 1978).

Evidence that cyanogens provide protection from mollusc feeding in clovers and trefoils has been obtained from both field and laboratory experiments. Field studies have been carried out over the course of 16 years in maritime populations of *L. corniculatus* by Jones *et al.* (1978). These have convincingly confirmed the tenet that the maintenance of high levels of cyanogenesis in certain local populations is due to the protection it affords from mollusc browsing. Laboratory studies have also shown that differential eating occurs on both *Lotus* and *Trifolium* plants. Of 13 species of slugs and snails examined, seven showed preferences for the acyanogenic forms, while the other six showed no selection. Some predator species are thus tolerant to cyanogenesis and have developed resistance to it by means of detoxification, a process which is now well understood (see Section C below).

Slugs capable of detoxifying cyanide, e.g. *Deroceras reticulatum*, nevertheless vary in their response to cyanogenesis according to the frequency of cyanogenesis in a given population (Burgess and Ennos, 1987). Slugs obtained from sites with a low frequency of cyanogenic clover tend to avoid eating cyanogenic forms, whereas slugs from sites where there is a high frequency of cyanogenic plants appear to have become adapted to the toxin and show much less discrimination. Thus the selective advantage that is enjoyed by a cyanogenic form under grazing by this slug will only be substantial where the frequency of cyanogenic plants is low (e.g. between 11 and 24%). There will be less protection when the frequency of the cyanogenic form is high (e.g. 65%).

Even heavily grazed plants, however, may not necessarily suffer lethal damage, since the cyanogenic glycosides are concentrated in the more vital organs (stem and cotyledon) rather than in the more expendable leaf tissue. Furthermore, the concentrations show a 33% increase in the seedlings, when comparing them at day 5 and day 35. Garden slugs, *Arion hortensis,* will graze and kill the youngest plants but will discriminate against grazing on the cyanogenic forms in older seedlings (Horrill and Richards, 1986).

One further factor in the interaction between slugs and clover, overlooked until recently, is the toxicity of the ketone released alongside the HCN (see Fig. 3.6). Linamarin releases acetone, while lotaustralin releases butanone and both these compounds are more deterrent to slug feeding than HCN (Jones, 1988). Insect feeding on cyanogenic plants (see below) is also probably limited by the repellency of the toxic aldehydes and ketones released at the same time as HCN is formed.

C. Other Protective Roles of Cyanogens

When snails and slugs feed on trefoil or clover, it is evident that many species are adapted to cyanide in the diet. Adaptation to HCN has been studied in farm animals and there is much evidence that detoxification occurs in sheep and cattle. Detoxification is by means of an enzyme rhodanese which converts the cyanide ion to thiocyanate:

$$CN^- + S \xrightarrow{\text{rhodanese}} CNS^-$$

The sulphur comes from β-mercaptopyruvic acid, $HSCH_2COCO_2H$, which in turn is converted to pyruvate. A similar process is involved in the clinical treatment of cyanide poisoning in man, when sodium thiosulphate is administered intravenously:

$$CN^- + Na_2S_2O_3 \rightarrow CNS^- + Na_2SO_3$$

Evidence that sheep are adapted to feeding on cyanogenic clover has been obtained by feeding them continuously with small amounts of the toxin in their diet. While unadapted sheep can be killed with a dose of 2.4 mg/kg body wt, animals that have become adapted can tolerate as much as 15–50 mg HCN/kg body wt. Sheep respond to the effects of mild HCN poisoning by ceasing to feed until the toxin clears from their systems.

It is thus clear that mammals are usually only killed by feeding on cyanogenic plants when presented with a single large dose at any one time. Such exposure occurs, for example, when tree stumps of the cyanogenic legume *Holocalyx glaziovii* produce young saplings just at the time when grass is scarce in Brazilian prairies. Cattle eat the saplings and die from the high intake of HCN. The protective function of cyanide to the plant is very clearly demonstrated in such instances (da Silva, 1940).

Humans are also subject to HCN poisoning, since cassava roots, used as a staple food in West Africa, contain significant amounts of cyanogenic glycoside, even after powdering and conversion to flour. It has been estimated that people living on cassava receive a daily dose of 35 mg HCN, half the lethal dose. Adaptation is clearly via rhodanese detoxification. However, high levels of thiocyanate may occur as a result and although this is not toxic, it has well-known goitrogenic properties and the long-term effects of living on cassava may be early mortality (see Evered and Harnett, 1988).

The ecological role of cyanogenic glycosides has been explored in other plants besides clover and *Lotus* and there is good evidence that cyanogenesis provides partial, if not complete, protection from predation by a wide spectrum of animal species. Common bracken, *Pteridium aquilinum*, a very successful weed, is another plant which is polymorphic for the cyanogenic character. Here, there is clear evidence of preferential feeding on the acyanogenic form by sheep and fallow deer living in Richmond Great Park, London (Cooper-Driver and Swain, 1976). There is also evidence that desert rodents avoid feeding on seeds of the jojoba plant, *Simmondsia chinensis*, because of the cyanoglucosides they contain (Sherbrooke, 1976).

In insects, the bitter taste of the glycosides appears to act as a deterrent, although the quality of the taste may be modified by other cellular constituents. How else can one explain the facts that only three of eight insect species avoided cyanogenic leaves of *Lotus*, when given a choice of cyanogenic and acyanogenic tissues, whereas four of four insect species avoided cyanogenic petals (Compton and Jones, 1985)? Since almost all insects are capable of detoxifying HCN via rhodanese (Beesley *et al.*, 1985), the protective value of cyanide may be due in insects to its further metabolism by β-cyanoalanine synthase to β-cyanoalanine, $NH_2CH(CO_2H)CH_2CN$, which also has toxic effects (Nahrstedt, 1985).

Finally, it may be observed that defence by HCN production is not confined to the plant kingdom. It is a protective device in animals, millipedes producing it to ward off attacks by ants. Some red warning-coloured moths also produce HCN at all stages in the life-cycle to make themselves unpalatable feeding to their predators. The protective role of cyanogenesis in insects is reviewed by Nahrstedt in Evered and Harnett (1988).

IV. Cardiac Glycosides, Milkweeds, Monarch Butterflies and Blue-Jays

What is now the classic example of plant–animal co-evolution, in which secondary plant compounds have a key role is the interaction between milkweeds, monarch butterflies and blue-jays. It is an interesting case where insects have capitalized on the plant toxins and used them in their defence against bird predators. It has the intriguing biological feature of warning coloration in that the insects involved are aposematic.

The various organisms—plants, insects, and birds—involved in this interaction are shown in Table 3.3. The chain of events linking these organisms has been worked out by the principal investigators, who include Brower (1969), Roeske *et al.* (1976), Rothschild (1972) and Reichstein and coworkers (1968). It is briefly as follows:

(1) The milkweed produces several cardiac glycosides within its tissues as a passive defence against insect feeding. The substances are both bitter tasting and toxic to higher animals.
(2) The monarch butterfly caterpillar learns to adapt to these toxins. They are sequestered during feeding and then stored safely within the insect body. This milkweed becomes the preferred food plant of the insect, since there are very few other competing feeders.

Table 3.3 Organisms of the milkweed–monarch butterfly–blue-jay interaction

Plants	Insects	Bird
Asclepias curassavica (milkweed) and other *Asclepias* spp. (Asclepiadaceae)	*Danaus plexippus* (monarch) and four other danaid butterflies	*Cyanocitta cristata bromia* (blue-jay)
Nerium oleander (Apocynaceae)	Other small insects	

(3) The adult butterfly flies away from the host plant with the protective cardiac glycosides stored within it.
(4) A naive blue-jay, in laboratory experiments, tries feeding on the butterfly. It receives a mouthful of bitter-tasting cardiac glycoside, causing it to vomit.
(5) Presented with a second monarch butterfly, the blue-jay turns away in distaste, since it has learned to associate the bright coloration of the butterfly with the bitter cardiac glycoside. The blue-jay avoids feeding on this butterfly and any others with the same warning coloration.

The situation, in practice, is more complex than this. The captive blue-jay represents a model predator; in the real world, birds occasionally succeed in overcoming this defence (see p. 97). Also, the fact that the blue-jay learns to associate the warning colour with the toxic compounds means that colour alone can allow the insect some protection. Indeed, Brower has calculated that only 50% of any given butterfly population need to carry toxins in order for the toxins to provide 100% protection from blue-jays. This biological device is presumably a safety factor to allow for the fact that the host plants may vary in the quantity and quality of toxins they synthesize and which are available to the insect in any one year. The concentration of glycoside in the insect will vary with the season and with the particular species of the host plant fed upon. Furthermore, not all cardiac glycosides are unpalatable, so that insects may occasionally feed on *Asclepias* without accumulating emetic compounds.

Plant-determined variations in the cardenolide content of adult butterflies can be evaluated by comparing the chromatographic separation of extracts from insects and from food plants (Fig. 3.10). There are some differences in pattern due to the metabolic conjugation in the insect of some of the plant components. Nevertheless, it is possible to recognize from these qualitative profiles which of several *Asclepias* species was fed upon by the butterfly at the caterpillar stage. In quantitative terms, the relationship between plant source and insect storage is more complex, since the larvae are able to adjust their intake according to the amount of cardenolide available in the food plant (Brower *et al.*, 1984).

The cardiac glycosides synthesized by the host plants include many different structures. Two examples are calotropin, one of the major glycosides of *Asclepias curassavica*, and oleandrin, from *Nerium oleander* (Fig. 3.11). All the plant toxins seem to be absorbed and passed through into the insect's body. Thus, no less than ten glycosides, including calotropin, have been identified in the body tissue of monarch butterflies reared on milkweed. Similarly, eleven

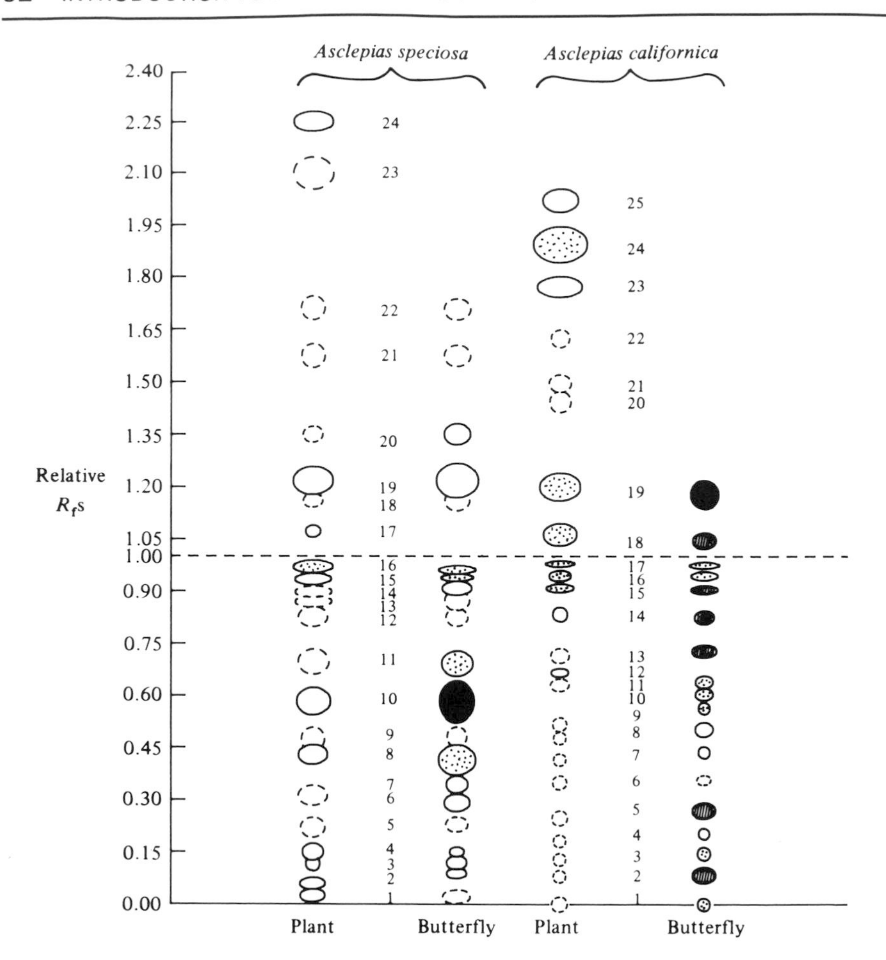

Figure 3.10 TLC separations of cardenolides from two *Asclepias* species and from monarch butterflies which have derived their toxins from these plants (R_fs are relative to digitoxin marker)

compounds have been found in the Lygaeid bug *Caenocoris nerii*, reared on oleander. Brower has calculated that the amount of cardiac glycoside stored on average in a single male butterfly is sufficient to cause five blue-jays to vomit 50% of the time! This odd statistic is a product of the particular method used for testing blue-jay feeding behaviour.

The protection cardiac glycosides afford to the monarch butterfly is also extended to other insects which live on milkweeds, namely four other aposematic butterflies, several Lygaeid bugs, Pyrgomorphid grasshoppers and beetles and one aphid. Additionally, yet other butterflies mimic the colour pattern of the monarch and gain aposematic protection, although they do not feed on *Asclepias* nor do they sequester toxins.

Finally, there is the question why the plants continue to synthesize toxins if the main benefactors are the insect feeders. The simple answer is that production of the toxins continues to provide the plant with protection not only from most insects but also from all

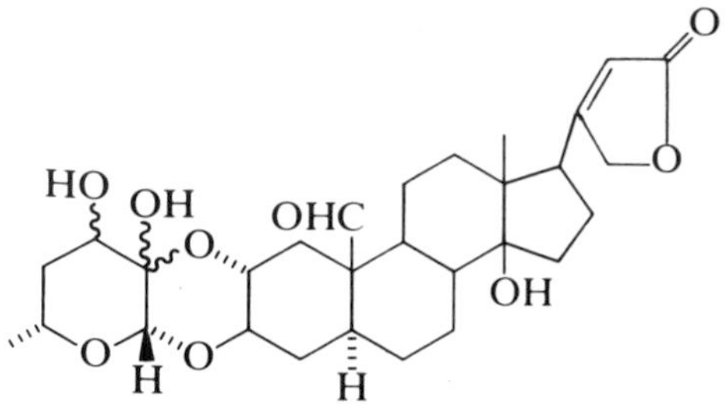

calotropin

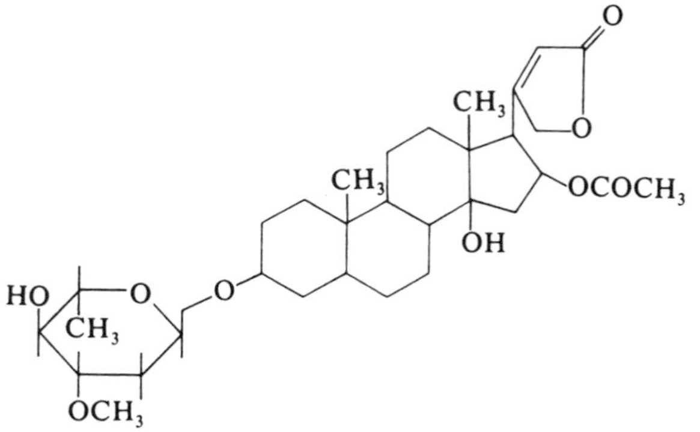

oleandrin

Figure 3.11 Structures of two cardiac glycosides

grazing animals; cattle, for example, will avoid eating such plants. The fatal dose of dry *Asclepias labriformis* leaf in a 100-pound sheep is as little as 0.8 ounces, so it is not surprising that cut fodder containing *Asclepias* material can often cause livestock deaths (Seiber *et al.*, 1984). Also the caterpillars feeding on the plant may accumulate the repellent odour of that plant and serve to reinforce its effectiveness against predators. In addition, the presence of these insects on the plant can act as a lure to pollinators, who will be drawn to the flower and will achieve its pollination.

V. Pyrrolizidine Alkaloids, Ragworts, Moths and Butterflies

A. Pyrrolizidine Alkaloids in Moths

A similar relationship to that linking milkweeds and butterflies in semitropical areas of North and Central America has been discovered in temperate climates, indeed in the meadows around the University of Oxford in England. In this case, the plants are the groundsel, *Senecio vulgaris*, and the ragwort, *S. jacobaea*, two very successful weeds of the Compositae family. These plants are powerfully protected from herbivores by containing in their leaves a series of alkaloids of the pyrrolizidine type. Many cases of cattle poisoning have been attributed to these *Senecio* alkaloids (Keeler, 1975). Nevertheless, the caterpillars of the tiger moth *Arctia caja* and the cinnabar moth *Tyria jacobaeae* feed with impunity on these two weeds and carry on their whole life-cycle on these same host plants.

Analysis of the caterpillars and adult moths shows, indeed, that all six pyrrolizidine alkaloids of the ragwort are sequestered and stored by these insects. There are changes in the proportions of the alkaloids in plant and in insect and there is evidence that the insect transforms one alkaloid of the host plant into another *in vivo*. The most remarkable demonstration of the non-toxicity of the alkaloids in the insect is the fact that the alkaloids are even present in the insect eggs. The protection of these moths from both their bird and other predators again involves warning coloration, both the caterpillars and the adults being brightly coloured and patterned.

The interaction between *Senecio* and moth is complicated by variations in alkaloid content of the host plant with season, and geographical site. The tiger moth is relatively polyphagous and also feeds on foxglove *Digitalis purpurea* and can even store its cardiac glycosides. Both classes of toxin have been found in the same insect (Rothschild *et al.*, 1979). Presumably, the caterpillar modifies its feeding habits in order to ensure receiving sufficient toxin of one type or another to protect itself.

While the cardiac glycoside–monarch butterfly and *Senecio* alkaloid–moth interactions are the only two to have been fully investigated, it is apparent that such interactions are common in aposematic insects. Rothschild (1973) has listed over 40 insect species (23 Lepidoptera, 1 Neuroptera, 7 Hemiptera, 5 Coleoptera, 1 Diptera and 6 Orthoptera) which have the ability to sequester and store plant toxins. Negative results were recorded for some 35 aposematic insects examined for toxins at the same time.

B. Pyrrolizidine Alkaloids in Butterflies

The ability of insects to feed on plants containing pyrrolizidine alkaloids and store them is remarkable in view of the high toxicity of these substances to other forms of life, particularly mammals. Mattocks (1972) suggests that the toxicity of pyrrolizidines lies not in their own structures, but in the major *in vivo* metabolites to which they are converted. It appears as if mammals, in attempting to detoxify the pyrrolizidine molecule by dehydrogenation, have accidentally produced much more dangerous compounds, with fatal consequences.

The pathway of metabolism in mammals (Fig. 3.12) involves hydrolysis of the ester groups of the alkaloid to give the main parent compound retronecine, and this is then dehydrogenated to a pyrrole, which is more toxic than retronecine and has the ability to bind to macromolecules like DNA in the liver. A more reactive metabolite, (*E*)-4-hydroxyhex-2-enal, is formed from the breakdown of this pyrrole *in vivo*, and this appears to be a major agent in pyrrolizidine poisoning (Segall *et al.*, 1985).

That some insects can safely store *Senecio* alkaloids without modification is exemplified by the two moths which feed on these plants in Oxfordshire meadows (see above). That other insects can metabolize them, without fatal consequences, has become apparent from

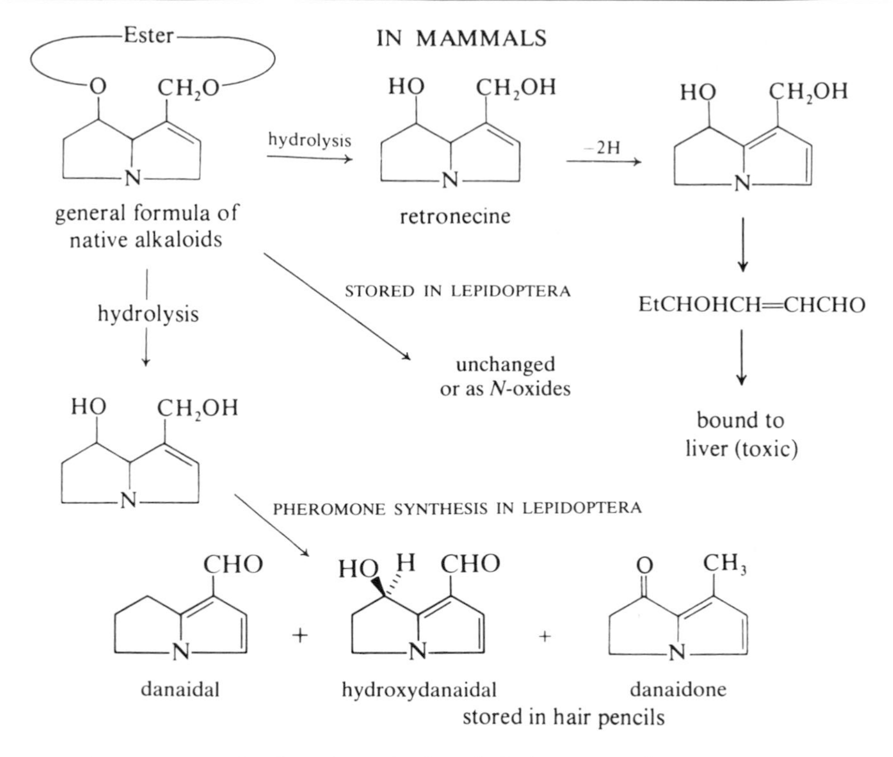

Figure 3.12 The fate of pyrrolizidine alkaloids in mammals and in insects

some recent observations on the *Asclepias*-feeding danaid butterflies, which were the main topic of an earlier section in this chapter.

It is clear that these butterflies have another secondary compound requirement, not provided for them by the milkweed (Edgar and Culvenor, 1974; Edgar *et al.*, 1974). This is for pyrrolizidine alkaloids, which are needed to manufacture aphrodisiacal substances which the male butterfly stores in its wing hair pencils and uses in its courtship display to attract the female. As the male hovers over the female, he sprinkles the hair pencil scent pheromones onto the female antennae as a 'love dust' to prepare her for mating. Three of the sex compounds are danaidal, hydroxydanaidal and danaidone (see Fig. 3.12) which are clearly different in their oxidation level from the pyrrole, the dangerous toxin in mammals. These pyrroles are clearly of dietary origin, derived from *Senecio* alkaloids. Indeed, one of the unchanged ester alkaloids of *Senecio* has been found in the insect hair pencils.

Male danaid butterflies have been observed to visit both borages and *Senecio* plants in order to satisfy the feeding requirement. Plants of the Boraginaceae are one of the few other natural sources of *Senecio*-type alkaloids. The adult butterflies are attracted to the withered leaves and are able to suck with their proboscis the alkaloid-containing exudation. They may also obtain their alkaloids from other plant secretions, e.g from *Senecio* nectar, from around the base of seed pods (as in *Crotalaria*) or even from the activities

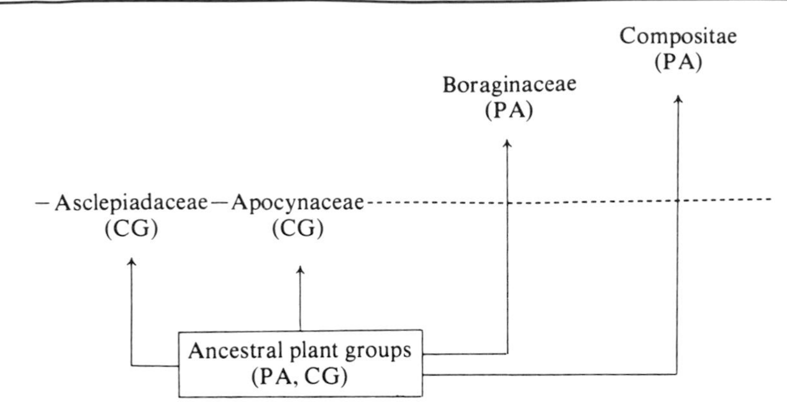

Figure 3.13 Hypothetical scheme for the evolution of plant families containing plant toxins

of other insects. They may enter into a casual symbiotic relationship with certain grasshoppers, which release the juice of plant leaves while feeding on *Heliotropium* plants. This very remarkable, indeed unique, behaviour of danaid butterflies has been carefully substantiated by both laboratory and field observations (cf. Bernays *et al.*, 1977).

One fascinating evolutionary feature of this unusual behaviour of danaid butterflies is their failure to obtain the courtship display compounds from their feeding as caterpillars. It is, indeed, very rare for adult butterflies to have any other requirements than nectar during their remaining life. It is clearly connected with their need for two different classes of plant substance—cardiac glycoside (for protection against bird predation) *and* pyrrolizidine alkaloid (for courtship display). One simple explanation of their odd present-day behaviour is that danaid butterflies originally fed as caterpillars on a plant containing *both* cardiac glycoside and alkaloid.

Going back in evolutionary time, one can envisage a scheme (Fig. 3.13) whereby some ancestral form of present-day Apocynaceae contained both types of toxin. Subsequent evolutionary pressures might have forced plants to choose between making either one or other of the two types of toxin, so that we reach the present-day situation, where the toxins are differentially distributed among Apocynaceae and Asclepiadaceae (cardiac glycosides), and Boraginaceae and Compositae (alkaloids). Such a theory would envisage that some primitive present-day Apocynaceae might conceivably still retain the capacity of synthesizing and accumulating both classes of toxin. A search for such a plant has, indeed, been successful and a species of *Parsonsia* contains both types of secondary compound (Edgar and Culvenor, 1975).

Other aspects of the role of cardiac glycosides and pyrrolizidine alkaloids in the life-style of danaid butterflies have come to light as a result of continuing research on these fascinating insects. It is now clear, for example, that the pyrrolizidine alkaloids are dual purpose: they are both protective and serve as essential pheromone precursors. This follows from regular observations that adult females, to a lesser extent than males, collect the alkaloid-rich exudates from withered borage or *Senecio* plants. Furthermore, many danaid species, including *Danaus plexippus* and *D. chrysippus*, have been found to store these alkaloids in their tissues during much of their adult lives (Edgar *et al.*, 1979).

If pyrrolizidine alkaloids are thus available to these butterflies for defence, one might question the value of sequestering cardiac glycosides as well. A simple answer would be that two types of chemical defence are better than one. A more subtle advantage of having cardiac glycosides, however, has been revealed in some feeding experiments of Smith (1978). He found that larvae of *D. chrysippus* fed on cardenolide-containing asclepiads were less subject to endoparasitic attack by flies than those fed on an asclepiad species lacking cardenolides. Since parasites are probably more important than predators in limiting danaid butterfly populations (Edmunds, 1976), the accumulation of cardenolides in the insect system may be just as essential as the acquisition of pyrrolizidine alkaloids.

The importance of cardenolides to the monarch butterfly has also been underlined by experiments in which larvae, fed on leaf-free artificial diets, were found to contain cardioactive substances in spite of their absence from the diet (Rothschild *et al.*, 1978). Indeed, it is possible that even in the wild where larvae are free to ingest cardenolides from their host plants, the adult butterfly may still contain its own cardioactive compounds, chemically differing from the plant cardenolides, in order to reinforce the toxic effectiveness of the dietary materials.

Although the monarch does not itself synthesize steroids recognizable as plant cardenolides, other insects are known to have this ability. Thus certain chrysomelid beetles do not have the opportunity of sequestering such compounds, since their host plants (e.g. various *Mentha* spp.) lack cardiac glycosides. The beetles nevertheless contain large amounts (100–200 μg/ml) of cardiac xylosides in their defensive secretions, which besides being toxic are intensely bitter (Pasteels *et al.*, 1979). The biosynthetic origin and effectiveness of insect defence substances are both large subjects and they will be considered in greater depth in Chapter 8.

The importance to the monarch butterfly of having more than one type of chemical toxin has been underlined by recent findings that two bird species are able to overcome their cardenolide defence among the dense, overwintering colonies of the butterfly that are found in central Mexico (Fink and Brower, 1981). These are insects which have migrated south during the autumn from their usual habitats in the central and eastern United States.

Predation is due to the black-backed oriole *Icterus abeillei* and the black-headed grosbeak *Pheuctius melanocephalus*. The success of these two birds is partly due to their lower sensitivity compared to other birds to the emetic effects of the toxins and partly due to the fact that the overwintering butterflies tend to contain less toxic cardenolides than normal. Mouse predation has also been observed in these sites (Glendinning and Brower, 1990). Indeed the ability of some birds and mice to predate on monarchs may lie in the fact that an increasing number of the larvae are feeding in the spring on milkweeds which only have low-potency cardenolides. Such plants, e.g. *Asclepias syriaca* and *A. speciosa*, have increased in abundance in the monarchs' natural habitat due to changes in agricultural and land use. In this instance, man may thus have triggered off a chain of events which has unwittingly altered the natural balance in butterfly populations in favour of their bird and mouse predators.

VI. Utilization of Plant Toxins by Animals

While the most dramatic examples of toxin utilization by animals are those involving cardiac glycosides and pyrrolizidine alkaloids (see Sections IV and V), these are not the only plant secondary constituents to be employed in this way. Continuing research has revealed at least ten other classes of plant toxin which can be sequestered from the diet and stored by insects for defence. These are listed in Table 3.4 (Harborne, 1987).

From these experiments, one may note the following six points.

(1) More than one class of toxin may be utilized by a given insect. Danaid butterflies not only obtain cardiac glycosides and pyrrolizidine alkaloids from plants but also apparently sequester a series of plant bases, called pyrazines, from larval food plants. Three such compounds have been identified in the warning odour secretions of both the monarch butterfly and the five-spot burnet moth *Zygaena lonicerae* (Rothschild *et al.*, 1984). Pyrazines represent yet another protective device in Lepidoptera against bird predators. They are musky odour components, which have other interesting effects, including the ability to sharpen the memory of the recipient.

(2) Different groups of insects vary in their utilization of toxins. Ithomiine butterflies, a sister group of the Danainae, utilize pyrrolizidine alkaloids as male sex pheromones after gathering them from withered plants in the adult stage. However, they differ from danaids in not acquiring any toxins, in the larval stage. This is surprising inasmuch as their food plants—members of the Solanaceae—are well endowed with alkaloids (Brown, 1984).

(3) The same toxins may be *both* sequestered from the diet *and* synthesized *de novo* by the insect. This is true in the case of the cyanogenic glycosides, linamarin and lotaustralin, which are synthesized by the burnet moth, *Zygaena trifolii* from valine and isoleucine (Wray *et al.*, 1983) and are also acquired from the food plant, *Lotus corniculatus* (Nahrstedt and Davis, 1986).

(4) Utilization of plant toxins may have unexpected consequences for an insect. Male adults of the Arctiid moth *Creatonotos gangis* differ considerably when raised on larval diets containing and lacking pyrrolizidine alkaloids. In the first case, the scent organs (coremata) grow to a large size and secrete up to 400 μg hydroxydanaidal per insect, while in the second case, the coremata are tiny and lack scent. Thus the plant alkaloid, or the pheromone derived from it, specifically regulates the growth of an organ which releases a chemical signal that is synthesized as the result of that regulation (Boppré, 1986).

(5) Some metabolism and/or conjugation of plant toxins commonly occur within the insect. This happens with the cardenolides and pyrrolizidine alkaloids in Lepidoptera, as already described. It is true also of salicin, which is sequestered from the food plant *Salix*, by the chrysomelid beetle, *Chrysomela aenicollis*. The beetle hydrolyses the sugar linkage and oxidizes the alcoholic group to an aldehyde (Fig. 3.14). The procduct, salicylaldehyde, becomes the principal component of the defensive secretion (Rowell-Rahier and Pasteels, 1982).

(6) Utilization of plant toxins is nearly always accompanied by warning coloration (cf.

Table 3.4 Classes of plant toxin sequestered by insects and stored for defence

Class of chemical	Typical structure	Plant source	Insect storing it
Aristolochic acids	Aristolochic acid	*Aristolochia* spp.	Butterfly, *Battus archidamus*
Cardiac glycosides	Calotropin	*Asclepias* spp.	Butterfly, *Danaus plexippus*
Cucurbitacins	Cucurbitacin D	*Cucurbita* spp.	Beetle, *Diabrotica balteata*
Cyanogenic glycosides	Linamarin	*Lotus corniculatus*	Moth, *Zygaena trifolii*
Glucosinolates	Sinigrin	*Brassica oleracea*	Butterfly, *Pieris brassicae*
Iridoids	Aucubin	*Plantago lanceolata*	Butterfly, *Euphydryas cynthia*
Methylazoxymethanols	Cycasin	*Zamia floridina*	Butterfly, *Eumaeus atala*
Phenols	Salicin	*Salix* spp.	Beetle, *Chrysomela aenicollis*
Pyrazines	3-Isopropyl-2-methoxypyrazine	*Asclepias curassavica*	Butterfly, *Danaus plexippus*
Pyrrolizidine alkaloids	Retronecine	*Senecio* spp.	Moth, *Arctia caja*
Quinolizidine alkaloids	Cytisine	*Cytisus scoparius*	Aphid, *Aphis cytisorum*
Polyhydroxy alkaloids	2,5-Dihydroxymethyl-3,4-dihydroxypyrrolidine	*Omphalea* spp.	Moth, *Urania fulgens*

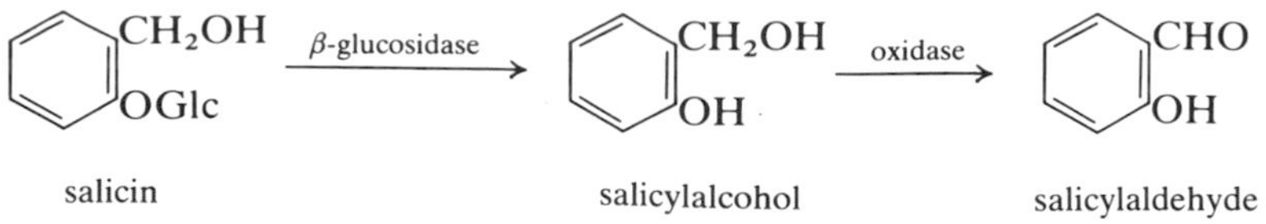

CH_2OH / $OGlc$ $\xrightarrow{\beta\text{-glucosidase}}$ CH_2OH / OH $\xrightarrow{\text{oxidase}}$ CHO / OH

salicin salicylalcohol salicylaldehyde

Figure 3.14 Conversion of ingested salicin to the defensive salicylaldehyde by a chrysomelid beetle

Rothschild, 1973). The most recent example is the rare hairstreak butterfly, *Eumaeus atala florida*, which absorbs the toxic glucoside, cycasin, from its cycad food plant. Both the red and yellow larvae and the striking orange-coloured adults are then protected by a cycasin content of 0.02% (larvae) and 1.0–1.8% (adult) (Rothschild *et al.*, 1986).

VII. Summary

In this chapter, the potentially toxic features of plants are considered as well as the responses of some animal herbivores to these toxins. There are many classes of toxin, the well-known alkaloids and cyclic peptides and the less familiar polyacetylenes and glucosinolates. It is suggested that the majority of plant species have the ability to harm their herbivores in some degree. Certain toxins, like the cyanogenic glycosides, are poisonous to all forms of life. Others, like the saponins, are specifically targetted at insects and molluscs, while yet others, such as the cardiac glycosides, specifically arrest the heartbeat of the mammalian grazer.

Many plant toxins exert their effects on animals in a relatively subtle way, sometimes over a long period of time. Some toxins, like the furanocoumarins, become more toxic on exposure to light. Others, like the quinolizidine alkaloids, have a teratogenic effect, causing skeletal damage in the offspring of grazing animals. Yet others, such as the pyrrolizidine alkaloids, are cumulative poisons, producing liver damage over a period of weeks or months.

The response of animals to plant poisons varies considerably. Some insects, like the Danaid butterflies and the Arctiid moths, can handle several classes of plant toxin with impunity, using them as pheromone precursors or as defence agents (see Sections IV and V). At least 12 classes of toxin can be borrowed from plants in this way and stored for protection against predation (Table 3.4). One insect, the bruchid beetle *Caryedes brasiliensis*, which feeds on legume seeds containing the non-protein amino acid canavanine, adapts the toxin to its own metabolism, converting it to useful nitrogen in the process (Section IIA).

Most animal feeders have a well-developed enzyme system to detoxify plant poisons and excrete them in water-soluble form (Section IIC). Detoxification can be costly and hence some animals may choose to discriminate in their feeding against toxic plants. This can be monitored in the clover plant, which exists in both cyanogenic and non-cyanogenic forms. Here the interaction is complex, especially when molluscs are the grazers (Section IIIB). Cyanogenic glycosides are distinctive in releasing two types of toxin on hydrolysis: hydrogen cyanide, a respiratory poison, and a toxic aldehyde or ketone. Although animal adaptation to the second type of toxin (e.g. benzaldehyde or acetone) has yet to be examined, it is already clear that cyanogenesis is a successful defence in plants against herbivory by several kinds of animal. The defensive role of other classes of plant toxin has also been surveyed in recent years and this will constitute much of the subject matter of Chapter 7. Cyanogenesis in clover is an example of a variable defence system and the biosynthetic cost of plant defence will also be considered in this later chapter.

Bibliography

Books and Review Articles

Bell, E. A. (1972). Toxic amino acids in the Leguminosae. In: Harborne, J. B. (ed.), 'Phytochemical Ecology', pp. 163–178. Academic Press, London.

Brattsten, L. B. (1992). Metabolic defenses against plant allelochemicals. In: Rosenthal, G. A. and Berenbaum, M. R. (eds.), 'Herbivores: their interactions with secondary metabolites', Vol. 2, 2nd edn, pp. 176–242. Academic Press, San Diego.

Brower, L. (1969). Ecological chemistry. *Scient. Am.* **220**, 22–29.

Dauterman, W. C. and Hodgson, E. (1978). Detoxification mechanisms in insects. In: Rockstein, M. (ed.), 'Biochemistry of Insects', pp. 541–577. Academic Press, New York.

Duffey, S. S. (1980). Sequestration of plant natural products by insects. *A. Rev. Entomol.* **25**, 447–477.

Evered, D. and Harnett, S. (eds.) (1988). 'Cyanide Compounds in Biology', John Wiley, Chichester.

Feeny, P. (1975). Biochemical coevolution between plants and their insect herbivores. In: Gilbert, L. E. and Raven, P. H. (eds.), 'Coevolution of Animals and Plants', pp. 3–19. Texas Univ. Press.

Fellows, L. E., Evans, S. V., Nash, R. J. and Bell, E. A. (1986). Polyhydroxy plant alkaloids as glycosidase inhibitors and their possible ecological role. In: Green, M. B. and Hedin, P. A. (eds.), 'Natural Resistance of Plants to Pests', pp. 72–78. American Chem. Soc., Washington.

Harborne, J. B. (1987). Chemical signals in the ecosystem. In: Dodge, J. D. (ed.), 'New Perspectives in Plant Science', pp. 39–58. Academic Press, London.

Janzen, D. H. (1969). Seed-eaters versus seed size, number, toxicity and dispersal. *Evolution* **23**, 1–27.

Jones, D. A. (1972). Cyanogenic glycosides and their function. In: Harborne, J. B. (ed.), 'Phytochemical Ecology', pp. 103–124. Academic Press, London.

Jones, D. A. (1988). Cyanogenesis in animal–plant interactions. In: Evered, D. and Harnett, S. (eds.), 'Cyanide Compounds in Biology', pp. 151–170. John Wiley, Chichester.

Keeler, R. F. (1975). Toxins and teratogens of higher plants. *Lloydia* **38**, 56–86.

Keeler, R. F. and Tu, A. T. (eds.) (1983). 'Handbook of Natural Toxins, Vol. 1, Plant and Fungal Toxins'. Marcel Dekker, New York.

Keeler, R. E., Van Kampen, K. R. and James, L. F. (eds.) (1978). 'Effects of Poisonous Plants on Livestock', 600 pp. Academic Press, New York.

Kuc, J. (1975). Teratogenic constituents of potatoes. *Recent Adv. Phytochem.* **9**, 139–150.
Mattocks, A. R. (1972). Toxicity and metabolism of *Senecio* alkaloids. In: Harborne, J. B. (ed.), 'Phytochemical Ecology', pp. 179–200. Academic Press, London.
Millburn, P. (1978). Biotransformation of xenobiotics by animals. In: Harborne, J. B. (ed.), 'Biochemical Aspects of Plant and Animal Coevolution', pp. 35–76. Academic Press, London.
Mitchell, J. and Rook, A. (1979). 'Botanical Dermatology', 787 pp. Greenglass, Vancouver.
Riechstein, T., von Euw, J., Parsons, J. A. and Rothschild, M. (1968). Heart poisons in the monarch butterfly. *Science, N.Y.* **161**, 861–866.
Rodriguez, E., Towers, G. H. N. and Mitchell, J. C. (1976). Biological activities of sesquiterpene lactones. *Phytochemistry* **15**, 1573–1580.
Roeske, C. N., Seiber, J. N., Brower, L. P. and Moffitt, C. M. (1976). Milkweed cardenolides and their comparative processing by monarch butterflies. *Recent Adv. Phytochem.* **10**, 93–167.
Rosenthal, G. A. (1982). 'Plant Nonprotein Amino and Imino Acids', 273 pp. Academic Press, New York.
Rothschild, M. (1972). Some observations on the relationship between plants, toxic insects and birds. In: Harborne, J. B. (ed.), 'Phytochemical Ecology', pp. 1–12. Academic Press, London.
Rothschild, M. (1973). Secondary plant substances and warning coloration in insects. In: van Emden, H. (ed.), 'Insect–Plant Interactions', pp. 59–83. Oxford Univ. Press.
Scheline, R. R. (1978). 'Mammalian Metabolism of Plant Xenobiotics', 502 pp. Academic Press, London.
Seiber, J. N., Lee, S. M. and Benson, J. M. (1984). Chemical characteristics and ecological significance of cardenolides of *Asclepias* species. In: Nes, W. D., Fuller, G. and Tsai, L. S. (eds.), 'Isopentenoids in Plants: Biochemistry and Function', pp. 563–588. Marcel Dekker, New York.
Siegler, D. S. (1975). Isolation and characterisation of naturally occurring cyanogenic compounds. *Phytochemistry* **14**, 9–30.
Smith, J. E. and Moss, M. O. (1985). ''Mycotoxins', 148 pp. John Wiley, Chichester.
Toms, G. C. and Western, A. (1971). Phytohaemagglutinins. In: Harborne, J. B., Boulter, D. and Turner, B. L. (eds.), 'Chemotaxonomy of the Leguminosae', pp. 367–462. Academic Press, London.
Towers, G. H. N. (1980). Photosensitizers from plants and their photodynamic action. *Prog. Phytochem.* **6**, 183–202.

Literature References

Beesley, S. G., Compton, S. G. and Jones, D. A. (1985). *J. Chem. Ecol.* **11**, 45–50.
Bell, E. A. (1980). *Prog. Phytochem.* **7**, 171–196.
Berenbaum, M. (1978). *Science, N.Y.* **201**, 532–533.
Bernays, E., Edgar, J. A. and Rothschild, M. (1977). *J. Zool., Lond.* **182**, 85–87.
Boppré, M. (1986). *Naturwissensch.* **73**, 17–36.
Brighton, F. and Horne, M. T. (1977). *Nature, Lond.* **265**, 437–438.
Brower, L. P., Seiber, J. N., Nelson, C. J., Lynch, S. P., Haggard, M. P. and Cohen, J. A. (1984). *J. Chem. Ecol.* **10**, 1823–1857.

Brown, K. S. (1984). *Nature, Lond.* **309**, 707–709.
Burgess, R. S. L. and Ennos, R A. (1987). *Oecologia* **73**, 432–435.
Burnett, W. C., Jones, S. B., Mabry, T. J. and Padolina, W. G. (1974). *Biochem. Syst. Ecol.* **2**, 25–30.
Compton, S. G. and Jones, D. (1985). *Biol. J. Linn. Soc.* **26**, 21–38.
Cooper-Driver, G. A. and Swain, T. (1976). *Nature, Lond.* **260**, 604.
Culvenor, C. C. J., Clark, M., Edgar, J. A., Frahn, J. L., Jago, M. V., Peterson, J. E. and Smith, L. W. (1980). *Experientia* **36**, 377–379.
Daday, H. (1954). *Heredity* **8**, 61–78; 377–384.
Edgar, J. A. and Culvenor, C. C. J. (1974). *Nature, Lond.* **248**, 614–616.
Edgar, J. A. and Culvenor, C. C. J. (1975). *Experientia* **31**, 393–394.
Edgar, J. A., Culvenor, C. C. J. and Pliske, T. E . (1974). *Nature, Lond.* **250**, 646–648.
Edgar, J. A., Boppré, M. and Schneider, D. (1979). *Experientia* **35**, 1447–1448.
Edmunds, M. (1976). *Zool. J. Linn. Soc.* **58**, 129–145.
Erickson, J. M. and Feeny, P. (1974). *Ecology* **55**, 103–111.
Fink, L. S. and Brower, L. P. (1981). *Nature, Lond.* **291**, 67–70.
Glendinning, J. I. and Brower, L. P. (1990). *J. Animal Ecol.* **59**, 1091–1112.
Horrill, J. C. and Richards, A. J. (1986). *Heredity* **56**, 277–281.
Janzen, D. H., Juster, H. B. and Liener, I. E. (1976). *Science, N.Y.* **192**, 795–796.
Jones, D. A. (1966). *Can. J. Genet. Cytol.* **8**, 556–567.
Jones, D. A., Keymer, R. J. and Ellis, W. M. (1978). In: Harborne, J. B. (ed.), 'Biochemical Aspects of Plant and Animal Coevolution', pp. 21–34. Academic Press, London.
McIndoo, N. E. (1945). *US Dept. Agr. Bur. Entom. Plant Quarantine* **ET 661**, 1–286.
Nahrstedt, A. (1985). *Plant Syst. Evol.* **150**, 35–47.
Nahrstedt, A. and Davis, R. H. (1986). *Phytochemistry* **25**, 2299–2302.
Pasteels, J. M., Daloze, D., Dorsser, W. V. and Roba, J. (1979). *Comp. Biochem. Physiol.* **63c**, 117–121.
Rehr, S. S., Bell, E. A., Janzen, D. H. and Feeny, P. P. (1973). *Biochem. Syst. Ecol.* **1**, 63–67.
Rosenthal, G. A., Dahlman, D. L. and Janzen, D. H. (1976). *Science, N.Y.* **192**, 256–257.
Rosenthal, G. A., Janzen, D. H. and Dahlman, D. L. (1977). *Science, N.Y.* **196**, 658–660.
Rothschild, M., Marsh, N. and Gardiner, B. (1978). *Nature, Lond.* **275**, 649–650.
Rothschild, M., Aplin, R. T., Cockrum, P. A., Edgar, J. A., Fairweather, P. and Lees, R. (1979). *Biol. J. Linn. Soc.* **23**, 305– 326.
Rothschild, M., Moore, B. P. and Brown, W. V. (1984). *Biol. J. Linn. Soc.* **23**, 375–380.
Rothschild, M., Nash, R. J. and Bell, E. A. (1986). *Phytochemistry* **25**, 1853–1854.
Rowell-Rahier, M. and Pasteels, J. M. (1982). In: Visser, J. H. and Minks, A. K. (eds.), 'Insect–Plant Relationships', pp. 73–79. Pudoc, Wageningen.
Segall, H. J., Wilson, D. W., Dallas, J. L. and Haddon, W. F. (1985). *Science, N.Y.* **229**, 472.
Shaver, T. N. and Lukefahr, M. J. (1969). *J. econ. Entomol.* **62**, 643–646.
Sherbrooke, W. C. (1976). *Ecology* **57**, 596–602.
da Silva, R. (1940). *Arq. Inst. Biologico (Sao Paulo)* **11**, 461–488.
Smith, D. A. S. (1978). *Experientia* **34**, 844–845.
Stermitz, F. R., Lowry, W. T., Norris, F. A., Buckeridge, F. A. and Williams, M. C. (1972). *Phytochemistry* **11**, 1117–1124.
Watson, W. C. (1964). *Biochem. J.* **90**, 3P.
Wray, V., Davis, R. H. and Nahrstedt, A. (1983). *Z. Naturforsch.* **38c**, 583.

4 Hormonal Interactions Between Plants and Animals

I. Introduction

The idea of hormonal interactions between plants and animals seems remote and in the realms of science fiction when one considers how different their hormonal systems are. In animals, hormones are usually manufactured in special endocrine glands and are then transported to the site of action through the circulatory system. Animal hormones are mainly steroidal or peptidal in chemical structure and fall into easily grouped classes according to their effects.

By contrast, in plants, the ability to synthesize hormones is present in many cells. The site of synthesis varies as the plant develops and the hormones may only translocate relatively short distances. Chemically, plant hormones are very diverse, being of several structurally different types, i.e. purine-based (cytokinins), amino acid based (auxins) or terpenoid-based (dormins, gibberellins). One of the major plant growth hormones is even a gas, the compound ethylene, which exerts its effects through the air spaces between the plant cells.

That hormonal interactions do occur between plants and animals will be shown in this chapter. They are possible at many levels and depend on the ability of physiologically active chemicals to interact between the different types of living organisms. In some cases, the animal is the dominant partner of the interaction; for example, when leaf-cutting ants add auxin hormone to the fungal colonies on which they feed in order to maintain their growth and vitality. More frequently, the plant is dominant, exerting its effects by synthesizing animal hormones and pheromones and thus influencing the life and survival of its animal predators.

The fact that the endocrine system of animals is essentially absent from plants led scientists to reject early reports during the 1930s that mammalian female sex hormones occur in plant tissues. However, recent analytical experiments leave no doubt that both human male and female hormones are present in plants. Their function is, of course, still open to speculation and it is possible that they have a natural role in plants in relation to growth, flowering or sexual expression. Such an argument is supported in part by the discovery of a steroidal hormone, antheridiol, as a chemotactic sex substance in the water fungus, *Achlya bisexualis* (Hendrix, 1970). An alternative suggestion is made here that they are actually synthesized to deter mammalian feeding. Since hormonal activity is delicately balanced and depends on the right amount of a series of compounds arriving in sequence at the right site at the right time, an exogenous dietary source taken at the wrong time could have serious consequences in the reproduction of the female. The feeding deterrent view is supported by the presence in plants of several different compounds which resemble female hormones in structural terms and have oestrogenic activity. This topic of animal and plant oestrogens will be discussed in Section II of this chapter.

Support for the view that hormonal interactions occur between plants and animals has, however, come mainly from entomological studies and the discovery that not one but two classes of insect hormone occur in plants. They occur in relatively large quantities and with a variety of chemical structures. Their function is still speculative but it is clearly possible that they are deliberately produced by plants to interfere with insect metamorphosis and, hence, reproduction. The occurrences in plants of insect moulting hormones and of juvenile hormones will be considered in Sections III and V. One particularly interesting interaction involving insect moulting hormones, fruit flies and cacti will be described in Section IV.

One further type of hormonal interaction is possible between plants and insects in relation to pheromones. This has already been briefly considered in the chapter on pollination ecology under flower scent (p. 56). The question considered here is the dietary origin of insect pheromones. It is possible that pheromones are synthesized *de novo* by the insect, are manufactured from plant substances of dietary origin or that plant compounds are sequestered and used directly without modification. One situation where all three possibilities are represented is the bark beetle–pine tree interaction and this will be presented in the final section of this chapter.

II. Plant Oestrogens

Reports of the presence of female sex hormones in seeds of the date palm and the pomegranate, which first appeared in the literature during the 1930s (e.g. Butenandt and Jacobi, 1933), were treated with scepticism. The methodology of chemical identification was relatively primitive at that time and this factor partly justified the critical attacks on these studies. However, in recent years, much more accurate analytical techniques have been applied to the same and other plant sources (Table 4.1) and the occurrence of both female and male sex hormones in plants is now almost beyond dispute (but see Van Rompuy and Zeevaart (1979) for an alternative view). They occur in only trace amounts and there is at present the possibility of considerable quantitative variation between plant samples. The occurrence of as much as 17 mg/kg oestrone in pomegranate seeds recorded by Heftmann *et al.*. (1966) has been disputed by Dean *et al.* (1971), who found much smaller

Table 4.1 Occurrence of human sex hormones in plants

Compound	Plant source	Concentration (mg/kg)
Oestrone	Date palm, *Phoenix dactylifera*	
	seeds	0.40
	pollen	3.3
Oestrone	Pomegranate, *Punica granatum*	
	seeds	17.0
Oestriol	Willow, *Salix*	
	flowers	0.11
Oestrone	Apple, *Malus pumila*	
	seeds	0.1
Oestradiol-17-β	French bean, *Phaseolus vulgaris*	
	seeds	
Testosterone	Scotch pine, *Pinus sylvestris*	
Androstenedione	pollen	0.08 and 0.59
Androstanetriol	Rayless golden rod, *Haplopappus heterophyllus*	—

For references, see Heftmann (1975) and Young *et al.* (1978).

amounts present. The latter authors, however, noted its presence in seed, flower, leaf and root.

Whether these reports (Table 4.1) indicate that animal sex hormones are widespread in plants in trace amounts remains to be determined. There are considerable practical difficulties in screening plants for them, because of the low levels present and the time-consuming methods needed to prove their occurrence. Their production in plants could be completely accidental, as by-products of pathways leading to functionally more important plant sterols. Alternatively, they may be involved in plant growth and development or even in the control of sexual expression in plants. Indeed, the effect of exogenous application of these hormones has been tested in a number of plant species. Among other effects reported are that: (1) oestrogens stimulate seed germination and growth; (2) oestrogens promote floral development; (3) both oestrogens and androgens increase the expression of femaleness in the cucumber; and (4) testosterone application to *Equisetum* increases the number of female prothalli (see Heftmann, 1975). These data are far from indicating that the sex hormones concerned have an endogenous role in plants. In none of these cases was either a male or a female hormone detected in the plant in question.

Reports of unidentified oestrogenic materials in many plant tissues, based on their effectiveness in upsetting the menstrual cycle in women, cows or ewes, suggest that oestrogens may be much more widespread than is indicated in Table 4.1. During the Second World War, for example, women in Holland correlated the eating of tulip bulbs, forced on them by food shortages, with menstrual upsets and ovulation failures. Among other food sources which have had effects on oestrus in women and cows are garlic, oats, barley, rye grass, coffee, sunflower, parsley and potato tubers. It is possible that some or all of these plant materials lack the hormones themselves but contain instead compounds which

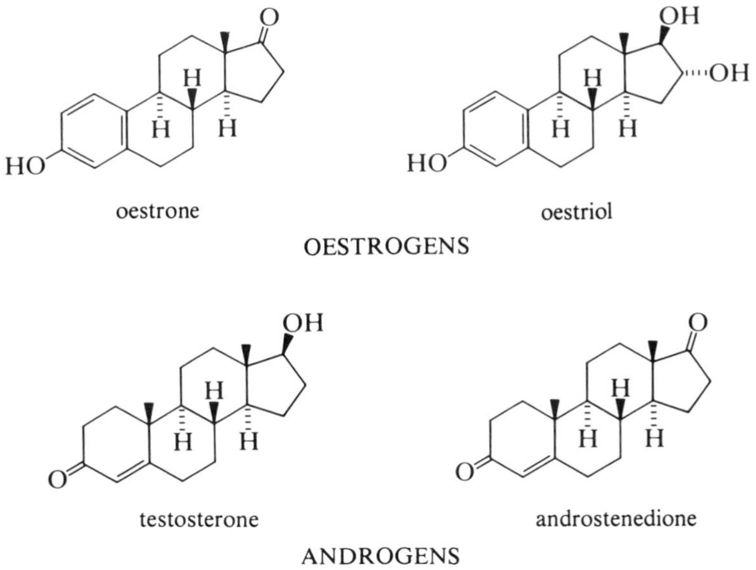

oestrone

oestriol

OESTROGENS

testosterone

androstenedione

ANDROGENS

Figure 4.1 Structures of human sex hormones found in plants

mimic their effect. This possibility is strengthened by the fact that one such steroidal mimic, miroestrol, has been isolated, following a deliberate study of a plant source of known oestrogenic potency. Furthermore, a series of aromatic oestrogens were discovered in plants of the Leguminosae, because of their effects on oestrus in sheep. Indeed, the general term 'phytoestrogen' has been coined to describe any plant compound which has this activity.

The compound miroestrol was isolated when scientists followed up a report that pregnant Burmese and Thai women used an extract of a legume tree root in order to bring on an abortion. The plant was identified as *Pueraria mirifica* and the active principle in the root characterized by Bounds and Pope (1960). Its structure, shown in Fig. 4.2, is remarkably close to that of the natural female hormone, oestrone. When given subcutaneously in multiple doses, miroestrol is as potent as 17-β-oestradiol. Its particular effectiveness as an abortifacient is because of its activity when taken orally. It is over three times as potent as

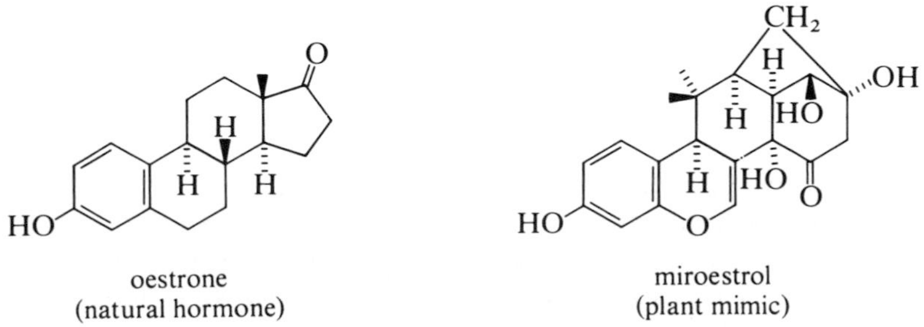

oestrone
(natural hormone)

miroestrol
(plant mimic)

Figure 4.2 Structural comparison between oestrone and miroestrol

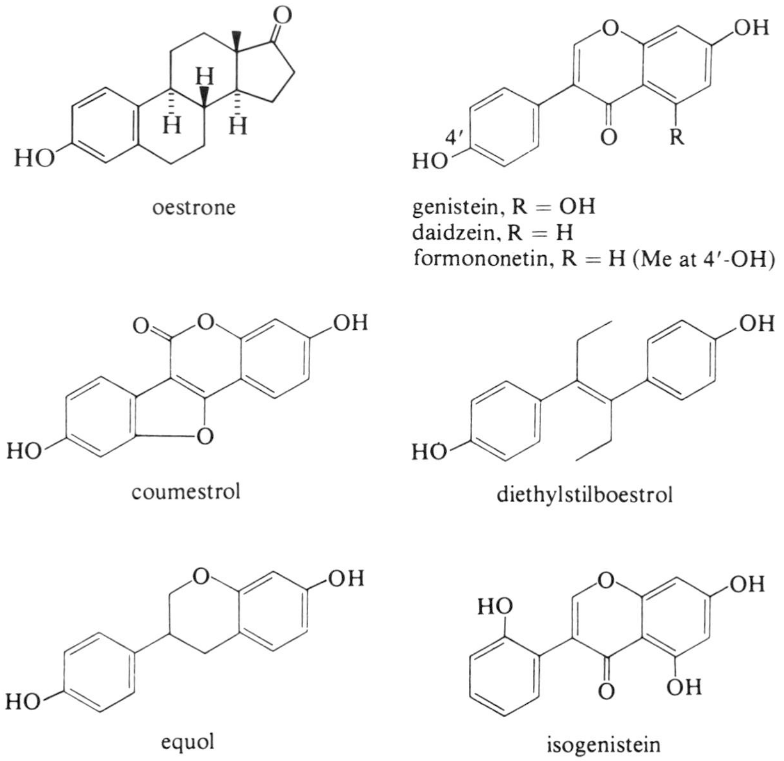

oestrone

genistein, R = OH
daidzein, R = H
formononetin, R = H (Me at 4′-OH)

coumestrol

diethylstilboestrol

equol

isogenistein

Figure 4.3 Isoflavonoids as oestrogenic mimics*

*Isoflavonoid structures have been turned through 180° on a vertical axis in order to emphasize the resemblance to the steroid, oestrone.

diethylstilboestrol, a synthetic compound used medicinally in place of oestrone because of its higher activity.

The discovery that isoflavonoids have oestrogenic activity in mammals was made during the 1940s when sheep in Australia were allowed to graze for longer periods than usual on pastures containing subterranean clover, *Trifolium subterraneum*. As a result of this practice, lambing percentages were seriously reduced (to less than 30%) and active material causing this infertility was traced to the clover plant. The principle was eventually isolated and identified as a mixture of two isoflavones, genistein and formononetin (Bradbury and White, 1954). Structural comparison (Fig. 4.3) with the hormone oestrone and the most active synthetic analogue, diethylstilboestrol, shows why these isoflavones are oestrogens—they mimic the steroidal nucleus of the natural female hormone. They are, in fact, rather weak oestrogens on a molar basis (Biggers, 1959) but are presumably effective because of the relatively large quantity (about 1% dry wt) in the clover fodder.

A more active substance, coumestrol, was later isolated by Bickoff (1968) from alfalfa, *Medicago sativa*, and ladino clover, *Trifolium repens*. Although coumestrol is 30 times more

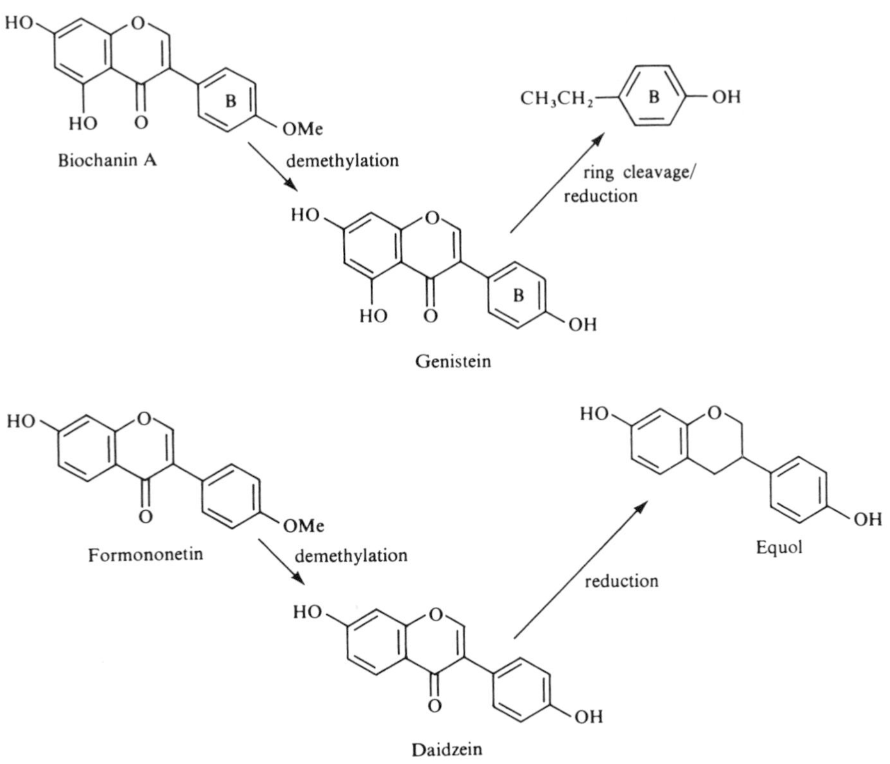

Figure 4.4 Degradative pathways of clover isoflavones by rumen microbes in sheep

active than genistein or formononetin, its concentration in legume fodder plants is generally much lower, so that it is probably less effective *in vivo* than the isoflavones. In fact, it has been shown (Shutt, 1976) that formononetin is the most important oestrogen in clover to sheep. This is because it is a pro-oestrogen, being converted (by demethylation and reduction) to a more active substance, the related isoflavan equol, within the animal body. This isoflavan was actually isolated from pregnant mares' urine as long ago as 1932, by Marrian and Haslewood. Comparative metabolic studies with genistein indicate that it is degraded in the rumen to inactive products, one of which is *p*-ethylphenol; a third isoflavone in clover, biochanin A, is similarly metabolized via genistein (Fig. 4.4).

All isoflavones are oestrogenic when given by parenteral injection to animals and much work has been done on structure–activity relationships in the isoflavonoid series using mice (Biggers, 1959). For example, it can be shown that the two *para*-substituted hydroxyl groups are needed for maximal activity: transfer of the 4′-hydroxyl in genistein to the 2′-position (to give isogenistein) reduces oestrogenic activity by 75% (Baker *et al.*, 1953).

From the agricultural viewpoint, the presence of isoflavonoids in clovers and other legume fodder plants is a hazard to farm animals, because of their effects on reproduction. Symptoms produced include difficult labour, infertility and lactation in unbred ewes. A survey of *Trifolium* has shown that 18 species have as high an isoflavone content as *T. subterraneum*, most other species having relatively small amounts. Plant breeding experiments have been carried out in the pasture clovers and strains with a safe, low isoflavone content are now available. Unfortunately, it is difficult to replace existing strains of subterranean clover in Australian pastures because they have become well adapted to their environment and have built up large reserves of seed in the soil. In spite of all attempts to reduce the feeding of breeding ewes on such pastures, it is estimated that each year one million Australian ewes fail to lamb because of 'clover' disease. Immunization procedures are being developed to overcome this problem (Shutt, 1976).

Isoflavones are more or less restricted in their distribution to the Leguminosae, so that if these compounds have a deterrent function in nature, such a function can only operate in this family. There is no reason, however, why a variety of quite different chemical structures should not have the same purpose in other plant families. Weak oestrogenic activity has been detected, for example, in some flavones and flavonols, two classes of compound which are widely present in the angiosperms.

The question remains—whether isoflavone synthesis is purely accidentally related to oestrogenic activity in mammals or whether these substances have been deliberately produced by the plant to interfere with the reproductive capacity of grazing animals. It may be significant that the isoflavone skeleton also provides the basis of disease resistance in legumes, since the phytoalexins formed in this family are nearly all reduced forms of genistein or formononetin (see Chapter 10).

That isoflavones do have an ecological role is supported by a report (Leopold *et al.*, 1976) that birds (quails) are also affected by the oestrogenic effects of pasture isoflavones. It appears that the birds feed on pastures rich in legume species and they use the presence of isoflavones as a form of population control. Thus in years of good rainfall, legumes that are eaten grow luxuriously and are relatively low in isoflavone on a fresh wt basis. There are no oestrogenic effects and egg laying is normal. However, in years of poor rainfall, the plants are less profuse in leaf and become richer in isoflavone on a fresh wt basis. An oestrogenic effect is exerted on the female quails, and egg laying is curtailed. Thus, there is a self-regulating system whereby the increase in population is kept at a low level when the food available to the birds is limited. Natural population limitation is a feature of many animal communities and it is possible that phytoestrogens have a role in other species besides quails.

Even simpler molecules than isoflavones may act as reproductive cues in certain animals. Thus Berger *et al.* (1977) have found that ferulic and *p*-coumaric acids and their vinyl analogues (Fig. 4.5) inhibit reproductive function in the herbivorous rodent, *Microtus montanus*, a native of North American grasslands. When fed on these substances, the female exhibited decreased uterine weight, inhibition of follicular development and cessation of breeding activity. Furthermore, the two phenolic acids are present in significant amount in the salt grass *Distichlis stricta*, which often represents over 90% of these rodents' diet. The amounts of these acids rise to high levels in the plant at flowering and fruiting, i.e. precisely when it is about to die back. It is suggested that these dietary constituents become a cue to these animals to cease reproductive effort in natural populations, just when their food supply becomes limiting at the end of the growing season.

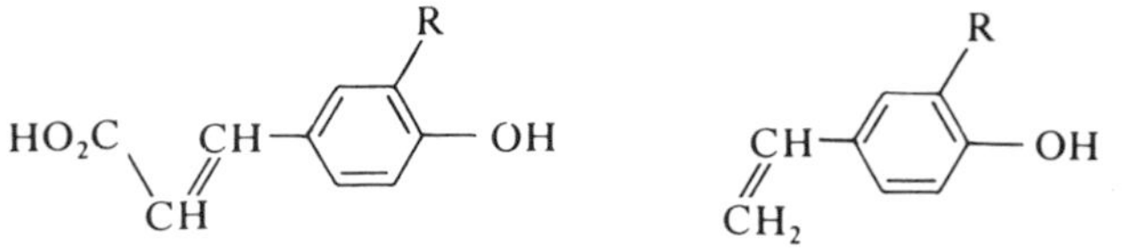

p-coumaric acid, R = H
ferulic acid, R = OMe

vinylphenol, R = H
vinylguaicol, R = OMe

Figure 4.5 Plant phenolics functioning as reproductive inhibitors in *Microtus montanus*

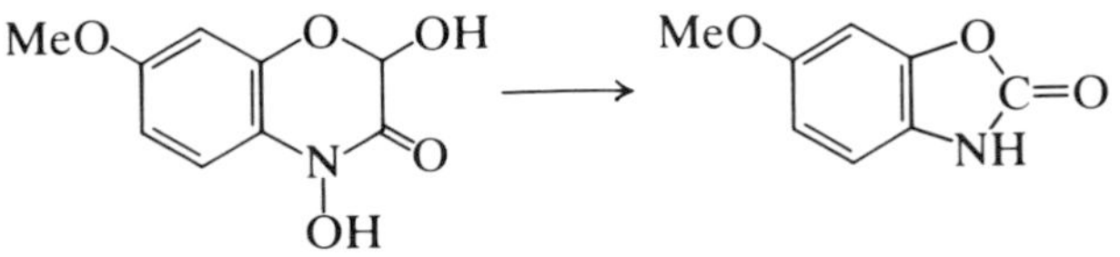

DIMBOA → 6-methoxybenzoxazolinone

Figure 4.6 Enzymic conversion of DIMBOA to 6-methoxybenzoxazolinone, a breeding stimulant in *Microtus montanus*

If dietary phenolics turn off breeding in this rodent during the autumn, then one might expect there to be a dietary cue for stimulating reproduction in the following spring. A search for such a cue has been successful (Saunders *et al.*, 1981) and another grass chemical, 6-methoxybenzoxazolinone, has been identified. This compound is derived after enzymic conversion from the hydroxamic acid DIMBOA present in many grasses (Fig. 4.6). DIMBOA itself is biologically active, since it has been identified as the feeding deterrent in wheat and maize to the aphid *Schizaphis graminum* (Argandona *et al.*, 1981).

III. Insect moulting hormones in plants

Before discussing the occurrence of insect hormones in plants, their role in the insect's life-cycle needs to be briefly stated. In the metamorphosis of insects, hormones are required to control the different stages in the life-cycle from larva to adult. They are required to initiate the changes in form that occur during growth, which in outline are as follows:

$$\underset{\text{1st stage}}{\text{Larva}} \xrightarrow{\text{MH/JH}} \underset{\text{2nd stage}}{\text{Larva}} \xrightarrow{\text{MH}} \text{Pupa} \xrightarrow{\text{JH}} \text{Imago}$$

the agents concerned being the juvenile hormone (JH) and the moulting hormone (MH). While the JH is required only for larval metamorphosis, MH, which literally controls the moulting of the outer case or skin at each stage, is needed at every step up to the emergence of the adult. In more advanced insects, the juvenile form differs so much from that of the adult that several larval stages are needed before pupation.

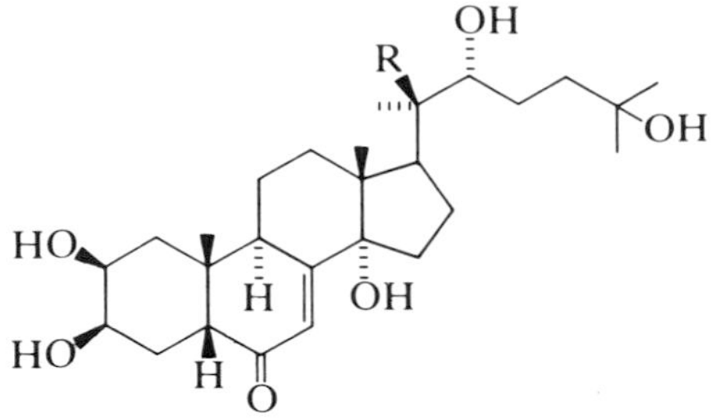

ecdysone, R = H
20-hydroxyecdysone, R = OH

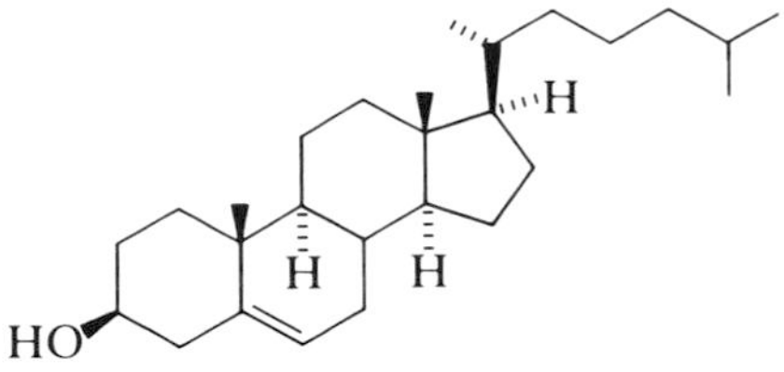

cholesterol

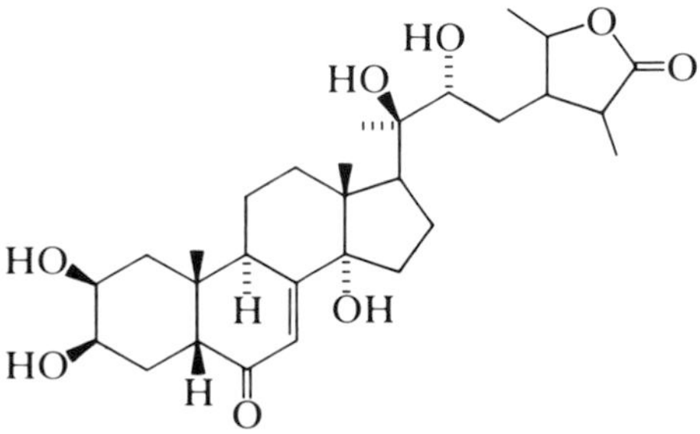

cyasterone

Figure 4.7 Structures of insect moulting hormones

For normal metamorphosis, then, these two hormones have a crucial role. They must be present in just the right amounts and at the right time in the cycle for development to proceed normally from larva to adulthood. While JH is synthesized in a pair of tiny cephalic glands, the corpora allata, MH is formed in the prothoracic glands located at the insect's anterior end.

That a moulting hormone was required in insect metamorphosis was first demonstrated over 40 years ago (see Wigglesworth, 1954). It was not, however, until 1954 that the pure hormone, named ecdysone, was isolated in sufficient amount for structural study. Butenandt and Karlson (1954) then isolated 25 mg of pure substance, starting from half a ton (500 kg) of silkworm pupae. It was not until 11 years later that the structure was determined by X-ray diffraction (Karlson *et al.*, 1965) as a hydroxysterol with a clear resemblance in structure to cholesterol (Fig. 4.7). A second compound, present in minor amount in silkworm, was identified as 20-hydroxyecdysone. Four further closely similar moulting hormones have subsequently been isolated from arthropods or crustacea, bringing the total number of zooecdysones to six.

So far, this account of insect hormones has been concerned solely with animals. A most amazing discovery, bringing this area of insect biochemistry and endocrinology into the realms of plant science, was made only a year or so after the structural elucidation of the two ecdysones. Takemoto *et al.* (1967) and Nakanishi (1968) jointly reported the presence of massive amounts of 20-hydroxyecdysone in the leaves of the common yew, *Taxus baccata*. The amount of this ecdysone (25 mg) produced by extracting half a ton of silkworms (i.e. 500 kg of insect tissue) was obtainable from as little as 25 g dried leaf or root of yew. An even richer source was found in the rhizomes of the common fern, *Polypodium vulgare*, 25 mg being obtained from only 2.5 g of rhizome.

The finding of insect hormone in such astonishing quantity in plants created a considerable sensation at the time. It was of immediate scientific value in providing an easy source of pure hormones for insect endocrine experiments. It was also of considerable practical interest, since such active materials or their analogues had potential in interfering with normal insect development and providing a means of pest control. What appeared to be a novel concept in insecticide chemistry, however, was, as Williams (1972) has put it, 'a strategy which appears to be an ancient art invented by certain plants and practiced by them for tens of millions of years'.

Plant surveys for ecdysones were rapidly initiated and they were found to occur regularly in ferns (in 22 of 43 spp. surveyed, especially in Polypodiaceae) and in gymnosperms (in 73 spp. of eight families, including Taxaceae and Podocarpaceae). Ecdysones do also occur in angiosperms, but less frequently. While 20-hydroxyecdysone is relatively ubiquitous in plants, the other zooecdysones are rare. Instead, a wide range of other structures are characteristic of plants. Over 30 phytoecdysones were characterized in the 4 years following the initial discovery in yew (Rees, 1971) and 70 more have subsequently been added to the list (Camps, 1991).

The most fascinating property of many phytoecdysones is their enormous hormonal activity, as compared to the zooecdysones. Some are up to 20 times more active. At such concentrations, they can have very damaging effects on insect development, partly because they resist the inactivation which zooecdysones undergo when applied to insects. For example, while ecdysone is 50% inactivated within 7 h of being fed to an insect, cyasterone, a phytohormone from *Cycas*, is only inactivated to the same extent after 32 h.

The biological effects of the plant hormones depend on whether they are applied cutaneously, by injection or by oral administration. Malformation, sterility and death is frequently the outcome of such administrations. The least effective route is via the mouth, probably because insects have developed methods of detoxification. Thus dietary hormones may undergo further hydroxylation, dehydrogenation at the 3-position, formation of conjugates or side-chain cleavage (Hikino *et al.*, 1975).

Among insects which are particularly sensitive to the effects of phytoecdysones are the silkworm, *Bombyx mori*, and the pink bollworm, *Pectinophora gossypiella*. The silkworm, for example, when so treated is unable to remove the old cuticle during the moulting process and may end up with two heads, with fatal consequences (Kubo *et al.*, 1983). The addition of crude extracts of other ecdysone-containing plants to the diet of a range of other insects, including the tsetse fly *Glossinia morsitans morsitans*, has also led to developmental anomalies and eventual mortality (Camps, 1991). In certain insects, exposure to dietary ecdysone can arrest feeding completely. Both ecdysone and 20-hydroxyecdysone at concentrations around 5 mg/kg fresh wt deter feeding in the cabbage white caterpillar *Pieris brassicae* (Camps, 1991).

Finally, the question may be posed: do phytoecdysones have any ecological function or is their presence in plants purely coincidental to their occurrence in insects? Are these hormones a major defence mechanism produced by plants to ward off insect attack? While conclusive proof for such a role is very difficult to produce, there are an increasing number of experiments which support such a role for these substances (see Williams, 1972).

Some of the points that might be argued in favour of an ecological function for plant ecdysones are as follows:

(1) Phytoecdysones occur mainly in two groups of relatively primitive plants—the ferns and gymnosperms—which still appear to be relatively (but not completely) free of insect predation (Hendrix, 1980). Thus they may have emerged as a defence mechanism at one particular phase in evolution before the advent of the angiosperms. Their role as feeding deterrents appears to have been taken over in these latter plants by alkaloids or ellagitannins.
(2) When administered to insects, phytoecdysones produce major abnormalities in growth, leading ultimately to sterility and early death.
(3) While it is possible to show that some insects detoxify orally administered ecdysone, there may be others which are not able to deactivate it in time. The structural variation present in the phytoecdysones would provide some protection against rapid deactivation. Insects are probably less able to cope with ecdysone entering through the cuticle and it is conceivable that some of the phytoecdysone may enter this way during insect feeding.
(4) Phytoecdysone synthesis is to be regarded as only one stage in a constant co-evolutionary situation between plants and insects. At the present time in evolutionary history, many insects might be expected to have partly or completely overcome this particular defence mechanism.
(5) Phytoecdysones do not necessarily have to be lethal to insects. Minor effects on metamorphosis or reproduction would probably be sufficient to reduce the fitness of the pest and keep its predation under reasonable control.

IV. The Fruit-Fly–Cactus Interaction

A most unusual example of insect–host plant specificity has been recorded by Kircher and Heed (1970) in the Sonoran desert of western United States. Here it has been found that four *Drosophila* species exhibit remarkable specificity in choosing to feed variously on rotting limbs of four cactus species growing in the area. The relationship between cactus and fruit fly is highly consistent, as shown by the emergence records of adult flies from larvae taken from the different cacti (Table 4.2). Each fly species feeds on a specific cactus, with very few exceptions.

Two types of interaction occur involving secondary chemistry, both feeding attractants and feeding repellents (see Chapter 5 for definitions of these terms) being present (Fig. 4.8). The feeding attractant is common to all the cacti and is a sterol which permits the insects to synthesize their moulting hormone. Since insects cannot make the sterol

Table 4.2 Emergence records of *Drosophila* species on rotting cacti in the Sonoran desert

Cactus	Associated *Drosophila* species	Host plant specificity[a]
Senita *Lophocereus schotti*	*D. pachea*	862 : 1
Saguaro *Carnegia gigantea*	*D. nigrospiracula*[b]	6803 : 1
Organ pipe *Lemairocereus thurberi*	*D. mojavensis*	28 : 1
Sina *Rathbunia alamosensis*	*D. arizonensis*	23 : 1

[a] Ratio of progeny of associated *Drosophila* sp. to number of progeny of non-associated species, i.e. on Senita, for every 862 *D. pachea* grubs emerging, one foreign species (*D. mojavensis*, etc.) was found. Data modified from Kircher and Heed (1970).

[b] *D. nigrospiracula* is unusual in that it will also live on two other host plants: *D. pachea* and *D. mojavensis* also each have one other possible host plant.

nucleus *de novo*, they are dependent on plants for their starting materials. The common starting material is sitosterol (see Fig. 4.9), which is the major sterol in plant tissues, and insects have to make a number of structural changes in order to convert this to ecdysone. These changes fall into three groups of reactions: (1) rearrangement of the double bond from the Δ5 to the Δ7 position; (2) removal of the 24-ethyl group (this is a structural feature which distinguishes almost all plant from animal sterols); and (3) various oxidations around the carbon skeleton to introduce three alcohol groups and one keto group.

While most insects have to carry out all these reactions, it is apparent that *Drosophila* species, by feeding on these cacti, can avoid carrying out the first step, the rearrangement of the Δ5 double bond. This is because the major sterol in Senita cactus is not sitosterol, but the compound schottenol (for formulae, see Fig. 4.9) which is similar in structure, except that the double bond in the second carbon ring is already in the right position

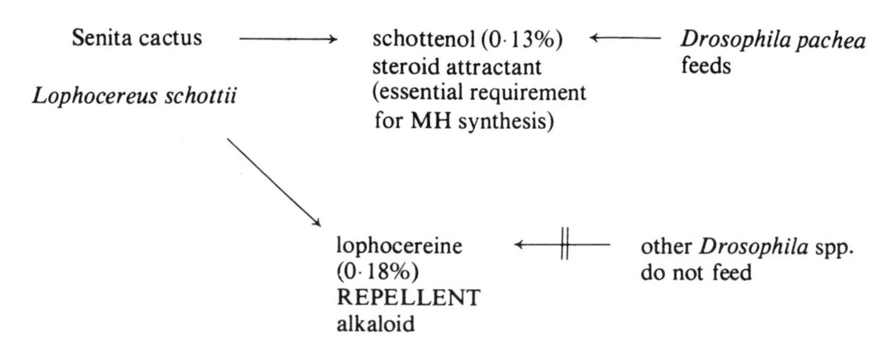

Figure 4.8 Interaction beween *Drosophila* and Senita cactus

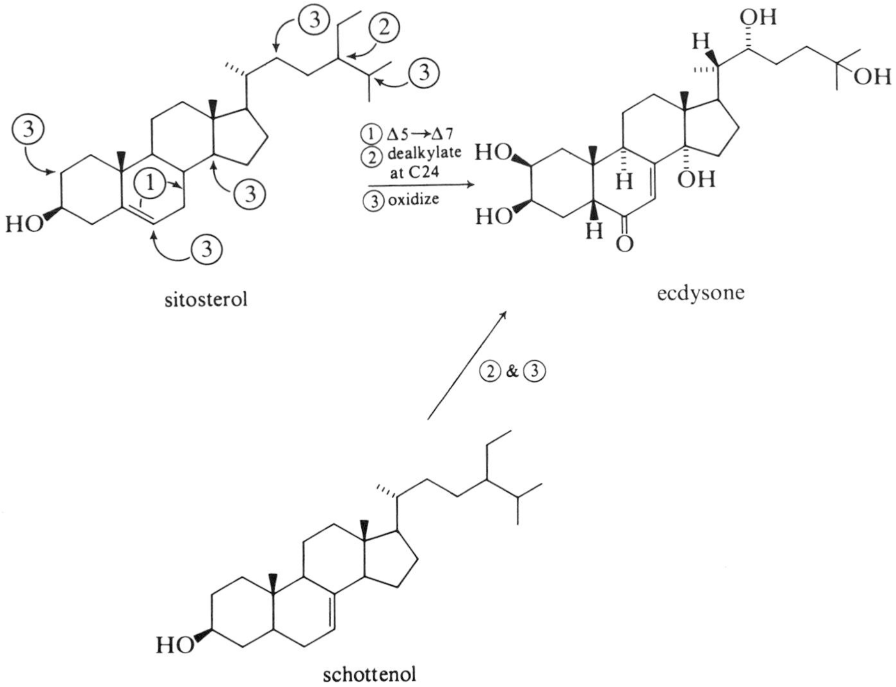

① Δ5→Δ7
② dealkylate at C24
③ oxidize

sitosterol

ecdysone

② & ③

schottenol

Figure 4.9 Pathway of synthesis of moulting hormone from sitosterol or schottenol

for ecdysone synthesis. By feeding on schottenol, the flies require less steps and less enzymes for synthesizing their moulting hormone. It is such an advantage to them that they have, indeed, become dependent on the host plant for this material. Laboratory-reared larvae fed on sitosterol are incapable of making ecdysone and thus do not develop into adult flies.

The second class of plant substances in this feeding syndrome are alkaloids. These apparently act as feeding deterrents to *Drosophila* species other than the one that is specifically associated with a particular cactus. Thus lophocereine occurs to the extent

lophocereine

repellent in Senita cactus to all except *D. pachea*

carnegeine

repellent in Saguaro cactus to all except *D. nigrospiracula*

Figure 4.10 Alkaloids of cacti

of 0.18% in Senita cactus; it is accompanied by a second alkaloid pilocereine (0.6%), a trimer of similar structure to lophocereine (Fig. 4.10). These two alkaloids act as repellents to other *Drosophila* feeders. Toxic effects can be shown in laboratory feeding experiments, only *D. pachea* being impervious to their effects. A second cactus, Saguaro, contains a different alkaloid, carnegeine, which is now repellent to *D. pachea* but not to *D. nigrospiracula*, the species which feeds on Saguaro in preference to any other host plant.

In this interesting example of co-evolution between plant and insect, the fruit-fly has reached the extreme of dependency on one or two species of host plant. Even a small change to a closely related plant is dangerous, causing death due to the presence of an unfamiliar alkaloid. The flies are also in a vulnerable position if their host plant disappears, since they will have to rapidly recover the ability to accomplish the $\Delta5 \rightarrow \Delta7$ double bond shift. The plants have evolved a complex secondary chemistry in terms of both sterols and alkaloids, which at least partly limits the predation of insects on them. Since the flies feed on the decaying parts of the cacti, insect feeding is relatively harmless and may even be beneficial in removing the dying limbs more rapidly than otherwise.

V. Insect Juvenile Hormones in Plants

The requirement for a juvenile hormone at any early stage of insect growth has already been mentioned in Section III (p. 111). It is required to control growth and the build-up of necessary material for metamorphosis. It may exert its effects for up to 6 weeks, as in the *Cecropia* moth larvae. The hormone is synthesized in the corpora allata glands which regulate its release into the blood.

From the first demonstration that a juvenile hormone was present in insects, it took 20 years before it was chemically characterized. In 1967, Roller and coworkers elucidated the structure of JH I (see Fig. 4.11), using less than 300 μg of material isolated from *Cecropia* moth. It has a sesquiterpenoid structure related to farnesol which itself has some activity. A second compound JH II of closely similar structure was also found in *Cecropia*. Subsequently, two further hormones, JH0 and JH III, were isolated from the tobacco hornworm *Manduca sexta*. From present information, it appears that these four compounds represent the juvenile hormone activity of the majority of insects. Many related structures have been synthesized and the relationship between structure and activity established (Pfiffner, 1971).

The story of the discovery that materials with JH activity occur in plants is one of those interesting accidents which illuminate the pages of the history of science. The detection of the so-called 'paper factor' arose when a Czechoslovak biologist, K. Sláma, was invited by C. M. Williams to Harvard University to culture his favourite experimental insect, the European bug *Pyrrhocoris apteris*. Mysteriously, all attempts to persuade the bugs to go through normal metamorphosis in the new surroundings failed. They obstinately remained in the fifth larval stage. A search of factors responsible for this failure in growth revealed that in moving to Harvard, Sláma had replaced the Whatman filter paper used in the petri dishes for growing the bugs on by U.S. paper towelling (Scott Brand 150). Substitution of the original filter paper led to normal growth and development.

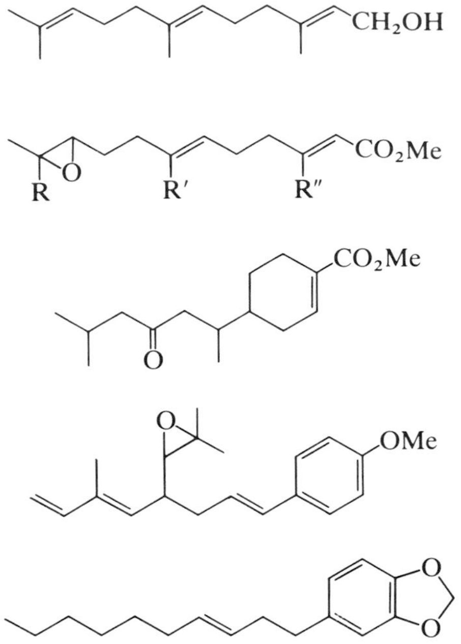

CH_2OH farnesol

CO_2Me R R′ R″

JH0 (R = R′ = R″ = Et)
JHI (R = R′ = Et, R″ = Me)
JHII (R = R′ = Me, R′ = Et)
JHIII (R = R′ = R″ = Me)

CO_2Me juvabione

OMe juvocimene 2

juvadecene

Figure 4.11 Structures of juvenile hormones and analogues

Clearly, some compound in the Scott Brand paper was continually providing juvenile hormone material to the bugs, thus arresting their normal metamorphosis.

A study of a range of paper products revealed that all U.S. newspapers and journals were highly active. By contrast, European and Japanese papers were completely inert. This difference in 'paper factor' was then traced to the fact that American paper is manufactured largely from balsam fir, *Abies balsamea* a tree which is not used in the European paper pulping industry. Thus some material present in the tree is carried through all the processes of paper manufacture and is still present on the printed page of the American journal *Science*. The corresponding English journal *Nature*, on the other hand, is completely free from this insect growth inhibitor, since different trees are used for making its paper.

Extraction of American paper eventually gave an active compound, juvabione, which was found to be a structural analogue of the natural insect hormone (Fig. 4.11). Juvabione, however, is only active for one family of insects, the Pyrrhocoridae, to which the European bug belongs. Treatment of the related Lygaeidae in the same group of insects, the Hemiptera, with juvabione gave no response. Plant juvenile hormones, therefore, seem to be rather selective in their effects.

The discovery of a second insect hormone in plant materials, following the phytoecdysone story (see Section III), led to interest and speculation about the possible ecological effects of such occurrences. Unlike the situation in the case of moulting hormones,

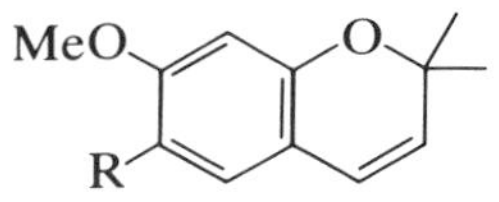

MeO O R

precocene 1 (R = H)
precocene 2 (R = OMe)
encecalin (R = COMe)

Figure 4.12 Anti-JH substances and a related chromene from plants

however, few other structures besides juvabione and its dehydro derivative appear to have been reported. There is, nevertheless, evidence that materials with JH activity are distributed in a range of plants. Bowers (1968) examined 52 species at random and found hormone activity in the *Tenebrio* assay in six—a frequency of 12%. Other surveys (e.g. Jacobson *et al.*, 1975) have also been successful in revealing such activity in further plant sources.

Two other juvenile hormone mimics isolated from angiosperm species are juvocimene 2 and juvadecene (see Fig. 4.11). Juvocimene 2 occurs in leaves of sweet basil, *Ocimum basilicum* (Labiatae), and is several orders of magnitude more active than JH I. Thus it will induce the formation of nymphal–adult intermediates in the milkweed bug, *Oncopeltus fasciatus*, at a concentration of 10 pg (Bowers and Nishida, 1980). Juvadecene, which occurs in the roots of *Macropiper excelsum* (Piperaceae), is likewise hormonally active. It induces supernumary metamorphosis, when applied to last-instar nymphs of the milkweed bug, at a concentration of 30 μg and is quite toxic at higher doses (Nishida *et al.*, 1983).

Since compounds with juvenile hormone activity occur in plants and are effective in arresting embryonic development in insects, sometimes with lethal side-effects, there seems little doubt that a similar ecological role can be put forward for them as for the plant-derived moulting hormones. These substances may be deliberately produced by plants as a sophisticated self-defence against insect predation. Preliminary evidence indicating that the free acid of juvabione (see Fig. 4.11) is indeed produced in response to balsam woolly aphid infestation in the wood of the coastal fir tree *Abies grandis* and then gives subsequent protection has been reported by Puritch and Nijholt (1974). The effectiveness of insect control through these hormonal derivatives can be seen in the fact that synthetic JH analogues have recently been marketed as insecticides against several agriculturally important pests.

Even more remarkable support for the viewpoint that secondary compounds provide hormonal defence in plants against insects is the recent discovery of anti-JH substances in plants. Two chromenes, called precocene 1 and 2 (see Fig. 4.12), have been isolated from the insect-resistant composite plant *Ageratum houstoniatum*. When added to insect diets, these two substances interfere with JH activity in such a way that precocious metamorphosis occurs. Thus, the nymphs of the milkweed bug miss out one or more larval states to become imperfect adults. The net result is usually sterility in the females (Bowers, 1991). The precocenes appear to inhibit JH synthesis only after they have been converted within the insect's glands to the corresponding epoxides (Brooks *et al.*, 1979).

A search of members of the Compositae, the family to which *Ageratum houstoniatum* belongs, for precocene-like activity has failed to yield any more potent substances and indeed, although more than 170 chromenes and related benzofurans have been isolated from these plants, hardly any have anti-JH activity (Proksch and Rodriguez, 1983). However, the related encecalin (see Fig. 4.12), from *Encelia farinosa*, turns out to be insecticidal to the milkweed bug and deters *Heliothis zea* from feeding. By contrast with these natural materials, a number of synthetic compounds such as fluoromevalonate are similar in action to the precocenes in inhibiting JH biosynthesis in insects. Several such agents have potential for insect control on crop plants, although none has yet been put on the market (Staal, 1986).

VI. Pheromonal Interactions and the Pine Bark Beetle

Infestation of pine trees by various bark beetles is a serious cause of damage and loss in these commercially important timbers. It has been estimated that 54% of natural deaths of all mature conifers in North America are due to beetle attack. Studies of the causes of these infestations have been carried out over a period of years, as part of a search for a means of control. One major tree affected in North America is the ponderosa pine, *Pinus ponderosa*, which has a range of pests including the Western pine beetle *Dendroctonus brevicomis* and the California five-spined ips, *Ips paraconfusus*. The interaction clearly involves the monoterpenes of the pine but is a very complicated one. It is difficult, even today, to be sure of all the steps of this interaction. The present account mainly is restricted to the work on *D. brevicomis* carried out by American entomologists (Wood, 1982; Byers *et al.*, 1984).

An outline of the different stages in the infestation by *D. brevicomis* is indicated in Fig. 4.13. Some of the chemical substances involved are shown in Figs. 4.14 and 4.15.

The oleoresin of pine bark is rich in volatile terpenes and traces of this vapour exude from the tree. This vapour, or certain of its components, attract the female beetle, which then settles to feed on the bark (stage 1).

Once the females are established, they begin to attract males to the site for reproductive purposes (stage 2). The first male beetles arriving release frontalin. This synergizes with the *exo*-brevicomin of the females and myrcene, sequestered directly from the oleoresin of the tree, to produce a rich pheromone bouquet, which elicits mass aggregation. The two bicyclic ketals, *exo*-brevicomin and frontalin, are synthesized *de novo* by the beetles.

The female beetles, originators of the infestation, appear to control the sex ratio within the population by varying the proportions of the sex pheromone mixture exuded at any given time (stage 3). This is apparent from experiments in which all three components have been added artificially to the 'frass' of a bark infestation. Such a mixture attracts males and females in a 1:1 ratio, whereas if frontalin is omitted, the ratio changes to 2:1 in favour of the males.

Finally, the area of infestation reaches its optimal size where there is only enough food for the existing beetles. At this stage, the female beetle produces (*E*)-verbenol which repels foraging females, while the male beetle produces verbenone and (+)-ipsdienol, which repel approaching males. The ketone verbenone is formed from one of the major bark terpenes, α-pinene, by a two-stage oxidation, via the alcohol (*E*)-verbenol (Fig. 4.15).

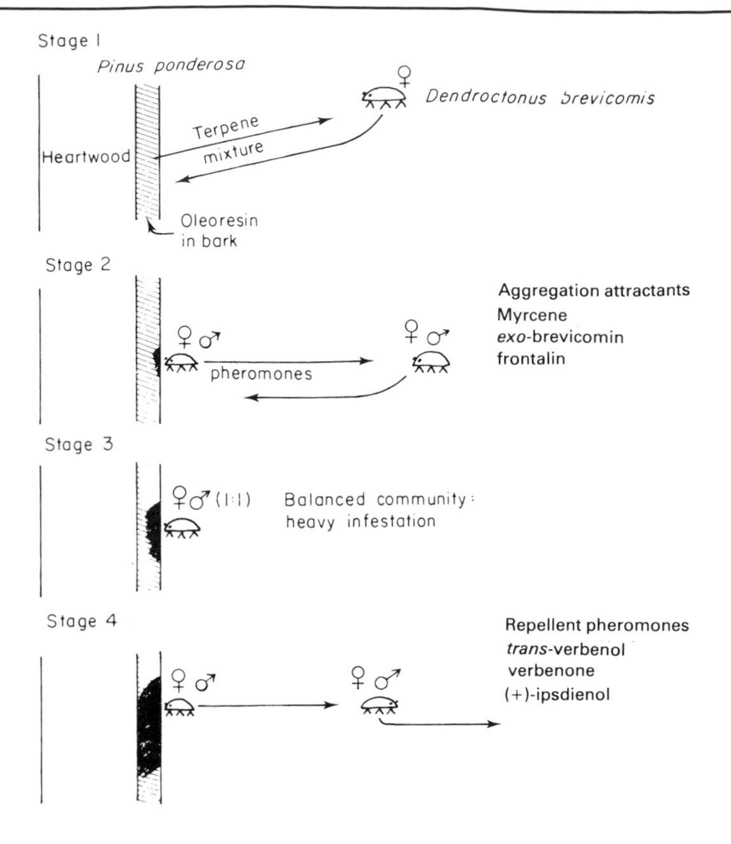

Figure 4.13 The plant bark beetle–ponderosa pine interaction

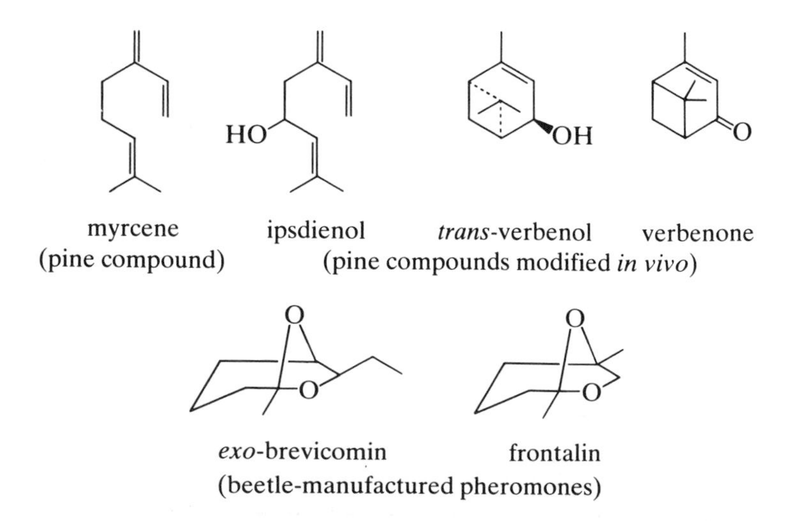

Figure 4.14 Pheromones of the pine bark beetle and their origins

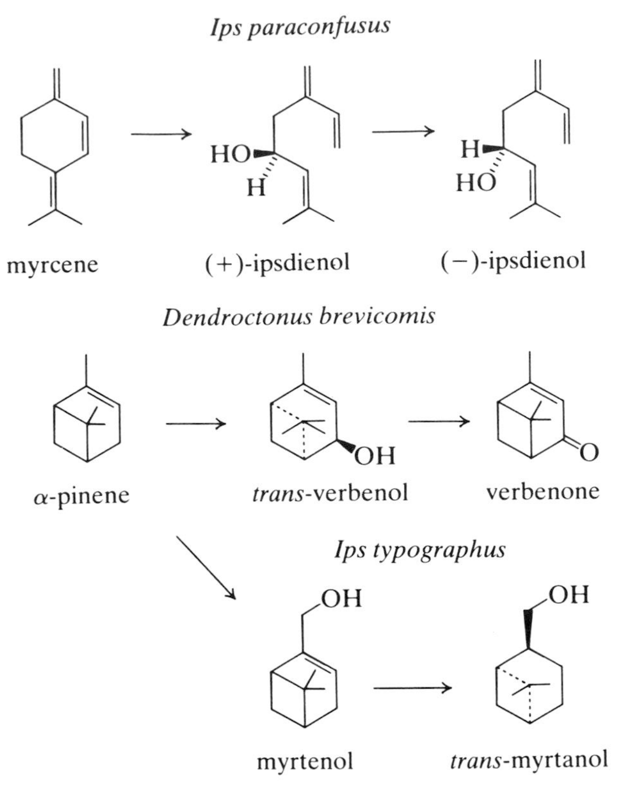

Figure 4.15 Production of beetle pheromones from plant terpenes via microbial oxidations

From the ecological viewpoint, it is particularly interesting that the beetle makes use of dietary compounds in at least two ways, either without structural alteration (myrcene) or with minor modification (verbenol, verbenone). Another interesting aspect is that of synergism (Table 4.3), whereby the male attractant must contain all three components for maximum effectiveness. Addition of myrcene to the frontalin-brevicomin mixture doubles the effectiveness of the attractant. Furthermore, myrcene is essential and it

Table 4.3 Synergistic effects of terpenoid mixtures on beetles

Pheromone mixture	Number males trapped in 24-h period
Brevicomin + frontalin	193
Brevicomin + frontalin + myrcene	389
Brevicomin + frontalin + limonene	196
Frontalin + 3-carene[a]	34

[a] This mixture is highly attractive to the females.

cannot be replaced by the closely related structure, limonene (Table 4.3). It may be noted that a mixture of frontalin and 3-carene, instead of attracting males, now is highly attractive to females. This mixture, of course, would be an ideal lure to attract females away from pine trees before they start infestations and it has been put forward as a means of control.

A further twist in this complex interaction may be due to micro-organisms, which are probably responsible for the *in vivo* conversion of α-pinene to verbenone within the beetle. Thus, a bacterium has been isolated from the gut of *Ips paraconfusus* which can oxidize α-pinene to verbenol, while a symbiotic fungus of *Dendroctonus* has been shown to be able to further oxidize verbenol to verbenone (Brand *et al.*, 1976). It is conceivable, therefore, that the development of micro-organisms in the host plant may be important in influencing the behaviour of the beetle on a successfully colonized tree.

The other bark beetle *Ips paraconfusus* that has been extensively studied is aptly named since its pheromones are confusingly similar in some points but different in others from those of *Dendroctonus*. For the sake of the student reader, they will not be elaborated here (but see Wood, 1982). One way that *Ips* is identical to *Dendroctonus* is in its utilization of plant terpenes for the production of certain pheromones. For example, the male beetle (but not the female), if exposed to vapours of myrcene, will absorb it into its hindguts and there oxidize it to the alcohols ipsdienol and ipsenol which are two male pheromones (Byers *et al.*, 1979). Convincing proof of this *in vivo* conversion within *Ips paraconfusus* has been obtained by feeding the beetle with myrcene deuterium labelled at the two methyls (see Fig. 4.15) and showing that the two pheromones produced are deuterated at the same two positions (Hendry *et al.*, 1980).

Males of a third bark beetle, *Ips typographus*, obtain their pheromones, myrtenol and (*E*)-myrtanol, by similar microbial oxidations, starting from α-pinene (Fig. 4.15) (Birgersson *et al.*, 1984).

It has recently been discovered that the various pheromones produced by the bark beetles are also important in restricting the feeding of the different beetle species to particular sites on the trees. Thus some of the pheromones emitted by *Dendroctonus brevicomis* will inhibit other species (e.g. *Ips paraconfusus*) from approaching and feeding at the same site; likewise, some of the pheromones of the other species will inhibit the approach of *D. brevicomis*. Thus these beetle pheromones are ecologically important in reducing interspecific competition for the host trees (Byers *et al.*, 1984).

One final point about this complex ecological interaction may be made—it is possible that the trees themselves may vary their oleoresin chemistry in order to develop resistance to beetle predation. Indeed, a study of populational variation in xylem resin terpenes has shown that directional selection is taking place in *Pinus ponderosa* for trees with high concentrations of limonene, while the other four main monoterpenes (α- and β-pinene, 3-carene and myrcene) remain very variable (Sturgeon, 1979). Such trees will be avoided because the limonene in quantity is toxic to the beetle and thus a feeding deterrent. Bark beetles prefer trees with low limonene, but with high levels of α-pinene, a pheromone precursor, and of myrcene, which is used directly as a pheromonal material (see Fig. 4.15). High β-pinene levels are acceptable because it is the least toxic monoterpene to *Dendroctonus brevicomis*.

Similar selective pressures have been observed in Douglas fir trees predated upon by the beetle *D. pseudotsugae*. These trees vary clinally in the western United States in the ratio of α- to β-pinene in the resins. In this case, β-pinene is repellent to the beetle, so that trees with a high β-/α-pinene ratio are less susceptible to insect attack (Heikkenen and Hrutfiord, 1965).

These data together suggest that bark beetles are exerting a frequency-dependent selection pressure on the chemically polymorphic populations of their host trees, the long-term effects of which would be to reduce the number of trees available for predation. The actual situation is obviously more complex than this; beetles may adapt to changing terpene patterns or else other beetle species with different terpene requirements and tastes may move in. Control of bark beetle predation, e.g. by pheromonal traps (see Chapter 8), still remains an essential requisite. For a discussion of the coevolutionary adaptations that can take place in conifer–bark beetle systems, see Raffa and Berryman (1987).

VII. Summary

That hormonal and pheromonal interactions occur between plants and animals is very clearly indicated in the examples discussed in this chapter (Table 4.4). Oestrogenic interactions are established in the case of humans, sheep, rodents and birds (quail). They are likely to occur in many other animals, since plant substances with oestrogenic effects are probably not uncommon. The fact that human sex hormones, both male and female, occur in trace amounts in a number of plants is now well founded (Section II), but there is as yet no explanation for these occurrences.

In the case of the insects, two major animal hormones—moulting and juvenile—are formed in plants. Furthermore, substances which mimic these hormones or which interfere with hormone synthesis are known to occur in plants. Insects are also affected by plant substances which are pheromonally active and they may also sequester and store plant substances to use them as pheromones or convert them to pheromones.

Table 4.4 Plant substances with hormonal or pheromonal effects on animals

Hormone/pheromone	Substance and source	Animals affected
Oestrogen	Isoflavones in *Trifolium*	Sheep, quail
	Miroestrol in *Pueraria*	Humans
	Phenolic acids in *Distichlis*	Mountain voles
Insect moulting	Phytoecdysones in *Taxus*	Lepidoptera
Insect juvenile	Juvabione in *Abies*	Lepidoptera
Sex	Myrcene in *Pinus*	Bark beetles
	(−)-γ-Cadinene in *Ophrys*	Solitary bees*
	Carvone epoxide in *Catasetum*	Euglossine bees*
	Methyleugenol in *Cassia*	Fruit-flies*

* For details, see Chapter 2.

One might assume that all these interactions are accidental or opportunistic and have little to do with the theory of co-evolution. Nevertheless, there is the striking fact that not one but two classes of insect hormone—juvenile and moulting—are produced in plants. The ability to produce ecdysones (the moulting hormones) as chemical signals is probably something that was established quite early in evolution. Thus, ecdysones have recently been found to occur in primitive helminth parasites (e.g. trematodes). Again, the steroids controlling gametogenesis in fungi (e.g. antheridiol, oogoniol) may be related to them, being derived with them from a common cholesterol-based precursor (Karlson, 1983). The synthesis and accumulation of these steroids as defence agents in many ferns and gymnosperms is therefore not so surprising after all. As will be seen later (Chapter 7), phytoecdysones probably represent only one of a variety of co-evolutionary adaptations in plants to herbivore attack.

Bibliography

Books and Review Articles

Bradbury, R. B. and White, D. (1954). Oestrogens and related substances in plants. *Vitamins and Hormones* **12**, 207–233.

Bowers, W. S. (1991). Insect hormones and antihormones in plants. In: Rosenthal, G. A. and Berenbaum M. R. (eds.), 'Herbivores: their Interactions with Secondary Plant Metabolites, Vol. 1, 2nd edn, pp. 431–456. Academic Press, San Diego.

Camps, F. (1991). Plant ecdysteroids and their interaction with insects. In: Harborne, J. B. and Barberan, F. A. T. (eds.), 'Ecological Chemistry and Biochemistry of Plant Terpenoids', pp. 264–286. Oxford Scientific Publications.

Heftmann. E. (1975). Functions of steroids in plants. *Phytochemistry* **14**, 891–902.

Hendrix, J. W. (1970) Sterols in growth and reproduction of fungi. *A. Rev. Phytopath.* **8**, 111–130.

Kircher, H. W. and Heed, W. B. (1970). Phytochemistry and host plant specificity in *Drosophila. Recent Adv. Phytochem.* **3**, 191–208.

Pfiffner, A. (1971). Juvenile hormones. In: Goodwin, T. W. (ed.), 'Aspects of Terpenoid Chemistry and Biochemistry', pp. 95–136. Academic Press, London.

Rees, H. H. (1971). Ecdysones. In: Goodwin, T. W. (ed.), 'Aspects of Terpenoid Chemistry and Biochemistry', pp . 181–222. Academic Press, London.

Shutt, D. A. (1976). The effects of plant oestrogens on animal reproduction. *Endeavour* **35**, 110–113.

Silverstein, R. M., Brownlee, R. G., Bellas, T. E., Wood, D. L. and Browne, L. E. (1968). Brevicomin: principal sex attractant in the frass of the female Western pine beetle. *Science, N.Y.* **159**, 889–890.

Staal, G. B. (1986). Antijuvenile hormone agents. *A. Rev. Entomol.* **31**, 391–429.

Williams, C. M. (1972). Hormonal interactions between plants and insects. In: Sondheimer, E. and Simeone, J. B. (eds.), 'Chemical Ecology', pp. 103–132. Academic Press, New York.

Wood, D. L. (1982). The role of pheromones, kairomones and allomones in the host selection and colonization behaviour of bark beetles. *A. Rev. Entomol.* **27**, 411–446.

Literature References

Argandona, V. H., Niemeyer, H. M. and Corcuera, L. J. (1981). *Phytochemistry* **20**, 665.
Baker, W., Harborne, J. B. and Ollis, W. D. (1953). *J. Chem. Soc.* 1859–1863.
Berger, P. J., Sanders, E. H., Gardner, P. D. and Negus, N. C. (1977). *Science, N. Y.* **195**, 575–577.
Bickoff, E. M, (1968). Rev. Ser. I. (1968). Commonwealth Bur. Pastures and Field Crops, pp. 1–39. Hurley, Berks.
Biggers, J. D. (1959). In: Fairbairn, J. W. (ed.), 'Pharmacology of Plant Phenolics', pp. 51–69. Academic Press, London.
Birgersson, G., Schlyter, F., Lofquist, J. and Bergstrom, G. (1984). *J. Chem. Ecol.* **10**, 1029.
Bounds, D. G. and Pope, G. S. (1960). *J. Chem. Soc.* 3696–3705.
Bowers, W. S. (1968). *Bio-Science* **18**, 791–799.
Bowers, W. S. and Nishida, R. (1980). *Science N. Y.* **209**, 1030–1031.
Brand, J. M., Bracke, J. W., Britton, L. N. and Markovetz, A. J. (1976). *J. Chem. Ecol.* **2**, 195–199.
Brooks, G. T., Pratt, G. E. and Jennings, R. C. (1979). *Nature, Lond.* **281**, 570–572.
Butenandt, A. and Jacobi, H. (1933) *Z. Physiol. Chem.* **218**, 104–112.
Butenandt, A. and Karlson, P. (1954). *Z. Naturforsch.* **96**, 389–391.
Byers, J. A., Wood, D. L., Browne, L. E., Fish, R. H., Piatek, B. and Hendry, L. B. (1979). *J. Insect Physiol.* **25**, 477–482.
Byers, J. A., Wood, D. L., Craig, J. and Hendry, L. B. (1984). *J. Chem. Ecol.* **10**, 861.
Dean, P. D. G., Exley, D. and Goodwin, T. W. (1971). *Phytochemistry* **10**, 2215–2216.
Heftmann, E., Ko, S. T. and Bennett, R. D. (1966). *Phytochemistry* **5**, 1337–1339.
Heikkenen, H. J. and Hrutfiord, B. F. (1965). *Science, N. Y.* **150**, 1457–1459.
Hendrix, S. D. (1980). *Amer. Nat.* **115**, 171–196.
Hendry, L. B., Piatek, B., Brownell, L. E., Wood, D. L., Byers, J. A., Fish, R. H. and Hicks, R. A. (1980). *Nature, Lond.* **284**, 485.
Hikino, J., Ohizuma, Y. and Takemoto, T. (1975). *J. Insect Physiol.* **21**, 1953–1963.
Jacobson, M., Redfern, R. E. and Mills, G. D. (1975). *Lloydia* **38**, 473–476.
Karlson, P. (1983). *Hoppe-Seyler's Z. Physiol. Chem.* **364**, 1067.
Karlson, P., Hoffmeister, H., Hummel, H., Hocks, P. and Spitelber, G. (1965). *Chem. Ber.* **98**, 2394–2402.
Kubo, I., Klacke, J. and Asano, S. (1983). *J. Insect Physiol.* **29**, 307–316.
Leopold, A. S., Erwin, M., Oh, J. and Browning, B. (1976). *Science, N. Y.* **191**, 98–99.
Marrian, G. F. and Haslewood, G. A. D. (1932). *Biochem. J.* **26**, 1227.
Nakanishi, K. (1968). *Bio-Science* **18**, 791–799.
Nishida, R., Bowers, W. S. and Evans, P. H. (1983). *Arch. Insect Biochem. Physiol.* **1**, 17–24.
Proksch, P. and Rodriguez, E. (1983). *Phytochemistry* **22**, 2335.
Puritch, G. S. and Nijholt, W. W. (1974). *Can. J. Bot.* **52**, 585–587.
Raffa, K. F. and Berryman, A. A. (1987). *Amer. Nat.* **129**, 234–262.
Roller, H., Dahm, K. H., Sweeley, C. C. and Trost, B. M. (1967). *Angew. Chem. Intern. Ed. English* **6**, 179–180.
Saunders, E. H., Gardner, P. D., Berger, P. J. and Negus, N. C. (1981). *Science, N. Y.* **214**, 67–68.
Sturgeon. K. B. (1979). *Evolution* **33**, 803–814.
Takemoto, T., Ogawa, S., Nishimoto, N., Arihari, S. and Bue, K. (1967). *Yakugaku Zasshi* **87**, 1414–1418.

Van Rompuy, L. L. L. and Zeevaart, J. A. D. (1979). *Phytochemistry* **18**, 863–865.
Wigglesworth, V. B. (1954). 'The Physiology of Insect Metamorphosis'. Cambridge Univ. Press, London.
Young, I. S., Hillman, J. R. and Knights, B. A. (1978). *Z. Pflanzenphys.* **90**, 45–50.

5 Insect Feeding Preferences

I. Introduction

Until recently, the role of secondary compounds in plants has remained largely obscure. Many plant physiologists have regarded them as waste products of primary metabolism and of no possible survival value to plants. This situation has been completely changed largely due to the attention paid to these substances by biologists interested in the complex and subtle interactions that take place between plants and insects. Fraenkel (1959), in what is now a classical article, was one of the first to voice the suggestion that secondary compounds are directly involved in the feeding behaviour of insects. However, it was not until the major review of Ehrlich and Raven (1964) on the probable factors controlling the co-evolution of butterflies and plants that secondary substances became the cornerstone of a new theory of biochemical co-evolution between animals and plants.

The conclusions from Ehrlich and Raven's article can best be presented here in their own words: 'A systematic evaluation of the kinds of plants fed upon by the larvae of certain subgroups of butterflies leads unambiguously to the conclusion that *secondary plant substances play the leading role* in determining patterns of utilization. This seems true not only for butterflies but for all phytophagous groups . . . In this context, the irregular distribution in plants of secondary compounds . . . is immediately explicable. Angiosperms have, through occasional mutations and recombinations, produced a series of chemical compounds not directly related to basic metabolism. Some of these compounds by chance serve to reduce or destroy the palatability of the plant in which they are produced. Such a plant protected from phytophagous animals enters a new adaptive zone. Phytophagous insects, however, can evolve in response to physiological obstacles . . . selection (from insect populations) could carry a recombinant or mutant (in turn) into a new adaptive zone. Here it would be free to diversify largely in the absence of competition from other feeders. Thus, the diversity of plants not only may tend to augment the diversity of phytophagous animals, the converse may also be true.'

Subsequent to the publication of Ehrlich and Raven's review, the role of such compounds as alkaloids, terpenoids and flavonoids has been extensively explored in this co-evolutionary situation of plants with their insect herbivores. Important reviews of this topic are those of Dethier (1972), Feeny (1975), Fraenkel (1969), Meeuse (1973) and Schoonhoven (1968, 1972). There are several symposium volumes, e.g. Hedin (1983) and Visser and Minks (1982), which should also be consulted.

The theory rests on a variety of observations but seven main points may be discerned.

(1) The theory explains why there are three major areas of *enormous diversity* in biology—in the angiosperms, in the insect kingdom and in secondary compound chemistry. In the case of angiosperms, well over a quarter of a million species are estimated to be present in the world. Estimates of the species diversity in insects vary between 2 and 5 million; the Lepidoptera alone has 15,000 species. Finally, with regard to secondary compounds, the total of known structures must be in the region of 50,000; at least 10,000 alkaloids and 20,000 terpenoids have been characterized. A much greater number of structures await recognition in the vast number of plants not yet sampled or chemically analysed.

(2) The theory explains a conspicuous non-event (Feeny, 1975), namely that the destructive potential of herbivorous insects has not prevented the green plants from dominating the earth, i.e. higher plants must possess effective defences against overpredation.

(3) Most herbivorous insects discriminate between plants in feeding and many feed on a small number of related species belonging to the same genus, tribe or family.

(4) The host plants of a given insect may share similar secondary compounds but be different in general morphology and anatomy.

(5) Many secondary compounds are highly toxic to insects. This applies not only to alkaloids, but to many terpenoids and oxygen heterocyclic compounds (see Chapter 3).

(6) Plants can arrive at the same solution to an ecological problem (e.g. animal predation) by a variety of routes, i.e. they practise chemical mimicry. For example, a repellent bitter taste in plants may be produced by the synthesis of an alkaloid (e.g. quinine), a saponin, a cardiac glycoside, a triterpenoid (e.g. cucurbitacin), a sesquiterpene lactone (e.g. lactupicrin) or a flavanone glycoside (e.g. naringin).

(7) All angiosperms tend to have at least one type of secondary compound in major concentration, i.e. they accumulate these substances in sufficient amount to be effective in controlling insect attack. The plant may have alkaloid *or* flavonoid *or* terpenoid; it is rare to find a plant rich in a variety of *different* classes of secondary substances.

In any theory involving the considerable time span taken for the angiosperms to evolve to the present day (some 135 m. years), a great deal rests on circumstantial evidence. However, as with the Darwinian theory of evolution, it is possible to devise experiments among present-day plants and insects to test these co-evolutionary ideas. There are, however, particular problems associated with testing this theory which have to be borne in mind in interpreting the data available to us. Thus, the toxicity of a secondary substance may be subtle, incomplete or difficult to measure accurately. Intake of the compound only needs to slightly reduce the fitness of the invading insect population to be ecologically significant; alternatively its effect may take the form of a hormonal interference (see Chapter 4).

In seeking a secondary substance involved in insect feeding, it is necessary to distinguish this function from other ecological roles, e.g. it may be an agent in plant–plant interactions (see Chapters 9 and 10). There are also problems in determining which of many related structures present in a particular plant may be active in co-evolution. Many trace components in plants may be biosynthetic intermediates and not relevant from the functional viewpoint. Finally, there are considerable problems in bioassay. Many oligo- or monophagous insects are reluctant to feed on an abnormal diet and it may be difficult to test the effectiveness of isolated plant fractions. Some insects may prefer to die from starvation rather than accept a diet lacking their normal feeding stimulant—this is true of the larvae of the cabbage white butterfly which live on crucifers (see p. 138).

In this chapter, it is planned to consider briefly the biochemical basis of insect feeding preferences and then provide a number of selected examples where secondary compounds act as attractants or deterrents to particular insects. Mention will be made of oviposition attractants since there is a clear chemical link with the processes of host plant searching and feeding. The question of the feeding preferences of higher animals, which have a similar biochemical basis, is reserved for Chapter 6. A critical examination of the evidence for plant defence, both static and dynamic, and for animal responses to these defence mechanisms will be reserved for Chapter 7.

II. Biochemical Basis of Plant Selection by Insects

A. Co-evolutionary Aspects

The choice of plants for feeding by present-day insect populations has to be considered within an overall evolutionary context. The situation observed today has been produced by evolutionary forces operating in the past and the interaction between plants and their insect predators is a dynamic one and likely to be subject to continual variation and change. At any one instance, the plant or the insect may appear to have the advantage. Both partners in the interaction, however, adapt themselves in different ways to the changing conditions. Other environmental pressures (e.g. climate, disease, etc.) also have their effect on the interaction:

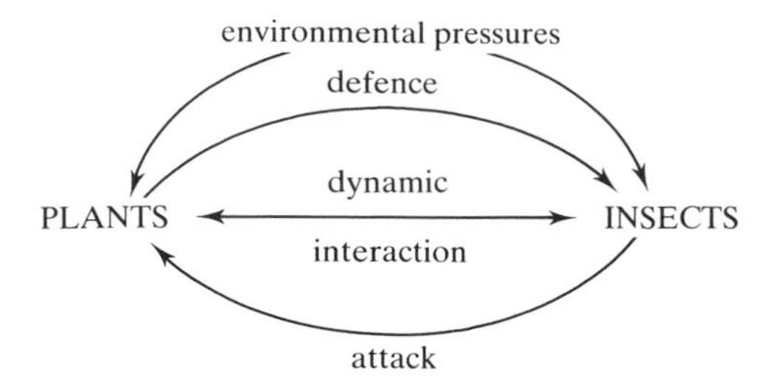

The plants that provide the source of food to insects—and this applies to practically all members of the Angiospermae–have evolved to avoid being overeaten. This can be achieved in a variety of ways. Morphological modifications, clearly produced in response to herbivore attack, include the armouring of the more attractive and accessible plant parts with spines, prickles, thorns or stinging hairs. Emigration via the seeds to new habitats represents another method by which plants can avoid overpredation. For example, a move by a plant from a mainland situation to an adjacent island may often be successful in 'throwing off' an animal pursuer.

Undoubtedly, though, the most significant type of defence a plant can erect is in the realms of chemical armoury. Effective and often drastic reduction in insect feeding may be achieved by altering the chemical components of the plant leaves. This can be done either by reducing the edibility or nutritive status of the leaves or more positively by introducing a toxin, an unpleasant taste or an offensive odour into the leaf tissue.

Insects evolve in this dynamic interaction by overcoming the plant's defences. Because insects are highly specialized organisms, they have a greater variety of responses available to them. They can adapt biochemically and anatomically to the digestion and assimilation of new plant foods. They can develop new feeding habits and new taste preferences. They are more mobile than plants and can move away to new pastures, when faced with an unpalatable or inedible species. Finally, they are adept at developing detoxification mechanisms and can neutralize the potency of a toxin, so that it no longer acts as a barrier to feeding. Such detoxification usually involves a chemical modification of the toxin *in vivo*, its conversion to a non-toxic conjugate or else its sequestration in special storage tissues within the insect.

B. Plant Chemicals as Defence Agents

As has already been mentioned in Chapter 3, the chemistry of plant defence is essentially the production of a variety of secondary metabolites. These compounds can be of many different structures. Toxicity to insects, for example, can be achieved by the synthesis of any one of a range of secondary compounds, be it an alkaloid, a terpenoid or a flavonoid. Frequently, mixtures of several closely related structures of the same class are produced by plants and it is likely that synergism occurs, one substance enhancing the effectiveness of a second substance as a feeding deterrent.

In order to repel insect feeding, a plant may not necessarily have to produce a substance that is highly toxic to the insect. It may be sufficient to produce a compound that is unpleasant or distasteful. Also, an effective barrier to most insect feeding can be erected by reducing the nutritional value of the plant. Thus it is apparent that the synthesis of some secondary compounds, especially of tannins, may have this effect on insect behaviour.

Two terms are widely used to refer to chemicals involved in insect feeding preferences: chemical attractants (or feeding stimulants) and chemical repellents (or feeding toxins). It can be confusing to the uninitiated that the same type of compound is often ascribed a role as an attractant to one insect *and* as a repellent to a second insect. This apparently contradictory situation can only be appreciated when the evolutionary aspects are taken into account. It is clear that *any* chemical substance implicated in insect–plant interactions has a defensive role, irrespective of whether it is a repellent or an attractant in any particular instance.

The hypothesis is that all secondary compounds affecting insects (and other animals) were first synthesized by plants (and continue to be synthesized) as a general defence against animal feeding. In this situation, one insect species evolves the means to detoxify the repellent (or toxin) and then begins to feed exclusively on that plant. The insect can do this because the repellent is no longer harmful to it. However, the original compound is still highly effective in repelling other insects from feeding so that it also has the enormous advantage that it does not have to compete with other species for its food.

To the successful insect, the toxin through its smell or taste becomes a valuable signal to guide it to its favoured host plant. Indeed, the substance becomes an attractant, because of its close association with this plant. In many cases, the insect becomes dependent on the presence of the attractant, so much so that it becomes an essential feeding stimulant to the insect. Moreover, a substance that is an attractant has its own hazards to the insect, because of this dependency, and the insect may become so 'hooked' on the substance that it is unable to feed on a diet lacking this component. Eventually, the size and success of the insect population may be closely controlled by the availability of the necessary host plant required for feeding.

To summarize, any chemical affecting insect feeding is basically a plant's defence against insect attack and even in situations where its role has been reversed into that of an attractant by one particular insect species, it still has its dangers to the successful predator.

C. Insect Feeding Requirements

Insects vary enormously in their responses to the plant world. Those that prey exclusively on other insects are clearly unaffected by plant chemistry but the vast majority of insects are phytophagous, i.e. are able to discriminate between plants through their chemical senses. Within this broad category, insects are divided into those that are polyphagous, oligophagous or monophagous.

Polyphagous insects are those that eat any plant which they are presented with. Locusts fall into this category, though even with these pestilential insects, it is possible to demonstrate some differential feeding habits (Chapman, 1976). Leaf-cutting ants are another group which utilize a wide variety of plant species for feeding purposes (see p. 148). The chrysanthemum leaf-miner *Liriomyza trifolii*, in spite of its common name, is another omnivore, feeding on over a 100 different host plants besides chrysanthemums.

Oligophagous insects, probably making up the majority of phytophagous insects, are those which feed on a relatively few related species belonging to one or only a few plant genera or families. They are selective in their feeding, being guided by other factors than purely nutritional requirements in their search for food. Of the many examples here, one may quote the danaid butterflies (see p. 90), which feed almost exclusively on plants of the Apocynaceae and Asclepiadaceae. Most aphids are oligophagous, generally feeding on plants in a single genus or single family (van Emden, 1972). Many other insect pests are also oligophagous, such as the oak leaf roller and the various pine bark beetles.

Finally, there are monophagous insects, which feed on a single plant species. The best known is the silkworm, *Bombyx mori*, which is restricted to mulberry leaves, *Morus nigra*. The condition of monophagy may be imposed on an oligophagous insect in certain habitats by a depauperate flora. The survival of some of the butterflies in the British fauna depends on particular host species being available to them. The swallowtail butterfly, *Papilio machaon* will only feed in England on the umbellifer, *Peucedanum palustre*, although in other European countries it will live on several related umbellifers. The close dependency of Lepidoptera on their preferred host plants is nicely illustrated in the book of Mansell and Newman (1968) on British butterflies, since each insect is shown against a background of one of its characteristic host plants.

There has been much debate in the past whether the preferential feeding behaviour of insects is solely determined by nutritional requirements or solely by their response to hostile chemicals within the leaf or other tissue. It is now generally agreed that both nutritional and other chemical agents guide insects to plants. The relative importance of nutrition versus secondary chemistry probably varies from insect to insect. However, secondary chemistry is usually the controlling factor, since all plants are relatively similar in nutritional value. Since the primary biochemical processes within the leaf are practically identical in all green plants, the relative amounts of sugar, lipid, polysaccharide, amino acid and protein are inevitably very similar. Again, physiological processes (e.g. senescence) are more likely to affect nutritional status in the plant than anything else. There is, in addition, little concrete evidence from insect feeding studies to suggest that certain plants are avoided because they are nutritionally inadequate (van Emden, 1973).

This is not to say that nutritional factors do not sometimes seriously affect insect predation on plants. The nitrogen content is often limiting, because of the relatively low levels (average leaf protein content is only 13%) and of the large variations with plant age. Nitrogen availability can thus be a major determinant in controlling the rates of growth and the reproduction levels in insect populations (McNeil and Southwood, 1978).

The main taste response of insects is to sweetness and free sugar, present universally in leaf tissue, is a major nutritional requirement. Other major requirements are for nitrogen (protein or free amino acid), vitamins and phospholipids. Trace elements are also important. Finally, there are the sterol requirements, which have already been discussed under insect metamorphosis in Chapter 4.

III. Secondary Compounds as Feeding Attractants

A. General

The role of secondary substances as feeding stimulants in insects has been widely studied and a wide range of plant structures have been implicated in such interactions. A selection from the many examples available in the primary literature is presented in Table 5.1. Further examples are discussed in the reviews of Dethier (1972) and Schoonhoven (1968, 1972).

Three general points may be noted. First, almost every class of secondary substance has been implicated in such interactions. Compounds representing at least eight different structural types are mentioned in the table. Second, compounds which are toxic or repellent generally are particularly prominent, e.g. the bitter cucurbitacins, the poisonous alkaloid sparteine, the acrid mustard oil allyl isothiocyanate and so on. Third, in the majority of cases, more than one compound has been implicated as feeding attractants. In one case, as many as 14 components are active, e.g. in the feeding of the catalpa sphinx moth.

Some of the above points will now be discussed in more detail, in relation to specific plant–insect interactions. The feeding attractant of the silkworm, the cabbage butterfly and the cabbage aphid will be particularly mentioned, since these have been most intensively studied; many of their behavioural responses are undoubtedly shown by many other phytophagous insects as well.

B. The Silkworm–Mulberry Interaction

Because of the valuable silk fibres it spins, the silkworm *Bombyx mori* has become one of the most beneficial insects to mankind. For this reason, more time and effort has been spent on the study of its feeding behaviour than on that of any other comparable insect. Scientifically, it is interesting in that it feeds exclusively on the leaves of the black and white mulberry trees, *Morus nigra* and *M. alba*. As a result of the experiments of Hamamura *et al.* (1962), it is apparent that a range of chemicals in the leaf are concerned in this very specific feeding behaviour. The substances can be divided into three groups: olfactory attractants, biting factors and swallowing factors (Table 5.2). Each group has a specific role in the insect feeding response. While some of the compounds listed are generally present in all plants, others are secondary constituents some at least of which are specifically associated with the host plant.

The olfactory attractants of mulberry leaf are a mixture of monoterpenes, which exert their primary effect by attracting the insect larvae to feed through their sense of smell. It has been shown that *Bombyx* larvae are sensitive to these monoterpene mixtures as soon as they approach within 3 cm of the leaf, their olfactory sense being relatively acute at this distance. The importance of smell as an attractant can be demonstrated by surgical removal of the insect's oral sense receptors. Insects so treated immediately lose their power of discrimination and they start biting into almost any plants which are presented to them.

The second group of substances involved in silkworm feeding, are the biting factors (Table 5.2). Three of the compounds included here—sucrose, inositol and sitosterol—

Table 5.1 Plant secondary compounds as feeding attractants

Insect class and species	Plant host	Chemical attractants	References
APHIDS			
Brevicoryne brassicae	*Brassica campestris* (cabbage)	Glucosinolate: sinigrin	van Emden, 1972
Acyrthrosiphon spartii	*Sarothamnus scoparius* (broom)	Alkaloid: sparteine	Smith, 1966
BEETLES			
Agasicles sp.	*Alternanthera phylloxeroides*	Flavone: 6-methoxyluteolin 7-rhamnoside	Zielske *et al.*, 1972
Diabrotica undecimpunctata	*Citrullus lanatus* (water-melon)	Triterpenoids: cucurbitacins	Chambliss and Jones, 1966
Scolytus mediterraneus	*Prunus* spp.	Flavonoids: taxifolin, pinocembrin, dihydrokaempferol	Levy *et al.*, 1974
S. multistriatus	*Ulmus europea* (elm)	Flavonoid: catechin 7-xyloside Triterpenoid: lupeyl cerotate	Doskotch *et al.*, 1973
BUTTERFLIES			
Papilio ajax	*Foeniculum vulgare* (fennel)	Essential oils: various	Dethier, 1941
Pieris brassicae	*Brassica campestris* (cabbage)	Glucosinolate	Schoonhoven, 1968
MOTHS			
Bombyx mori	*Morus nigra* (mulberry)	Flavonoids and essential oils (see Table 5.2)	Hamamura *et al.*, 1962
Ceratomia catalpae	*Catalpa* spp.	Iridoid glycosides: various	Nayer and Fraenkel 1963
Serrodes partita	*Pappea capensis* (wild plum)	Quebrachitol	Hewitt *et al.*, 1969
WEEVILS			
Sitonia cylindricollis	*Melilotus alba*	Coumarin	Akeson *et al.*, 1969

Table 5.2 Chemical factors of mulberry leaves associated with silkworm feeding

Attractants	Biting factors	Swallowing factors
ESSENTIAL OILS:	FLAVONOIDS:	INORGANIC ELEMENTS:
Citral	Isoquercitrin	Silicate
Terpinyl acetate	Morin	Phosphate
Linalyl acetate	TERPENOID:	CELL WALL COMPONENT:
Linalol	Sitosterol	Cellulose
β,γ-Hexenol	SUGARS:	
	Sucrose	
	Inositol	

From Hamamura *et al.* (1962).

are essential dietary requirements and they must all act as general feeding stimulants to most insects. The other two substances—morin and isoquercitrin—are much more restricted in their natural distribution and clearly have a different role, since they are not important dietary constituents. These two compounds, together with the essential oils in the leaf, must provide the specific basis for the attraction of the silkworm to its favoured food source. While one of these two flavonoids, isoquercitrin (or quercetin 3-glucoside) is relatively common in angiosperm leaves, the other, morin, is almost completely exclusive in its occurrence to this particular plant.

The structural requirements of these two flavonoids seem to be absolute in that substitution of closely related structures in the insect diet is ineffective in producing a feeding response. Indeed, if quercetin 3-glucoside is replaced by either the 3-rhamnoside or the 3-rutinoside (see Fig. 5.1 for formulae), a receptor site sensitive to repellent chemicals is triggered and the insect is put off feeding. It is noteworthy that simple substitution of one sugar (rhamnose) for another (glucose) in the flavonoid molecule can have such a dramatic effect on insect feeding. However, other examples are known in the flavonoid series where a change in the nature of the substituted sugar can dramatically alter taste properties (see Horowitz, 1964). Reversal of taste properties can also occur with other types of simple molecules. Thus, the amino acid L-alanine, $CH_3CH(NH_2)CO_2H$, is a feeding stimulant to the European corn borer while the structurally related isomeric β-alanine, $NH_2CH_2CH_2CO_2H$, is a deterrent (Beck, 1960).

The final stage in the silkworm feeding is the act of swallowing and here relatively common chemicals provide the necessary stimulus. These are the cellulose, richly present in the cell wall of this plant, and the mineral elements, silicate and phosphate. These substances provide the necessary 'bulk' to the insect alimentary canal in just the same way as 'roughage' is a needed requirement in mammalian diets.

The monophagy exhibited by the silkworm in its feeding choice of mulberry leaf indirectly suggests that most other plants and hence their secondary constituents are likely to be hostile to it. A dramatic illustration of the susceptibility of the silkworm to other secondary compounds is the finding by Jones *et al.* (1981) that 2-furaldehyde, one of the volatiles of the bald cypress *Taxodium distichum*, is growth inhibitory and toxic at

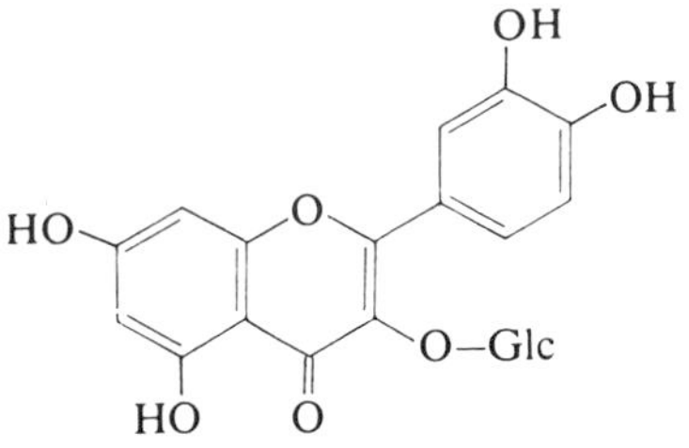

isoquercitrin
(*Bombyx mori*)

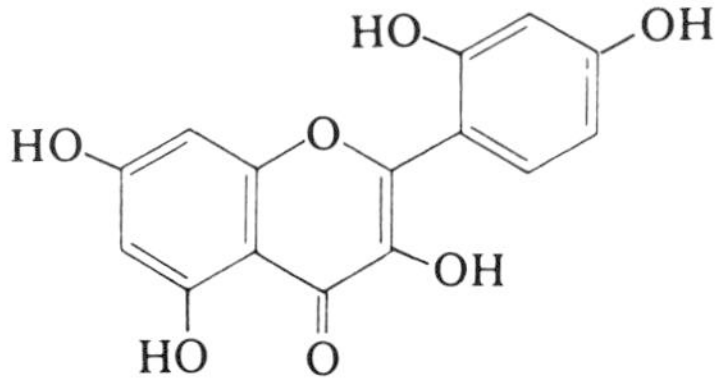

morin
(*Bombyx mori*)

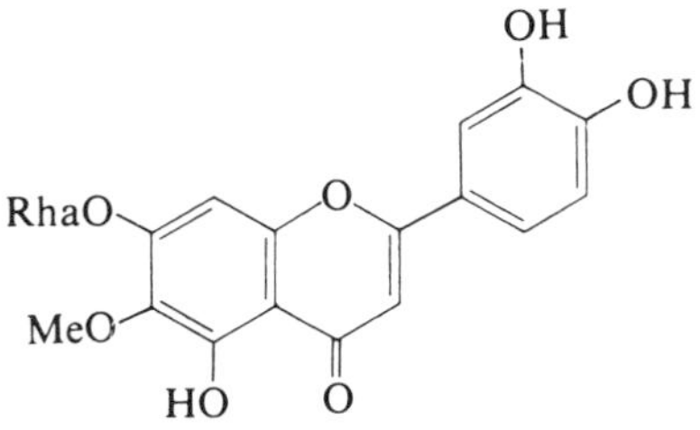

6-methoxyluteolin
7-rhamnoside (*Agasicles*)

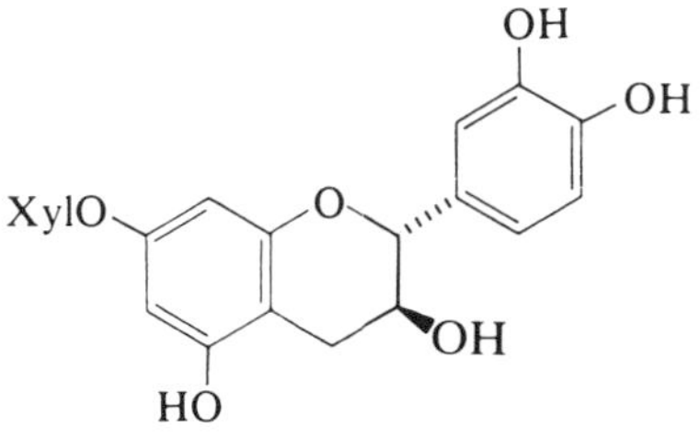

catechin 7-xyloside
(*Scolytus multistriatus*)

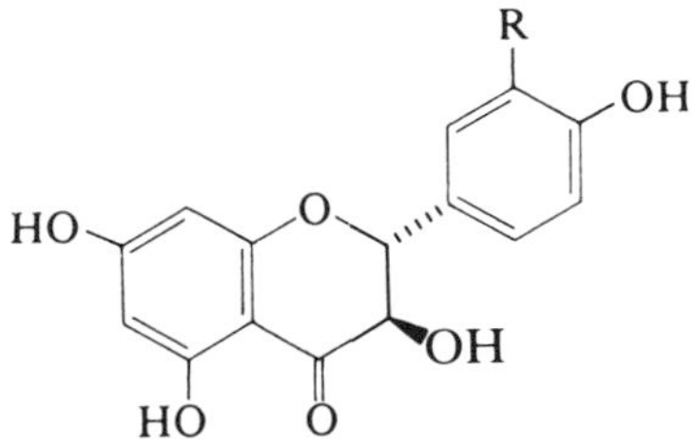

dihydrokaempferol (R = H)
taxifolin (R = OH)
(*Scolytus mediterraneus*)

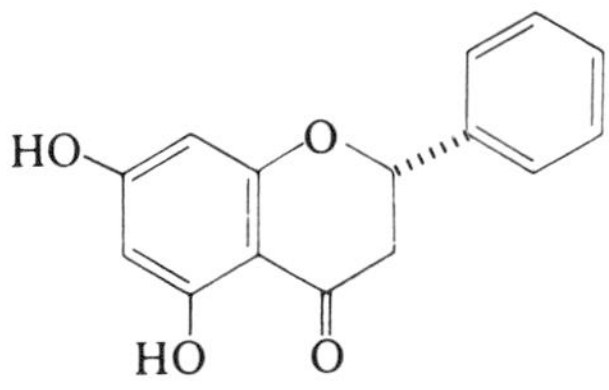

pinocembrin
(*Scolytus mediterraneus*)

Figure 5.1 Flavonoids as feeding attractants to insects

a concentration as low as 1 ppm. This discovery neatly explains retrospectively the demise of a local silk industry set up by American colonists in Georgia during the eighteenth century. Silk production was satisfactory until the larvae were transferred to a new rearing house built of bald cypress wood, when the industry went into a decline from which it never recovered.

The inhibition caused by 2-furaldehyde in silkworm larvae at such low concentrations appears to be due in part to an indirect effect on the silkworm enteric microflora. Indeed, 2-furaldehyde is known to be highly bactericidal. The growth inhibitory effects on the animal in this case appear to be due to a reduced microbial contribution to its nutrition and digestion. This phenomenon could occur in other insects so that endosymbiotic microflora may have to be taken into account whenever evaluating the deterrence of plant chemicals on insect feeding (see also Section IV).

C. Glucosinolates as Feeding Attractants in the Cruciferae

The best documented case of repellent substances becoming attractants is in the feeding behaviour of crucifer insect pests. Several particular insects have been studied in detail—the cabbage butterfly *Pieris brassicae*, the cabbage aphid *Brevicoryne brassicae* and the flea beetles *Phyllotreta* species—but what applies to these insects is undoubtedly also true for other predators on members of the cabbage family, the Cruciferae. The repellent substances, whose role has been dramatically reversed by these insects, are the acrid-smelling mustard oils released by these plants. The volatile oils occur in the plants in bound form as the thioglucosides (called glucosinolates) and are released enzymically by the action of the enzyme myrosinase which co-occurs with the glucosides in crucifer leaves. Although a range of different glucosinolates occur within the Cruciferae, the major compound of the cabbage and the one upon which most work has been done is sinigrin. This releases allyl isothiocyanate, its mustard oil, following enzymic hydrolysis and chemical rearrangement according to the reaction shown in Fig. 5.2. The corresponding thiocyanate and nitrile may be formed as minor products of this reaction.

Allyl isothiocyanate is the acrid, sharp-tasting principle of mustard and, although taken by man in small amounts as a table condiment, it is generally a feeding repellent to most animals. That it is not only a repellent but can actually be toxic to insects has been demonstrated by Erickson and Feeny (1974). Since insects will usually refuse sinigrin in the diet, these authors fed the substance to larvae of the black swallowtail butterfly *Papilio polyxenes* by infiltrating a 0.1% solution (based on leaf fresh wt) of sinigrin into one of its normal food plants, namely celery. This was sufficient to cause 100% mortality, the concentration of sinigrin in the celery being close to that present in crucifier plants. The authors argue from this that sinigrin in cabbage plants has a clear defensive function against insects not normally feeding on this family of plants.

The fact that this same sinigrin is a positive feeding stimulus to both cabbage butterflies and aphids has been demonstrated in a number of experiments (see Dethier, 1972). Thus larvae of the cabbage butterfly can generally be persuaded to feed on an artificial diet but it is essential that mustard oil (or the glucosinolate) be added to it. It is possible to take newly hatched larvae and get them to feed on a diet devoid of mustard oil, but even with these insects addition of glucosinolate to the diet immediately increases feeding intake by 20%. The dependency of the larvae on this feeding stimulant is most significantly demonstrated in the case of insects fed from the start on cabbage leaves. When

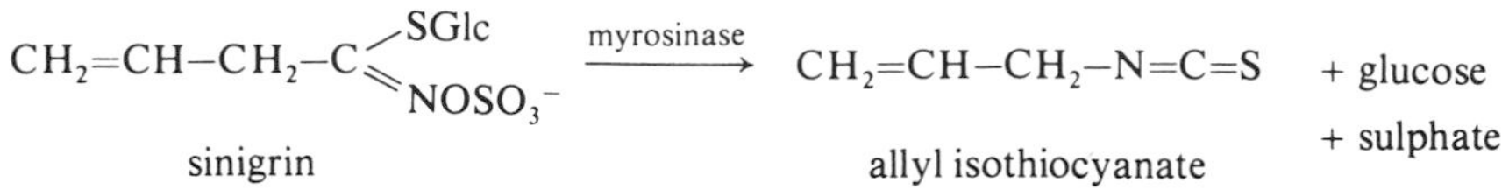

$$CH_2{=}CH{-}CH_2{-}C\begin{matrix}\diagup SGlc \\ \diagdown\!\!\diagdown NOSO_3^-\end{matrix} \xrightarrow{\text{myrosinase}} CH_2{=}CH{-}CH_2{-}N{=}C{=}S \quad + \text{glucose} \quad + \text{sulphate}$$

sinigrin — allyl isothiocyanate

Figure 5.2 Enzymic release of allyl isothiocyanate from sinigrin

these larvae are transferred to an artificial diet lacking sinigrin, they refuse to eat and, in fact, prefer to die rather than accept food lacking what has become an essential attractant. The importance of sinigrin to *Pieris brassicae* is shown in other aspects of their life-cycle. The adult female, for example, uses the same substance as an oviposition stimulant. Indeed, its effectiveness is apparent in that the butterfly can be fooled into laying its eggs on a piece of filter paper, as long as the paper has been previously soaked in a solution of sinigrin (see Section V).

The cabbage aphid is similarly attracted to the same host plant by the presence of sinigrin. While the aphid is highly specific to crucifers, it can be persuaded to feed on a non-host plant such as the broad bean *Vicia faba* by the simple expedient of infiltrating the bean leaves with a solution of sinigrin. Factors controlling the feeding of aphids on cabbage leaves have been studied by van Emden (1972) who concludes that there are two major chemical controls in aphid behaviour—the concentration of sinigrin and the free amino acid balance in the leaves.

The aphids use the presence of sinigrin as a sensitive guide to the host plant when they arrive at a new site for infestation. They may alight on non-host plants, but once they have inserted their sucking stylets and found sinigrin to be absent, they immediately fly off in search of a plant containing the essential stimulant. Once settled on a host plant, they are also guided to the most suitable site for feeding by the sinigrin content. Thus, young cabbage leaves have a very high sinigrin concentration, so much so that aphids avoid feeding on these tissues. It is the more mature leaves with medium sinigrin levels that they prefer to feed on. They also avoid senescent plants; here it is not a drop in sinigrin content which turns them off feeding but a change in the amino acid balance. In particular, older leaves contain *inter alia* more than average amounts of γ-aminobutyric acid and variation in this and other amino acids is apparently detected by aphids.

Aphids are thus highly sensitive to plant chemistry: to the feeding stimulant, its absolute concentration, and to nutritional factors, especially the amino acid balance. van Emden's reference (1972) to 'aphids as true phytochemists' is a very apt description for these highly perceptive insect feeders.

Many other insects that feed on crucifers are also stimulated to feed by the presence of glucosinolates in their food plants. Nearly all, however, show feeding preferences for certain species within the family—some even exhibit monophagy—so that there must be other factors controlling feeding choice. Variation in glucosinolate chemistry—which is quite marked in the family—could be partly responsible but it is now abundantly clear that other secondary components in the Cruciferae also exert an effect on feeding preferences. Glucosinolates may interact with other feeding stimulants or else with substances hostile to feeding, namely deterrents. Both situations are nicely illustrated by recent experiments carried out on flea beetles of the genus *Phyllotreta* which feed on crucifers and are often agricultural pests.

The first situation occurs with the monophagous horseradish flea beetle, *Phyllotreta armoracea*, which feeds only on the horseradish plant, *Armoracia rusticana*. This beetle appears to be controlled in its feeding by the combination of a general stimulant sinigrin with two species-specific stimulants, the flavonol glucosides, kaempferol and quercetin 3-xylosylgalactosides (Nielsen *et al.*, 1979). It is the disaccharide unit of these flavonols which is important in determining feeding, since other glycosides of these flavonols are less effective stimulants. Comparable effects of different flavonol glycosides on feeding behaviour have been recorded in the case of *Bombyx mori* larvae (see previous section). These 3-xylosylgalactosides appear to be characteristic of the horseradish plant and do not occur in any related potential host crucifer. Their efficiency as stimulants was confirmed in laboratory experiments when mixing them with sinigrin had a synergistic effect on *P. armoracea* feeding.

The second situation occurs with oligophagous *Phyllotreta nemorum*, which feeds happily on radish or turnip leaves but totally avoids eating leaves of *Iberis amara*. Not surprisingly, rejection can be related to the presence of feeding inhibitors, identified as cucurbitacins in this plant (Nielsen *et al.*, 1977). Cucurbitacins are well known as bitter-tasting toxins in the cucumber family Cucurbitaceae, but are rare in the Cruciferae; insects feeding on cucurbits treat them as feeding attractants (see Section IVD). Clearly then, cucurbitacins in the Cucurbitaceae and glucosinolates in the Cruciferae serve much the same function: they are inhibitors for non-adapted species and stimulants for adapted species. A plant species in either family with both toxins, such as *Iberis amara*, will thus be at a selective advantage to avoid attack by adapted species.

In this situation, one might predict that other crucifer species may have 'borrowed' feeding inhibitors from other angiosperm families. Indeed, this is so since cardiac glycosides—characteristic bitter-tasting toxins of the Asclepiadaceae and Apocynaceae (see Chapter 3)—are present in two crucifer genera, *Chieranthus* and *Erysimum*. Yet other beetle species, *Phyllotreta undulata* and *P. tetrastigma*, can be shown to refuse feeding on these plants because of the cardiac glycosides in their tissues (Nielsen, 1978). These inhibitors only provide limited protection, since *P. nemorum*, while rejecting *Iberis amara* (see above), will readily feed on *Cheiranthus* plants both in the laboratory and in the field. These complex interactions between flea beetles and their host plants recall the hackneyed aphorism: one man's meat is another man's poison. It might be more appropriate to summarize as follows: while all these flea beetles enjoy mustard with their meat, they clearly discriminate very precisely between one bitter taste and another.

D. Other Feeding Attractants

As discussed above, the major attractants of the silkworm are essential oils and flavonoids, while that of the cabbage butterfly and aphid is the glucosinolate sinigrin. These are not exceptional cases and it is apparent (Table 5.1) that practically every type of secondary constituent has been assigned a role as a feeding stimulant in a particular plant–insect interaction. The poisonous alkaloid sparteine, for example, acts as a stimulant to feeding of the broom aphid on *Sarothamnus scoparius*. Here, the aphid alters its feeding site according to where the highest concentration of sparteine occurs within the plants. Thus, it begins feeding on the young shoots in the spring and eventually moves during the summer months onto the flower buds and fruit pods as the sparteine content varies with the life-cycle of the plant (Smith, 1966). Again, it proved possible in this case (see also above, p. 139) to make these aphids transfer to a non-host plant *Vicia faba* by infiltrating the leaves with a sparteine solution.

ALKALOID

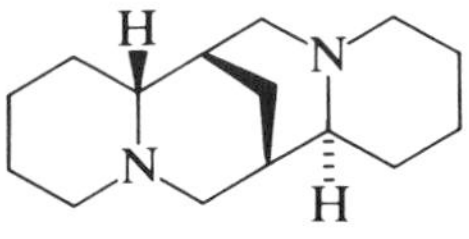

sparteine
(*Acyrthrosiphon spartii*)

IRIDOID

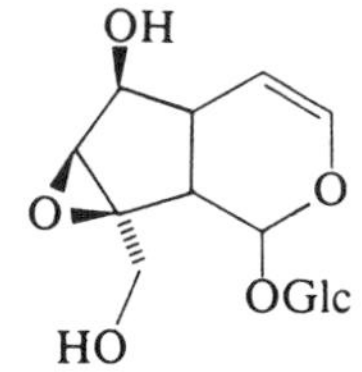

catalpol
(*Ceratomia catalpae*)

PHENYLPROPANOIDS

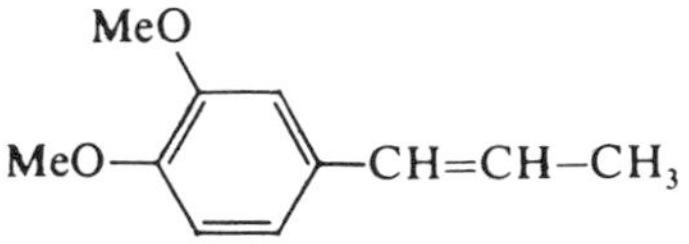

methylisoeugenol
(*Psila rosae*)

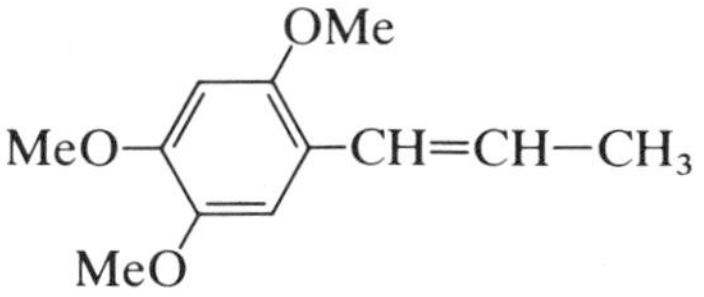

(*E*)-asarone
(*Psila rosae*)

Figure 5.3 Structures of some characteristic insect attractants

What is true of the broom aphid, *Acyrthrosiphon spartii*, may not hold true for other species. Thus *Aphis cytisorum*, which also feeds on several broom plants, actually sequesters and stores an alkaloid, cytisine, from the diet for defence purposes. However, if given a choice of plants with varying alkaloid levels, it prefers to feed on a plant low in alkaloid (Wink and Witte, 1985) so that the alkaloid content is only a limited attractant. Generalist aphids would be expected to avoid feeding on alkaloid-containing plants. This is apparent in the case of *Acyrthosiphon pisum*, since it was deterred from feeding on artificial diets to which different alkaloids had been added in turn (Dreyer *et al.*, 1985).

Essential oils, which act as olfactory stimulants to silkworm feeding on mulberry leaves, probably provide a signal to many other oligophagous insects that they are approaching their favoured host plant. Experiments with larvae of the swallowtail butterfly, which feed exclusively on members of the Umbelliferae, show them to be sensitive to at least eight different components in umbellifer essential oils.

Again, field studies with the aphid *Cavariella aegopodii*, which feeds on umbellifers in the summer, show that it is attracted to land on wild parsnip by the monoterpene attractant, carvone, which is present in its scent (Chapman *et al.*, 1981).

Besides containing mono- and sesquiterpenes, the oil fractions of umbellifer species often have aromatic phenylpropanoids present, which are volatile and physiologically active. Two of them, methylisoeugenol and asarone (see Fig. 5.3), occur together in carrot leaf and they act as feeding attractants to the carrot root fly (Guerin *et al.*, 1983).

Many other terpenoids can be active in feeding behaviour. One complex interaction is that of the catalpa sphinx moth feeding on *Catalpa* leaves (Bignoniaceae) since a mixture of 15 iridoid glycosides, including catalpol (see Fig. 5.3), are concerned in feeding attraction. It is interesting that a related monoterpene lactone, nepetalactone, occurring in a related family in *Nepeta cataria* (Labiatae), is an olfactory attractant to the domestic cat *Felis domestica*. Thus, the same type of molecule can be an olfactory stimulant in both plant–insect and plant–mammal interactions.

Here, a word may be said about flavonoids as feeding attractants in plants other than the mulberry. A range of other structures have been found to be active (Fig. 5.1). Thus, the *Agasicles* beetle is attracted to feed on alligator weed *Alternanthera phylloxeroides* by the presence of a particular flavone, 6-methoxyluteolin 7-rhamnoside. On the other hand, the elm bark beetle is attracted to elm bark by a flavanol, (+)-catechin 7-xyloside. This vacuole-soluble compound operates in conjunction with a triterpenoid, lupeyl cerotate, which presumably occurs on the bark surface. Interestingly, flavones have been implicated as feeding attractants to another beetle, which feeds on the bark of fruit trees belonging to the genus *Prunus* (see Table 5.1). In this case, three compounds are active: taxifolin, pinocembrin and dihydrokaempferol. A fourth flavanone in the bark, naringenin (5,7,4′-trihydroxyflavanone), is curiously inactive.

Finally, it is worth noting that willow-feeding beetles, e.g. *Chrysomela vigintipunctata*, are stimulated to feed by a combination of a common leaf flavonoid, luteolin 7-glucoside, with the more specific willow constituents, salicin and populin (Matsuda and Matsuo, 1985).

IV. Secondary Compounds as Feeding Deterrents

A. The Winter Moth and Oak Leaf Tannins

A few typical examples of secondary substances acting as deterrents to insect feeding are shown in Table 5.3. Some other cases have already been mentioned earlier in this chapter. The types of compound, as with the attractants, span the whole range of structural types from terpenoids and alkaloids to quinones and flavonoids. It is among this latter class of secondary compound that the most important barrier is provided in the angiosperms against herbivore feeding. These particular substances are the plant tannins, which occur very widely in relatively high concentration in the leaves of woody plants. There are, in fact, two groups of tannin present: the hydrolysable and the condensed tannins. The hydrolysable tannins are derivatives of simple phenolic acids such as gallic acid and its dimeric form, hexahydroxydiphenic acid, combined with the sugar, glucose. The condensed tannins have a higher molecular weight and are oligomers formed by condensation of two or more hydroxyflavanol units. Some typical structures are illustrated in Fig. 5.4.

Tannins, by definition, have the ability to tan animal hide to form leather, i.e. they combine with protein, often irreversibly, by forming bonds with the peptides and other functional groups. Such bonding prevents proteins from being attacked by trypsin and other digestive enzymes, the significance of which will be apparent later. The other property of tannins, linked with their tanning ability, is their taste—an astringency which

Table 5.3 Plant secondary compounds as feeding deterrents

Insect class and species	Plant host	Chemical repellent	Reference
ANT			
Atta cephalotes	*Astronium graveolens*	Monoterpene: β-ocimene	Hubbell *et al.*, 1983
BEETLES			
Leptinotarsa decemlineata	*Solanum demissum*	Alkaloid: demissine	Sturchkow, 1959
Monochamus alternatus	*Pinus densiflora*	Hydrocarbon: ethane	Sumimoto *et al.*, 1975
Scolytus multistriatus	*Carya ovata*	Quinone: juglone	Gilbert *et al.*, 1967
BOLLWORM			
Heliothis zea	*Gossypium barbadense* (cotton)	Terpenoid: gossypol Flavonoids: quercetin glycosides	Shaver and Lukefahr, 1969
MOTH			
Operophtera brumata	*Quercus robur* (oak)	Flavonoids: tannins	Feeny, 1970
Spodoptera ornithogallii	*Vernonia glauca*	Sesquiterpene lactone: glaucolide-A	Burnett *et al.*, 1974

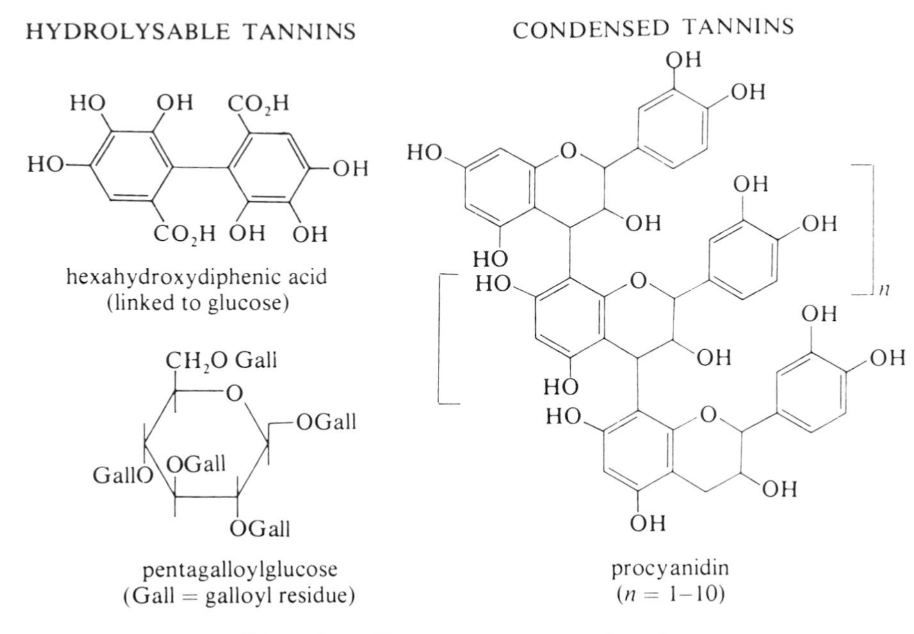

Figure 5.4 Characteristic tannins of the oak

causes the puckering of the tongue. This astringency is repellent to higher animals, birds, reptiles and probably also insects (see p. 145).

The importance of tannins in controlling the feeding of winter moth larvae on oak trees has been established by the work of Feeny (1970). He was interested in the curious behaviour of these caterpillars, which feed happily on oak leaves in the spring but abruptly cease feeding on oak in mid-June, turning to other tree species for sustenance. In seeking an explanation for this odd feeding behaviour, Feeny compared oak leaves growing in the spring with those growing in June in terms of nutritional status but could find no obvious differences. The odd behaviour was not explicable in terms of other environmental parameters, e.g. an increase in bird predation on these insects. It was only when he measured tannin content that real differences became apparent (Fig. 5.5), a sharp increase occurring just at the time when insect feeding ceased.

Chromatograms of leaf tannin extracts made on the two dates showed not only qualitative but also quantitative differences, leaf maturity being associated with an increasing number of tannin components. While the hydrolysable tannins were present equally in April and June leaves, the condensed tannins only appeared in significant amount in the older leaves. The actual concentration of condensed tannin increases from 0.5 to 5% dry weight during the season.

What is true for the winter moth probably applies to the other insect predators feeding on oak. While no less than 110 species of Lepidoptera have been recorded in oak trees in early June, only 65 of these species can be found on the same trees in mid-August. Furthermore, the density of insects on oaks undergoes a dramatic decline about the

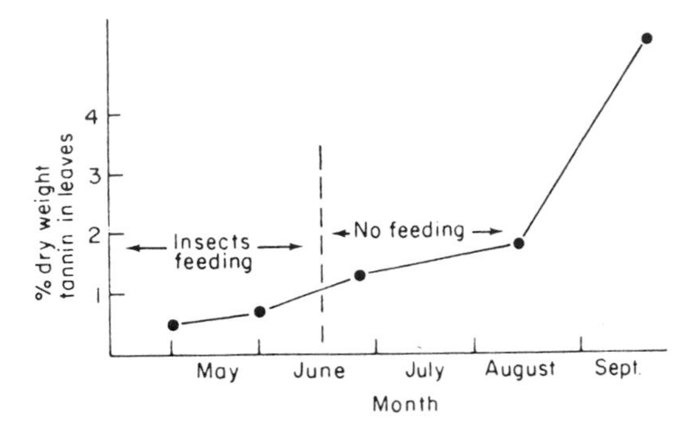

Figure 5.5 Correlation between tannin content and insect feeding behaviour on oak

same time. While the tannin increase is the major determinant in this interaction, increasing toughness of the leaf is also a factor. Clearly, the tree, in response to insect predation, produces increasing amounts of tannin to avoid complete defoliation. The trees also respond to insect attack by producing 'lammas' shoots and the photosynthetic area after the attack is largely restored by late June. During the summer months, then, the oak can carry out sufficient photosynthesis to build up energy to store for the winter months ahead.

The immediate reason why the oak moth larvae stop feeding in mid-June may be the repellent taste of the leaf, as the tannin levels build up. The insect is simply avoiding a high intake of tannin. Originally, it was thought that the deleterious effects of a high tannin diet were anti-nutritional. Thus, complexing of the plant protein with the tannin released by insect feeding might reduce the availability of the protein nitrogen to the insect. It is now clear that this is unlikely. Instead, plant tannins are probably directly toxic in unadapted insects due to tannin binding to insect tissue within the gut.

Evidence for this second view comes from studies on two swallowtail butterflies, one of which is adapted to feeding on tannin-containing plants and the other is not. When the non-adapted species, *Papilio polyxenes,* was artificially fed tannin in its diet, it died and postmortem examination revealed damaging lesions in the midgut. No such lesions developed in the adapted species *P. glaucus,* which survived without harm on a tannin diet. Survival could be due to its ability to secrete surfactant lipids, which then line the gut and prevent tannin complexing with structural protein at the gut wall (Steinly and Berenbaum, 1985; Martin and Martin, 1984).

That tannin concentration in the plant is a key feature in determining the feeding behaviour of insects is well illustrated with our experiments at Reading on the aphid *Aphis craccivora,* which lives on the groundnut *Arachis hypogaea.* In this plant, the condensed tannin, procyanidin, is unusually concentrated just in the petiole, the point at which the aphid lands to feed. Groundnut cultivars with low tannin in the phloem (0.1–0.2% wet weight) were susceptible to aphid attack, but cultivars with higher tannin (0.3–0.7%) were avoided as long as there was a choice of low tannin plants available. By contrast, aphids forced to feed on the high-tannin plants continued to do so but they became less fecund as a result; the average number of nymphs produced in the first five days of reproduction fell from 38 in the control to 19. The consequence of high dietary intake of tannin in the phloem-feeding insect is to reduce its reproductive capacity (Grayer *et al.*, 1992).

So far, the insect feeding deterrency of tannins relates almost entirely to the condensed tannins of plants. Little is known about the hydrolysable tannins, although those of the oak leaf are known to be toxic to farm animals. However, there is one experiment with geraniin, the hydrolysable tannin of *Geranium* leaves, which shows it to be a growth inhibitor in the larvae of *Heliothis virescens*. Geraniin releases ellagic acid within the caterpillar and this interferes with growth by chelating some of the essential metals needed for normal development (Klocke *et al.*, 1986).

Other insects besides Lepidoptera and aphids have been shown to be affected adversely by dietary tannins and, in summary, it would appear that tannins represent a major barrier in plants to insect feeding. The effectiveness of the barrier is directly related to the concentrations present and tissues low in tannin will be vulnerable to herbivory. The avoidance of high-tannin plants is probably due to the toxic effects of the tannin within the insect gut and not to any effects on nutritional quality of the plant protein. Undoubtedly, insects can become adapted to feeding on high-tannin leaves (as can vertebrates, see Chapter 7, Section IVC) but the mechanism of adaptation has yet to be fully understood.

B. The Colorado Beetle and *Solanum* alkaloids

The Colorado beetle *Leptinotarsa decemlineata*, with its brightly coloured yellow and black marking, is a familiar and serious pest of the potato crop. Although it has been excluded from Great Britain, it is prevalent both in North America and in continental Europe. It can produce serious crop loss due to leaf damage. A search for potatoes resistant to Colorado beetle attack has, therefore, been an important goal in potato breeding. Resistance was first discovered not among cultivated varieties but among related wild tuber-bearing *Solanum* species, native to South America. In particular, resistance was recognized in *S. demissum* and the source of resistance was traced to the major steroidal alkaloid of the leaf, a substance called demissine. Although demissine has a structure closely related to that of solanine, the major alkaloid of the cultivated potato *S. tuberosum*, it is apparently sufficiently different to repel beetle attack (Fig. 5.6).

A number of other *Solanum* alkaloids have been tested for activity and the relationship between repellency and molecular structure has been at least partly established. The three key features in the structure of demissine which are important for deterrence are: (1) the presence of a tetrasaccharide sugar at the 3-position; (2) the presence of xylose as one of the four sugar moieties; and (3) the absence of a $\Delta 5$-double bond. As soon as the 3-sugar is reduced to a trisaccharide, xylose is lost and the $\Delta 5$-double bond introduced, as in solanine, all repellency disappears. Tomatine, a major alkaloid in the tomato and also in several potato relatives, has a tetrasaccharide containing xylose and lacks a $\Delta 5$-double bond. It is, indeed, as repellent as demissine. Infiltration of potato leaves with tomatine solution, at a concentration of 2 mmol/kg causes 50% reduction in beetle feeding while a concentration of 3 mmol/kg leaf results in 100% larval mortality. Thus like most other feeding deterrents, it can have toxic consequences to the insect invader.

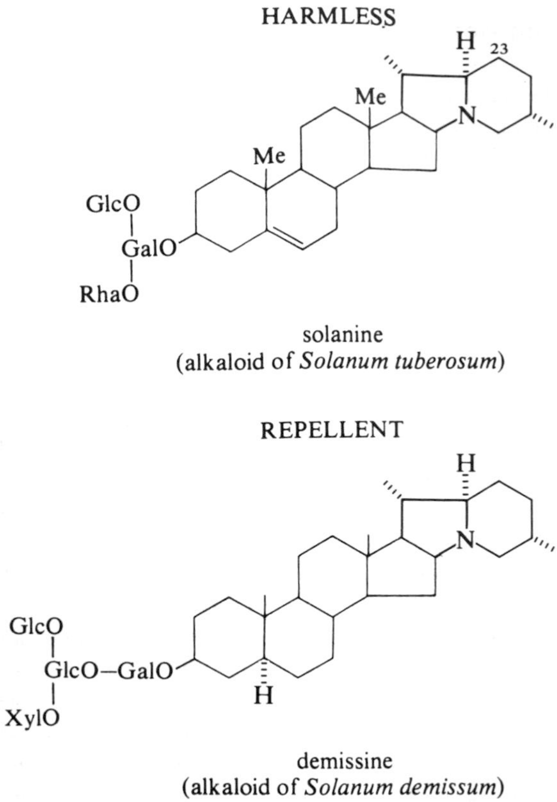

HARMLESS

solanine
(alkaloid of *Solanum tuberosum*)

REPELLENT

demissine
(alkaloid of *Solanum demissum*)

Figure 5.6 Structure of *Solanum* alkaloids affecting Colorado beetle feeding

It is interesting that deterrence to beetle attack is closely dependent on chemical structure and small changes in part of the deterrent molecule can completely abolish the deterrence value. This may be because the deterrent acts on the insect at the membrane level, possibly interfering with the absorption of the phytosterols of the potato leaf, which are required by the beetle for ecdysone synthesis (see p. 115). Since alkaloids of the *Solanum* are in fact steroidal molecules, it is also possible that they have a direct effect in blocking ecdysone biosynthesis.

Information on the deterrent properties of demissine is clearly of practical value, since breeding experiments with *S. demissum* and *S.tuberosum* might yield a potato resistant to beetle attack. In fact, such a breeding programme has been developed and potato plants resistant to the potato beetle have been obtained. Another source of resistance to Colorado beetle attack has been identified in some strains of the wild *Solanum chacoense* (Sinden *et al.*, 1980). Here resistance is due to the replacement of solanine as one of the chief alkaloids by the 23-*O*-acetyl derivative (called leptine II) (see Fig. 5.6). Again, there is the possibility of using this resistance to breed new potato varieties free from beetle infestation. Such breeding programmes are important, since Colorado beetles eventually become habituated after several generations to 'so-called' resistant plants.

C. Leaf-cutting Ants

The relationship between leaf-cutting, fungus-growing ants and plants is an unusual one, since the ants do not feed directly on the plants they cut. Instead, they feed the cut pieces of leaf to the fungal colony in the nest and feed themselves on the mycelial mat. Chemical signals are important in the social behaviour of these, as with most ant groups, and it is known that scent trails are laid to foragers to lead others to the cutting areas (Chapter 8, Section IIB). Additionally, cuts in the leaves are highly attractive to other ants and this induces further cutting in the vicinity (Cherrett, 1972).

The choice of food plant by leaf-cutting ants is extremely catholic and few plants seem to be immune to such attack. These ants are essentially impervious to most of the secondary compounds which normally repel phytophagous insects. Latex production, as in Euphorbiaceae, is one obvious barrier to leaf-cutting ants and there is evidence from the field that plant species containing latex systems are significantly less attacked than species without (Stradling, 1978). Nevertheless, one might suspect that all higher plants without latex are unprotected from leaf-cutting ants. However, recent investigation indicates that these insects are sensitive to chemical cues and some plant species have apparently evolved a defence chemistry inimicable to ant cutting.

Through the monitoring of plant extracts with a special bioassay with *Atta cephalotes*, it has been shown that a number of tropical plant species are immune to attack through the presence of specific terpenoids in their tissues (Hubbell *et al.*, 1983). While some of these compounds (Table 5.4) are volatile (e.g. β-ocimene and caryophyllene epoxide), others are non-volatile (e.g. the triterpenoid, jacquinonic acid) so that deterrence presumably requires tactile as well as odour foraging by the ant. The protective secretions in some species are quite simple, being based on a single chemical (e.g. jacquinonic acid in *Jacquinia pungens*), while in others, mixtures are present (e.g. six triterpenoids in the leaf of *Cordia alliodora*).

Structure–activity relationships are not obvious, since a range of unrelated terpenoid derivatives are active (Fig. 5.7). There is, however, a degree of specificity in that the great majority of known terpenoids are completely inactive. This is true of the terpenoids present in any given immune plant. For example, in *Eupatorium quadrangulare*, there are five sesquiterpene lactones present in the leaf, but only two of these are active.

The effects of the volatile repellents on the foraging behaviour of the ant can be quite dramatic. Thus, β-ocimene is a repellent in both field and laboratory experiments. When ants are offered sugar-coated oat flakes in a controlled laboratory experiment, they remove 59 pieces in a 60-min period. If then offered sugar-coated oat flakes treated with β-ocimene, they only remove 2 pieces in the same time period. It appears that the ants reject plants containing these particular chemicals because the compounds harm their fungal food. Indeed, when caryophyllene epoxide (Fig. 5.7) the repellent of *Melampodium divaricatum* leaves, was tested on the fungal symbiont of *Atta cephalotes* in the nest, the fungus shrivelled up and collapsed within two days (Hubbell *et al.*, 1983). Thus, the plant gains protection from leaf cutting by the synthesis of an antifungal agent, rather than an ant repellent. It would be interesting to know whether the antifungal compounds known to be produced by plants specifically as a protection against infection have a protective effect against the devastating raids of leaf-cutting ants.

Table 5.4 Terpenoids that are repellent to the leaf-cutting ant *Atta cephalotes*

Plant species	Terpenoids
Hymenaea courbaril (Leguminosae)	Caryophyllene epoxide
Lasianthaea fruticosa (Compositae)	Lasidiol angelate
Cordia alliodora (Boraginaceae)	Six triterpenoids
Astronium graveolens (Anacardiaceae)	*trans*-β-Ocimene
Melampodium divaricatum (Compositae)	Caryophyllene epoxide; spathulenol, sesquiterpene and kolavenol
Eupatorium quadrangulare (Compositae)	Two sesquiterpene lactones
Jacquinia pungens (Theophrastaceae)	Jacquinonic acid

D. Other Feeding Deterrents

As described above, the simple monoterpene, β-ocimene, is an important repellent to leaf-cutting ants. How far monoterpenes can act as feeding repellents generally is not clear, although, as has been mentioned earlier (Chapter 4, Section VI), changes in the α-/β-pinene ratios in pine oleoresins can apparently limit the feeding of some pine bark

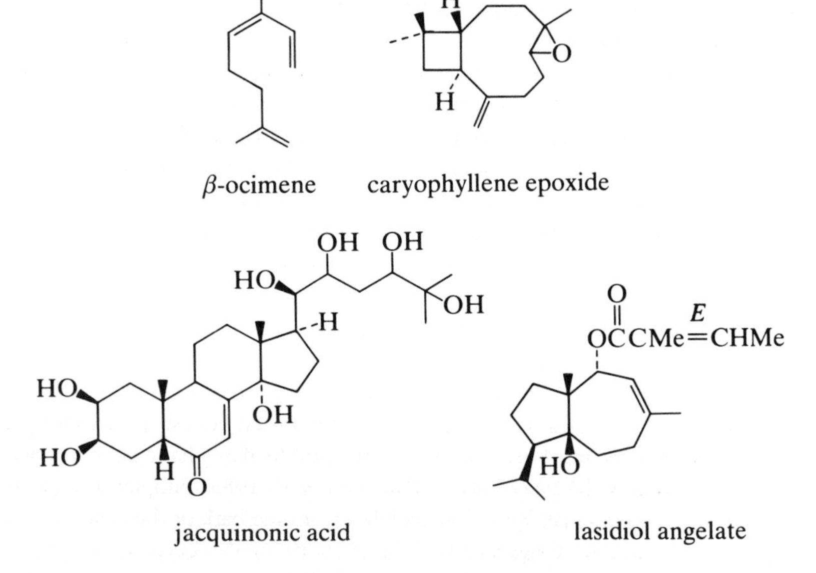

Figure 5.7 Terpenoids that protect plants from leaf-cutting ants

beetles. However, not all beetles are affected by these olfactory compounds. In the case of *Monochamus alternatus,* which feeds on pine needles, ethane gas is the active repellent and *not* the five monoterpenes which are present in the needle vapour. It is remarkable that such a simple compound as ethane, C_2H_6, can be effective in this way. As little as 0.60 μl of the gas are produced from 1 g of pine needles, but this is sufficient to cause feeding to cease (Sumimoto *et al.*, 1975).

Pine beetles are also sensitive to the presence of ethanol, which may be formed from sugar by anaerobic fermentation in the stumps and crevices of damaged trees. For example, the pine shoot beetle, *Tomiscus piniperda,* is attracted to decaying Scots pine, *Pinus sylvestris,* by the distinctive bouquet of the terpenoids, terpinolene and α-pinene, mixed with alcohol. However, the amount of alcohol released at the site is critical to acceptance; while low levels are an attractant, high concentrations repel the beetle from feeding (Klimetzek *et al.*, 1986).

Higher terpenoids (see Fig. 5.8) are undoubtedly significant deterrents in many instances. Glaucolide-A, sesquiterpene lactone of *Vernonia* plants, has been shown to deter not only the yellow striped army-worm (see Table 5.3) but also other insects which try to feed on these composite species. Glaucolide-A is effective in feeding diets at concentrations between 0.1 and 1%. Other sesquiterpene lactones, such as 8-deoxylactucin and lactupicrin from chicory, *Cichorium intybus,* have also been shown to be insect antifeedants (Rees and Harborne, 1985). There is growing evidence that non-volatile terpenoids in plants that stop insects feeding are also inhibitors of larval growth and development. For example, Elliger *et al.* (1976) found that two diterpenes, kaurenoic and trachylobanoic acids, present in the sunflower *Helianthus annuus* florets were inhibitory and larvicidal to several Lepidoptera; in addition, there was a correlation between resistance to attack by the sunflower moth *Homeosoma electellum* and varietal variation in diterpene acid content. In laboratory experiments, the deleterious effects of the two acids on larval growth were relieved in part by the addition of cholesterol to the diet, suggesting that the effect of the diterpene acids is to upset the normal hormonal processes, especially perhaps ecdysone synthesis (see Chapter 4).

The most bitter and distasteful triterpenoids in plants are probably the cucurbitacins, a series of 20 tetracyclic triterpenes occurring in the cucumber and other members of the Cucurbitaceae. While serving as attractants to the cucumber beetles, they are repellent to most other insects. For example, cucumber beetles, offered a choice of non-bitter and bitter fruits to eat, will feed almost exclusively (in a ratio of 11:1) on the bitter fruits. By contrast, the honey bee, *Apis mellifera,* faced with the same choice, not surprisingly opts for the fruit lacking cucurbitacin (in a ratio of 1:7).

One other notable triterpenoid antifeedant is the compound azadirachtin, which was discovered following repeated observations that the neem tree *Azadirachta indica* (Meliaceae), which grows in Africa, is never eaten by the desert locust *Schistocerca gregaria.* While azadirachtin is the most active, several other antifeedants have been characterized from neem and the related *Melia azedarach* (Nakatani *et al.*, 1985). Simpler sesquiterpenoid antifeedants, e.g. warburganal, have been isolated from the bark of the East African trees *Warburgia stuhlmannii* and *W. ugandensis* (Kubo *et al.*, 1976). Warburganal is a less general repellent than azadirachtin, since while active against larvae of army-worm, it does not

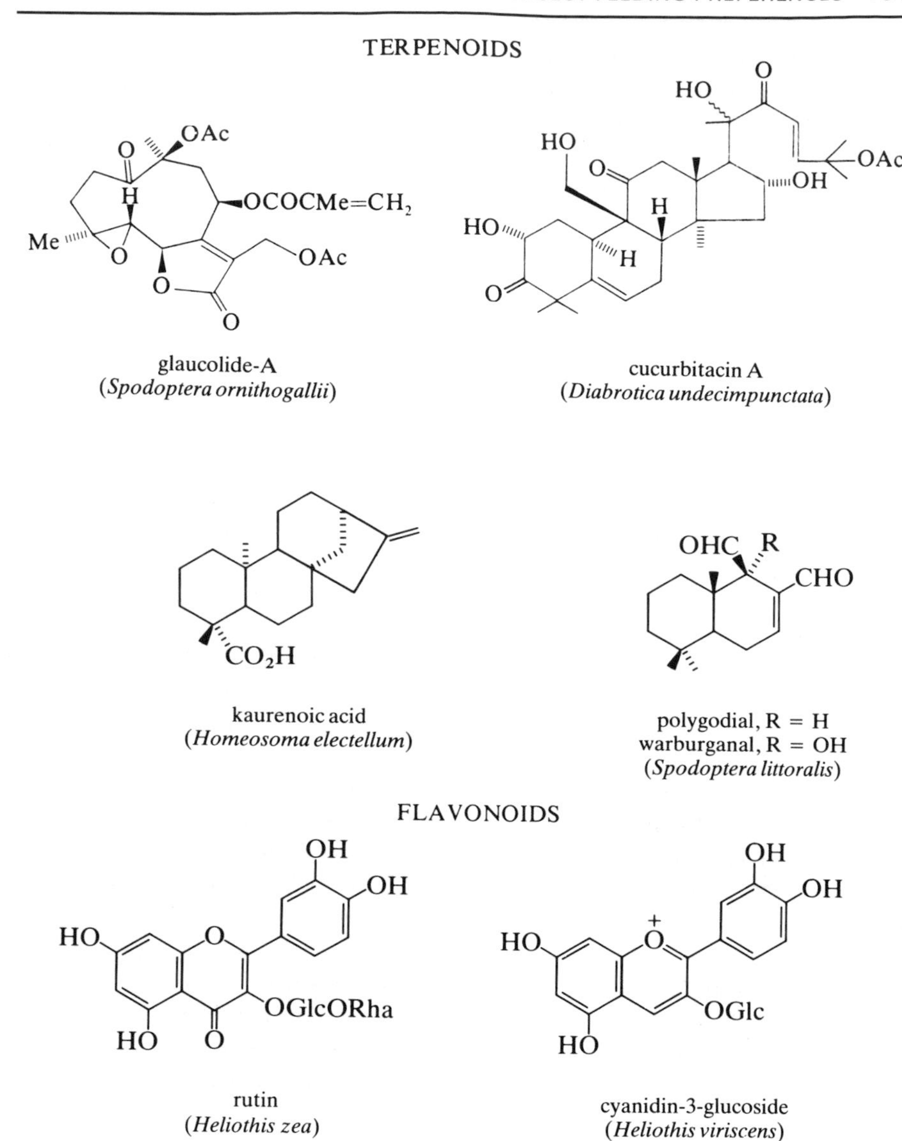

Figure 5.8 Some insect feeding deterrents

have any deterrent effect on locust feeding. Interestingly, the antifeeding compounds in *Warburgia* taste 'hot' to humans: whether insects can detect this taste and are thus deterred from feeding remains for future study.

One group of secondary compounds which may be important deterrents are the various flavone and flavonol glycosides which accumulate in leaves of most angiosperms. They accumulate particularly in herbaceous species and appear to replace, at least in part, the condensed tannins which occur so characteristically in leaves of woody angiosperms (Harborne, 1979). The toxicity of the common flavonol glycosides such as rutin and isoquercitrin to several insects feeding on cotton or tobacco plants has already been referred to (see Table 5.3). A major effect of dietary flavonoids on Lepidopteran pest species is a dramatic reduction in larval development. Larvae of the tobacco budworm, *Heliothis virescens,* are reduced to less than 10% of the growth of a control group by the addition of isoquercitrin or the related anthocyanin pigment, cyanidin 3-glucoside, to the diet. This tobacco pest also attacks the cotton crop and it has been proposed that breeding varieties of cotton with increased anthocyanin in the leaves should effectively provide resistant strains (Hedin *et al.,* 1983). Other *Heliothis* species such as *H. zea* (see below) also seem to be sensitive to the presence of flavonoids such as rutin in the diet.

Simple phenolic glycosides, which occur in willow and poplar leaves, may also deter certain Lepidoptera and beetles that feed on these plants. The compound salicortin, for example, which occurs in *Salix* spp. causes larvae of the large willow beetle, *Phratora vulgatissima,* to fail to pupate (Kelly and Curry, 1991).

One more specialized type of plant defence against insects is that provided by concentrations of secondary compounds stored in glandular hairs or trichomes on upper leaf surfaces. These toxins may be released either when the insect arrives on the plant or when it starts to eat; in either case, feeding is effectively prevented. One plant in which trichomes are well developed on the leaf is the cultivated tomato and one of the deterrent molecules present is the ketohydrocarbon 2-tridecanone, $CH_3(CH_2)_{10}COCH_3$. This has been shown to be toxic to several potential pests, including the fruitworm, tobacco hornworm and an aphid. The same ketone has, incidentally, been detected in defence secretions of termites and caterpillars so that it is clearly a repellent used in both plant and insect defence. Interestingly, the concentration of 2-tridecanone in the cultivated tomato trichomes is low, but very much larger amounts occur in trichomes of the wild tomato relative, *Lycopersicon hirsutum* f. *glabratum* (Williams *et al.,* 1980). It appears, therefore, that this may be another example where a useful insect defence present in a plant has been accidentally diminished during the process of domestication and plant breeding. The sesquiterpene zingiberene replaces 2-tridecanone as the trichome defence of another wild tomato species, *L. hirsutum* f. *hirsutum.* It is very poisonous to Colorado beetle larvae, killing them when applied at a concentration of 12–25 μg (Carter *et al.,* 1989).

Another major trichome component in tomato is the flavonol glycoside, rutin. Remarkably, a third of the total phenolic content of the leaf is stored in the trichomes. Furthermore, larval growth of the fruitworm, a tomato pest, is significantly inhibited by the addition of rutin to an artificial diet (Isman and Duffey, 1982). Thus, there is evidence here that by storing a variety of toxins in the trichomes, this plant is able to deter feeding of a range of insect groups.

Other examples of insect feeding deterrents can be found in the reviews of Beck and Reese (1976) and of Rhoades and Cates (1976). The book of Rosenthal and Janzen (1979) on herbivores and secondary plant metabolites should also be consulted.

V. Feeding of Slugs and Snails

The molluscs are another large group of animals which, like the insects, cause considerable damage to plants through feeding. While the common slug, *Arion hortensis*, is a notorious garden pest in temperate climes, there are many lesser known species of slugs and snails which feed fairly widely on plant tissues. Deterrent chemicals undoubtedly play a role in limiting the herbivory of these animals, just as they do with insects. The role of cyanogenic glycosides in deterring feeding by some mollusc species on *Lotus corniculatus* has already been referred to in Chapter 3 (Section IIIB). Essential oils are also significant deterrents, as has been established in the case of the slug *Ariolimax dolichophallus* feeding on the labiate, *Satureja douglasii*. This plant, which grows along the Pacific seaboard of North America, varies in its leaf terpenoids and several chemical races can be recognized. Types containing high proportions of camphene or camphor were more palatable in laboratory experiments than those high in pulegone or carvone. These differences in essential oil composition were reflected in the differential herbivory of the slug on populations in the field (Rice *et al.*, 1978).

Perhaps more general deterrents in plant tissues to common slugs and snails are not the terpenoids mentioned above but are the phenolics. This is the conclusion of Mølgaard (1986) from a series of controlled experiments on which plants were preferred by *Helix pomatia* and *Arion ater* for eating. Plants rich in caffeic acid esters (e.g., *Plantago* species) were usually avoided. This effect of phenolic acids also applies to molluscs which feed on decaying plant material, e.g. the detritus feeders *Melampus bidentatus* and *Orchestia grillus*. Here, the two closely related ferulic and *p*-coumaric acids turned out to be most distasteful and these acids accumulate in the detritus of many common grasses (Valiela *et al.*, 1979).

The effects of plant alkaloids on mollusc feeding are complex, with immature animals being the more sensitive to them. Younger snails of *Arianta arbustorum* avoid feeding on the leaves of *Adenostyles alliariae*, a plant rich in pyrrolizidine alkaloids, while older snails eat them regularly. A laboratory feeding trial with the alkaloids and the sesquiterpenes of this plant indicated that the former were toxic to the snail, while the latter caused feeding aversion (Speiser *et al.*, 1992).

The effects of slugs feeding on mosses and lichens have also been evaluated in recent years. Lawrey (1983) found that slugs were deterred from feeding on many lichen species; the toxins causing avoidance were soluble in acetone and were presumed to be secondary compounds. It is interesting that those lichen species which should have been the most alluring to the slugs because of the higher nutritional levels present were, in fact, the most distasteful because of their heavy loads of secondary substances. Experiments at Reading University (Davidson *et al.*, 1990) have similarly shown that slugs avoid feeding on the shoots of most moss species that are available to them. However, they do eat the capsules which are nutritionally rich and lack any obvious protection. Slug avoidance of mosses is probably due to a combination of phenolic chemistry with the indigestibility of the particular kind of cell wall that is present in these lower plants.

Many different kinds of mollusc live in the marine environment and there is evidence here too of the feeding behaviour being determined by secondary chemistry. For example, oligomers and polymers of phloroglucinol (see Figure 5.9), in the brown algae *Fucus vesiculosus* and *Ascophyllum nodosum* effectively limit the grazing of the marine snail *Littorina littorea* (Geiselman and McConnell, 1981). With the brown alga *Alaria marginata*,

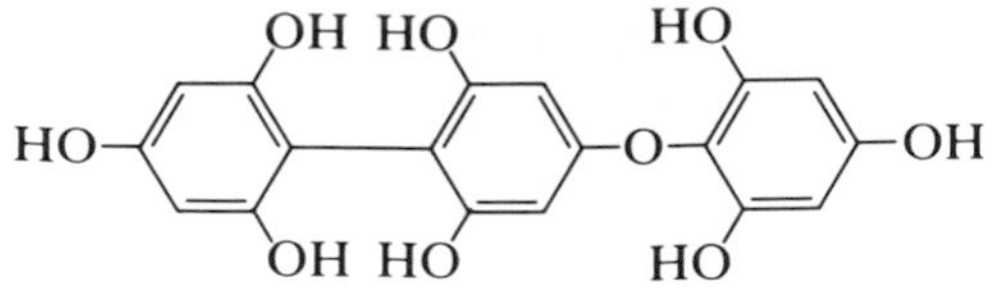

phloroglucinol trimer
(*Fucus vesiculosus*)

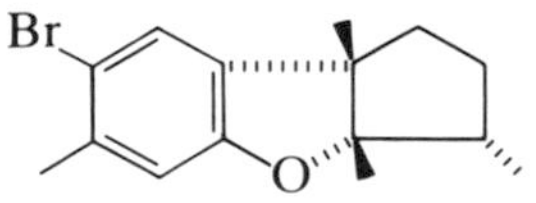

aplysin
(*Laurencia pacifica*)

Figure 5.9 Chemical defences in algae to mollusc feeding

these polyphenols are concentrated in the reproductive fronds, which allows the snail *Tegula funebralis* to selectively feed on the vegetative parts (Steinberg, 1984). The same phenomenon of selective feeding has been observed in the case of the snail *Lacuna vincta* grazing on the canopy-forming kelp *Laminaria longicruris* (Johnson and Mann, 1986). Polyphenols are concentrated in the intercalary meristem, which is thus protected, and the snail feeds on the stipe and holdfast. These polyphenols are not an absolute defence since sea urchins are able to consume the whole seaweed. In this kelp, there appears to be a balance between growth rate and reproductive output versus chemical defence, the limited expenditure on the latter allowing the alga to recover rapidly if predated upon by either herbivore.

The red alga *Laurencia pacifica*, which grows in warmer seas than *Laminaria*, produces several halogenated sesquiterpenoids such as aplysin (Fig. 5.9) which protect it from most molluscs. However, the sea hare, *Aplysia californica*, is able to graze on it and while doing so, concentrates the toxins in its digestive glands and then enjoys protection from its predators (Naylor *et al.*, 1983). Some molluscs thus have the ability to sequester algal toxins in the same way that aposematic insects sequester and store plant toxins (see Chapter 3). Interestingly and remarkably, another red alga *Erythrocystis saccata*, a semi-parasite on *L. pacifica* appears to do the same thing. It takes up these sesquiterpenes from its host and is protected from grazing by the sequestered defence materials (Crews and Selover, 1986).

Freshwater habitats also have their slugs and snails and here, most attention has been given to the feeding of *Schistosoma glabrata*. This is one of several snails directly implicated in the transmission of the debilitating human parasitic disease schistosomiasis. The snail is quite harmless but it is a vector of the schistosome organism causing the disease and by living in irrigation canals and stagnant water holes, it can spread the disease in those parts of Africa and South America where it is endemic. This snail is relatively susceptible to plant secondary chemicals and many efforts have been made to control it by application of molluscicidal plant extracts. Saponins are particularly effective, partly because they are readily soluble in water. *Phytolacca dodecandra* is a promising control plant, since its saponins can kill *S. glabrata* snails within 24 h at a concentration as low as 2 ppm. The saponins of this plant are based on oleanolic acid and the lethal dose varies between 1.5 and 32 ppm according to the nature of the sugar side chain (Marston and Hostettmann, 1985).

A more effective approach to the control of snails carrying the schistosome has been proposed: the release of a bound molluscicide together with specific attractants or phagostimulants to selectively remove the target animal. The chemoreception of *S. glabrata* has therefore been examined and it has been found that certain protein amino acids (e.g. phenylalanine and tyrosine) and propionic acid are effective feeding attractants (Thomas *et al.*, 1980). Why the amino acids are active is not entirely clear, but propionic acid is probably a snail pheromone; it has been detected in the mucus deposited by the foot glands of the creature.

VI. Oviposition Stimulants

Plant chemicals play an important role, together with visual cues (Prokopy and Owens, 1983), in attracting phytophagous insects to their chosen host plants for both feeding (usually in the larval state) and oviposition (adult female). Therefore, one might expect the same chemicals to be involved in some cases and this is certainly true for the cabbage white butterfly. It is, however, difficult to generalize at the present time since we know much less about the chemistry of oviposition stimulation than we do about feeding attractants (Ahmad, 1983). Most of the limited data available are collected in Table 5.5. One might expect oviposition to be cued largely by volatile chemicals (Visser, 1986) but this is not so; non-volatile chemicals are equally well represented.

The one interaction that has been studied in most detail is that of the carrot fly oviposition on the carrot plant. In this interaction, a mixture of six surface chemicals present in the leaf wax produce the maximum response. One of these, the polyacetylene falcarindiol, not only occurs in the greatest amount but also is far the most active component. Nevertheless, its activity is significantly synergized by the other five components. What is taking place here is a carrot specialist insect locating its host plant by the 'chemical signature' of the leaf wax. All these compounds do occur in other umbellifer species, but this particular quantitative profile is entirely characteristic of the carrot leaf. It may finally be noted in this case that two of the oviposition stimulants, (*E*)-asarone and methyleugenol, are also feeding attractants for the carrot fly (see Section IIID).

Chemicals that stimulate one specialist insect to oviposit on its host plant may not necessarily have any effect on a second one. Thus the female of the black swallowtail butterfly also lays its eggs on carrot leaves, but the stimulant is a mixture of water-soluble components, a hydroxycinnamic acid, a flavone glycoside and two organic bases (Feeny *et al.*, 1983; Feeny, 1987). The citrus-feeding swallowtail also depends on water-soluble chemical cues; a mixture of four flavonoids, together with adenosine and a tryptamine derivative, has been identified in the oviposition-stimulating fraction of the citrus host plant (Table 5.5).

In the case of these female butterflies, it is not entirely clear how they are able to detect water-soluble stimulants which must be located in the vacuoles of epidermal leaf cells. They do characteristically 'drum' the leaf with their forelegs before egg laying and in doing so must presumably release sufficient of the stimulant for tarsal chemoreception.

Table 5.5 Oviposition stimulants from plants

Insect and host plant	Chemicals	Reference
Carrot fly (*Psila rosae*) on carrot (*Daucus carota*)	Methyleugenol (+),* *t*-asarone (+),* osthol (++), bergapten (++), xanthotoxin (+), falcarindiol (++++)	Städler and Buser, 1984
Black swallowtail (*Papilio polyxenes*) on carrot	Luteolin 7-malonyl-glucoside, *t*-chlorogenic acid	Feeny *et al.*, 1983
Papilio xuthus on *Citrus unshiu*	Vicenin-2, narirutin, hesperidin, rutin, *N*-methylserotonin, bufotenine, adenosine	Nishida *et al.*, 1987
Rice grain weevil (*Sitophilus zeamais*) on rice grains	Sterols, steryl ferulates, diglycerides	Maeshima *et al.*, 1985
Cabbage white (*Pieris brassicae*) on cabbage	Sinigrin* and glucobrassicin	David and Gardiner, 1962; Traynier and Truscott, 1991
Diamond-back moth (*Plutella maculipennis*)	Allyl isothiocyanate*	Gupta and Thorsteinson, 1960
Codling moth (*Laspeyresia pomonella*)	α-Farnesene	Wearing and Hutching, 1973

* Also recorded as feeding attractants for these insects.

Volatile signals are involved in other cases, e.g. the cabbage-feeding diamond-back moth is stimulated by the acrid-smelling allyl isothiocyanate, but it is notable that the cabbage white butterfly responds mainly to the non-volatile glucosinolate precursors, sinigrin and glucobrassicin, for oviposition.

Plant chemicals in non-host plants can turn off adult females from ovipositing. The cabbage white butterfly, for example, will not lay her eggs on the leaves of the Siberian wallflower, *Cheiranthus* × *allionii*, although they are rich in the chemical cue of glucosinolate. She sensibly avoids the plant, because the leaves also contain cardenolides based on strophanthidin, which would be poisonous to the young larvae. These cardenolides occur on the leaf surface as well as within the leaf and they produce a strong deterrent response in the adult butterfly (Rothschild *et al.*, 1988). Some insects (e.g. female fruit-flies) deliberately mark fruit after egg laying in order to avoid accidentally laying two eggs in the same fruit. The nature of the oviposition deterrent in the faeces, used for this purpose, has been identified for the European cherry fruit-fly. It turns out to be a hydroxy fatty acid conjugated with glucose and taurine. It would thus seem to be a normal insect metabolite which has been adapted for practical use in this way as a chemical signal for preventing oviposition (Hurter *et al.*, 1987).

VII. Summary

In this chapter, the chemical basis of insect–plant interactions is established and many examples are given where feeding and oviposition of phytophagous insects are dependent on plant secondary chemistry. Most of the major classes of natural product are included, namely alkaloids, flavonoids and terpenoids (Table 5.6). Although one particular structure has often been highlighted in a given plant–insect interaction, in practice synergism with other related structures may well be involved. The amount of the chemical in the plant is very important and a substance (e.g. sinigrin) which may be attractive at one concentration can deter at another concentration.

Other chemical components of the plant leaf interact with secondary chemistry in attracting or repelling insect feeders. Sweetness, one of the nutritional characteristics of all plants, is usually an attractant. The natural green odour of plant leaves, a mixture of leaf alcohol and leaf aldehyde, can affect the behaviour of some insects, such as the Colorado beetle; plant fermentation products, e.g. ethanol, can affect others. Overall, each given plant species will have an odoriferous or tactile chemical profile which the phytophagous specialist insect can recognize and exploit.

While plants may be generally protected from insect feeding by a particular series of toxins in their tissues, the switch from deterrence to attraction can happen relatively easily. Plant–insect interactions are not immutable and insects adapt to new host plants. This can be seen to operate whenever plant crops are introduced into new geographical areas; new insect pests emerge from the surrounding habitat to attack these plants. Even in well-established agricultural practice, harmless insects can sometimes move in and become a problem. This happened ten years ago in California when the anise swallowtail butterfly, *Papilio zelicaon*, which normally feeds on fennel, began to attack the orange orchards. The chemical link in this case lay in the presence of the same three feeding attractants, anethole, methylchavicol and anisaldehyde, in the essential oils of members of two related families, the Umbelliferae (fennel) and the Rutaceae

Table 5.6 Some of the major chemicals controlling feeding and/or oviposition in phytophagous insects

Chemical type and examples		Insects affected
Alkaloid	Demissine	Colorado potato beetle
	Sparteine	Broom aphid
Flavonoid	Catechin 7-xyloside	Elm bark beetle
	Kaempferol glycosides	Horseradish flea beetle
	Tannin	Winter oak moth
Glucosinolate	Sinigrin	Cabbage aphid Cabbage white butterfly
Phenylpropene	Methyleugenol	Carrot fly
Phenolic	Salicortin	Large willow beetle
Monoterpenoid	Citral	Silkworm
	β-Ocimene	Leaf-cutting ant
Diterpenoid	Kaurenoic acid	Sunflower moth

(orange). This illustrates the fact that there is much practical incentive, as well as theoretical reason, for continued studies of those chemicals in plants which determine insect feeding behaviour.

Bibliography

Books and Review Articles

Ahmad, S. (ed.) (1983). 'Herbivorous Insects: Host-seeking Behaviour and Mechanisms', pp. 257. Academic Press, New,York.

Beck, S. D. and Reese, J. C. (1976). Insect–plant interactions: nutrition and metabolism. *Recent Adv. Phytochem.* **10**, 41–92.

Cherrett, J. M. (1972). Chemical aspects of plant attack by leaf-cutting ants. In: Harborne, J. B. (ed.), 'Phytochemical Ecology', pp. 13–24. Academic Press, London.

Dethier, V. G. (1972). Chemical interactions between plants and insects. In: Sondheimer, E. and Simeone, J. B. (eds.), 'Chemical Ecology', pp. 83–102. Academic Press, New York.

Ehrlich, P. R. and Raven, P. H. (1964). Butterflies and plants: a study in co-evolution. *Evolution* **18**, 586–608.

Feeny, P. (1975). Biochemical co-evolution between plants and their insect herbivores. In: Gilbert, L. E. and Raven, P. H. (eds.), 'Co-evolution of Animals and Plants', pp. 3–19. Univ. Texas Press, Austin, Texas.

Feeny, P. (1976). Plant apparency and chemical defense. *Recent Adv. Phytochem.* **10**, 1–40.

Fraenkel, G. (1959). The *raison d'etre* of secondary plant substances. *Science, N. Y.* **129**, 1466–1470.

Fraenkel, G. (1969). Evaluation of our thoughts on secondary plant substances. *Entomol. expl. appl.* **12**, 474–486.

Harborne, J. B. (1979). Flavonoid pigments. In: Rosenthal, G. A. and Janzen, D. H. (eds.), 'Herbivores: their Interaction with Secondary Plant Metabolites', pp. 619–656. Academic Press, New York.

Hedin, P. A. (ed.) (1983). 'Plant Resistance to Insects', p. 375. Amer. Chem. Soc., Washington, DC.

Meeuse, A. D. J. (1973). Co-evolution of plant hosts and their parasites as a taxonomic tool. In: Heywood, V. H. (ed.) 'Taxonomy and Ecology', pp. 289–316. Academic Press, London.

Prokopy, R. J. and Owens, E. D. (1983). Visual detection of plants by herbivorous insects. *A. Rev. Entomol.* **28**, 337–364.

Rhoades, D. F. and Cates, R. G. (1976). A general theory of plant herbivore chemistry. *Recent Adv. Phytochem.* **10**, 168–213.

Rosenthal, G. A. and Janzen, D. H. (eds.) (1979). 'Herbivores: their Interaction with Secondary Plant Metabolites', 718 pp. Academic Press, New York.

Schoonhoven, L. M. (1968). Chemosensory bases of host plant selection. *A. Rev. Entomol.* **13**, 115–136.

Schoonhoven, L. M. (1972) Secondary plant substances and insects. *Recent Adv. Phytochem.* **5**, 197–224.

van Emden, H. F. (1972) Aphids as phytochemists. In: Harborne, J. B. (ed.), 'Phytochemical Ecology', pp. 25–44. Academic Press, London.
van Emden, H. F. (ed.) (1973). 'Insect–Plant Relationships'. Blackwell Scientific Pub., Oxford.
Visser, J. H. (1986). Host odour perception in phytophagous insects. *A. Rev. Entomol.* **31**, 121–144.
Visser, J. H. and Minks, A. K. (eds.) (1982). 'Proceedings of the Fifth International Symposium Insect Plant Relationships', 464 pp. Pudoc, Wageningen.

Literature References

Akeson, W. R., Haskins, F. A. and Gorz, H. J. (1969). *Science, N. Y.* **163**, 293–294.
Beck, S. D. (1960). *Ann. Entomol. Soc. Amer.* **53**, 206–212.
Burnett, W. C., Jays, S. B., Mabry, T. J. and Padolina, W. G. (1974). *Biochem. Syst. Ecol.* **2**, 25–29.
Carter, C. D. Gianfugna, T. J. and Sacalis, J. N. (1989). *J. Agr. Fd. Chem.* **37**, 1425–1428.
Chambliss, O. and Jones, C. M. (1966). *Science, N. Y.* **153**, 1392–1393.
Chapman, R. F., Bernays, E. A. and Simpson, S. J. (1981). *J. Chem. Ecol.* **7**, 881–888.
Chapman, R. F. (1976). 'A Biology of Locusts'. Studies in Biology, No. 71. Edward Arnold, London.
Crews, P. and Selover, S. J. (1986). *Phytochemistry* **25**, 1847–1852.
David, W. A. L. and Gardiner, B. O. C. (1962) *Bull. Entomol. Res.* **53**, 91–109.
Davidson, A. J., Longton, R. E. and Harborne, J. B. (1990). *Bot. J. Linn. Soc.* **104**, 99–113.
Dethier, V. G. (1941) *Amer. Nat.* **75**, 61–73.
Doskotch R. W., Mikhail, A. A. and Chatterji, S. K. (1973). *Phytochemistry* **12**, 1153–1156.
Dreyer, D. L., Jones, K. C. and Molyneux, R. J. (1985). *J. Chem. Ecol.* **11**, 1045–1052.
Elliger, C. A. Zinkel, D. F., Chan, B. G. and Waiss, A. C. (1976). *Experientia* **32**, 1364–1365.
Erickson, J. M. and Feeny, P. (1974). *Ecology* **55**, 103–111.
Feeny, P. (1970). *Ecology* **51**, 565–581.
Feeny, P. (1987). In: Labeyrie, V., Fabres, G. and Lachaise, D. (eds.), 'Insects–Plants', pp. 353–354. Junk, Dordrecht, Netherlands.
Feeny, P., Rosenberry, L. and Carter, M. (1983). In: Ahmad, S. (ed.), 'Herbivorous Insects: Host-seeking Behaviour and Mechanisms', pp. 27–76. Academic Press, New York.
Geiselman, J. A. and McConnell, O. J. (1981). *J. Chem. Ecol.* **7**, 1115–1133.
Gilbert, B. L., Baker, J. E. and Norris, D. M. (1967). *J. Insect Physiol.* **13**, 1453–1459.
Grayer, R. J., Kimmons, F. M., Padgham, D. E., Harborne, J. B. and Ranga Rao, D. V. (1992). *Phytochemistry* **31**, 3795–3800.
Guerin, P. M., Stadler, E. and Buser, H. R. (1983). *J. Chem. Ecol.* **9**, 843–861.
Gupta, P. D. and Thorsteinson, A. J. (1960). *Entomol. expl. appl.* **3**, 305–314.
Hamamura, Y., Hayashiya, K., Naito, K., Matsuura, K. and Nishida, J. (1962). *Nature, Lond.* **194**, 754–755.
Hedin, P. A., Jenkins, J. N., Collum, D. H., White, W. H. and Parrott, W. L. (1983). In: Hedin, P. A. (ed.), 'Plant Resistance to Insects', pp. 347–366. Amer. Chem. Soc., Washington, DC.
Hewitt, P. H., Whitehead, V. B. and Read, J. S. (1969). *J. Insect Physiol.* **15**, 1929–1934.
Horowitz, R. M. (1964). In: Harborne, J. B. (ed.), 'Biochemistry of Phenolic Compounds', pp. 545–572. Academic Press, London.

Hubbell, S. P., Wiemer, D. F. and Adejore, A. (1983). *Oecologia* **60**, 321–327.
Hurter, J. and 14 others (1987). *Experientia* **43** 157–164.
Isman, M. B. and Duffey, S. S. (1982). *Entomol. expl. appl.* **31**, 370–376.
Johnson, C. R. and Mann, K. H. (1986). *J. expl. Mar. Biol. Ecol.* **97**, 231–267.
Jones, C. G., Aldrich, J. R. and Blum, M. S. (1981). *J. Chem. Ecol.* **7**, 89–114.
Kawazu, K., Nakajima, S. and Ariwa, M. (1979). *Experientia* **35**, 1294–1295.
Kelly, M. T. and Curry, J. P. (1991). *Entomol. Exp. Appl.* **61**, 25–32.
Klimetzek, D., Kohler, J., Vite, J. P. and Kohnle, U. (1986). *Naturwissensch.* **73**, 270–271.
Klocke, J. A., van Wagenen, B. and Balandrin, M. F. (1986). *Phytochemistry* **25**, 85–91.
Kubo, I., Lee, Y. W., Pettei, M., Pilkiewicz, F. and Nakanishi, K. (1976). *J. Chem. Soc. Chem. Commun.* 1013–1014.
Lawrey, D. J. (1983). *Am. J. Bot.* **70**, 1188–1194.
Levy, E. C., Ishaaya, I., Gurevitz, E., Cooper, R. and Lavie, D. (1974). *J. Agric. Fd Chem.* **22**, 376–382.
Maeshima, K., Hayashi, N., Murakama, T., Takahashi, F. and Komae, H. (1985). *J. Chem. Ecol.* **11**, 1–10.
Mansell, E. and Newman, L. H. (1968). 'The Complete British Butterflies in Colour'. Edbury Press & Michael Joseph, London.
Marston, A. and Hostettmann, K. (1985). *Phytochemistry* **24**, 639–652.
Martin, M. M. and Martin, J. S. (1984). *Oecologia* **61**, 342–345.
Matsuda, K. and Matsuo, H. (1985). *Appl. Entomol. Zool.* **20**, 305–313.
McNeil, S. and Southwood, T. R. E. (1978). In: Harborne, J. B. (ed.), 'Biochemical Aspects of Plant and Animal Coevolution', pp. 77–99. Academic Press, London.
Mølgaard, P. (1986). *Biochem. Syst. Ecol.* **14**, 113–121.
Nakatani, M., Takao, H., Miura, I., and Hase, T. (1985). *Phytochemistry* **24**, 1945–1948.
Nayer, J. K. and Fraenkel, G. (1963). *Ann. Entomol. Soc. Amer.* **56**, 119–122.
Naylor, S., Hanke, F. J., Manes, L. V. and Crews, P. (1983). *Prog. Chem. Org. Nat. Prod.* **44**, 189.
Nielsen, J. K. (1978). *Entomol. expl. appl.* **24**, 562–569.
Nielsen, J. K., Larsen, L. M. and Sorensen, H. (1977). *Phytochemistry* **16**, 1519–1522.
Nielsen, J. K., Larsen, L. M. and Sorensen, H. (1979). *Entomol. expl. appl.* **26**, 40–48.
Nishida, R., Ohsugi, T., Kokubo, S. and Fukami, H. (1987). *Experientia* **43**, 342–344.
Rees, S. B. and Harborne, J. B. (1985). *Phytochemistry* **24**, 2225–2231.
Rice, R. L., Lincoln, D. E. and Langenheim, J. (1978). *Biochem. Syst. Ecol.* **6**, 45–53.
Rothschild, M., Alborn,. H., Stenhagen, G. and Schoonhoven, L. M. (1988). *Phytochemistry* **27**, 101–108.
Shaver, T. N. and Lukefahr, M. J. (1969). *J. Econ. Entomol.* **62**, 643–646.
Sinden, S. L., Sanford, L. L. and Osman, S. F. (1980). *Am. Potato J.* **57**, 331–343.
Smith, P. (1966). *Nature, Lond.* **212**, 213–214.
Speisser, B., Harmatha, J. and Rowell-Rahier, M. (1992). *Oecologia* **92**, 257–265.
Städler, E. and Buser, H. R. (1984) *Experientia* **40**, 1157–1159.
Steinberg, P. D. (1984). *Science, N. Y.* **223**, 405–407.
Steinly, B. A. and Berenbaum, M. (1985). *Entomol. Exp. Appl.* **39**, 3–9.
Stradling, D. J. (1978). *J. Animal Ecol.* **47**, 173–188.
Sturchkow, B. (1959). *Z. Vergl. Physiol.* **42**, 255–302.
Sumimoto, M. Shiraga, M. and Kondo, T. (1975) *J. Insect Physiol.* **21**, 713–722.
Thomas, J. D., Assefa, B., Cowley, C. and Ofuso-Barko, J. (1980). *Comp. Biochem. Physiol.* **66c**, 17–27.

Traynier, R. M. M. and Truscott, R. J. W. (1991). *J. Chem. Ecol.* **17**, 1371–1380.
Valiela, I., Koumjian, L., Swain, T., Teal, J. M. and Hobbie, J. E. (1979). *Nature, Lond.* **280**, 55–56.
Wearing, C. H. and Hutching, R. F. N. (1973). *J. Insect Physiol.* **19**, 1251–1256.
Williams, W. G., Kennedy, G. G., Yamamoto, R. T., Thacker, J. D. and Bordner, J. (1980). *Science, N. Y.* **207**, 888–889.
Wink, M. and Witte, L. (1985). *Phytochemistry* **24**, 2567–2568.
Zielske, A. G., Simons, J. N. and Silverstein, R. M. (1972). *Phytochemistry* **11**, 393–396.

6 Feeding Preferences of Vertebrates, Including Man

I. Introduction

Apart from nutritional considerations, the selection of plant species as foods by vertebrates is based on taste and aroma, a complex response from tongue and nostril. Rejection of otherwise nutritious plants may, therefore, be due to either the absence of an agreeable flavour (i.e. no positive response to the species) or the presence of disagreeable chemicals or toxins. The experimental problems of determining food selection by higher animals are, however, very considerable, compared to those with insects. It is, for example, highly expensive to test the effects of plant alkaloids on feeding behaviour in large farm animals, simply because of the number of fatalities that would inevitably arise in a statistically significant sampling. Again, to examine the feeding preferences of wild deer is complicated by the facts that they feed over wide areas and that there are practical problems in observing the behaviour of such shy creatures at close quarters. Much of the information available, therefore, is circumstantial or based on limited observations of the natural history type.

The very fact that chemical substances can play a role in feeding, social and reproductive behaviour of mammals has taken some time to be generally accepted. Animals undoubtedly live in a world linked by chemical communication systems. While chemical signals are important in influencing food selection, as will be shown in this chapter, they are also involved in social and reproductive behaviour. Although unsuspected for many years, there are now known to be pheromonal interactions in higher apes and man; odour compounds in the sweat have been implicated in male–female recognition encounters (see Chapter 8).

The chemicals of taste and smell are detected in man by receptor sites in the mouth and nose and much is known about the physiology of these responses (Moncrieff, 1967; Harper *et al.*, 1968a, b). In the case of taste buds these are grouped in papillae on the surface of the tongue and most papillae appear to be sensitive to more than one taste. However, there are regions of distribution of the four main types of receptor; the sweet taste is more easily sensed on the tips of the tongue, the bitter taste at the back, the sour taste at the edges and the saltiness on the tip and at the edges. These taste buds are highly sensitive to dilute solutions of appropriate chemicals. The threshold of salty taste is, for example, at a sodium chloride concentration of about 0.05% and that for bitterness is at a concentration of brucine of 0.0001%.

The olfactory receptors of man number over a million and are situated in a small 5 cm^2 region at the top and towards the rear of the nose. These receptors are tightly packed in this region and are protected from direct contact with the exterior environment by means of a series of folds. Attached to each olfactory receptor are a number of short and long cilia or hair-like filaments, which lie within a mucoid material bathing the receptor region. Receptor cells connect directly with the olfactory bulb in the brain and are capable of transmitting some 10^8 bits of information per second. One of the many remarkable features of the olfactory stimulus is that only a small number of molecules are required for perception to take place. For methyl mercaptan, for example, it has been estimated that a minimum of 40 molecules distributed over several receptors is sufficient for perception.

With both taste and smell, individuals vary in their response and there is a degree to which genetic factors control the ability to detect different chemicals. Blindness to the bitter taste of various phenylthiocarbamides and thioureas has been extensively explored by geneticists and it has shown to be inherited as a recessive character in a simple Mendelian fashion. The parent compound phenylthiocarbamide is non-bitter to some 5% of humans. Individuals who cannot detect a certain smell are similarly described as being blind to that smell or 'anosmic'. The frequency of anosmia within human populations varies with the smell. Thus, the skunk-like odour of *n*-butylmercaptan, which is obnoxious at a concentration of 0.0075% in 90% methanol, cannot be detected by about one person in a thousand (Harper *et al.*, 1968a). By contrast, trimethylamine, the fishy primary odour, is not apparent to as many as 7% of human subjects (Amoore and Forrester, 1976).

Selection of food by man and other mammals is inevitably a complex matter involving flavour, taste, palatability, colour and odour. It is often difficult to determine the controlling factor in situations where the detection system is so highly sophisticated. Although human tastes are classified within four groups—salty, sweet, bitter, sour—the palate can respond to and recognize a variety of shades within these groups and similarly the human nose can differentiate between many scents and odours (see e.g. Table 6.1). However, as with insects, a major attractant in mammals is undoubtedly sweetness. By contrast to the attraction of sweetness, repellents are sharpness, bitterness and astringency. In man, whose taste responses are highly developed, a balance between sweetness and acidity or astringency is often necessary and both tastes are commonly present in foods and drinks; otherwise there is insipidity (Bate-Smith, 1972).

Table 6.1 Some characteristic odours and chemicals that can represent them

Odour	Chemical[a]	Odour	Chemical[a]
Almond-like	Nitrobenzene	Metallic	*n*-Nonyl acetate
Aromatic	Benzaldehyde	Minty	Menthol
Burnt	Methyl benzoate	Musk-like	Exaltolide
Cool	Camphor	Onion-like	Dimethylsulphide
Disinfectant	Phenol	Petrol-like	Benzene
Faecal	Skatole	Rancid	Valeric acid
Fishy	Trimethylamine	Soapy	Stearic acid
Floral	Hydroxycitronellal	Sour	Acetic acid
Fragrant	β-Ionone	Spermous	1-Pyrroline
Fruity	Benzyl acetate	Spicy	Eugenol
Heavy	Coumarin	Sweaty	*Iso*-Valeric acid
Malty	Isobutyraldehyde	Sweet	Vanillin

[a] Based on the opinion of seven odour experts (Harper, 1975). Assessment of the odour of a particular chemical is subjective and the same compound can represent different odours to different people. In addition, with most odours, more than one individual chemical may be chosen to represent it.

While we know a lot about the taste responses of man and can infer much about the taste responses of insect herbivores, it is difficult to obtain much information about the enormous range of animals in between. Can we assume that there is a common basis for taste responses in animals generally? Evidence that tastes are indeed basically similar in hedgehogs, rats and man has been obtained in a rather odd test, devised because of wartime shortage of food, on the palatability of wild birds' eggs to these three groups of feeders (Bate-Smith, 1972). The results showed that all three groups responded to bitterness in the eggs to the same degree. Furthermore all three classes of subject were equally repelled by this bitterness. One should not, of course, conclude from this experiment that all animals respond to bitterness to the same degree. Swain (1976) has shown that reptiles, and especially tortoises, are relatively insensitive to the bitterness of quinine in solution, their sensitivity being only one-tenth that of man himself.

In the present chapter, it is intended to consider in turn the available information on feeding preferences of domestic animals, wild animals and man. Much of the data discussed has been derived from the reviews of Arnold and Hill (1972), Bate-Smith (1972) and Rohan (1972), which should be consulted for most of the references in this field. More recent work on mammalian herbivory is reviewed in the book of Palo and Robbins (1991).

II. Domestic Animals

A. Responses to Individual Chemicals

In choosing animal species in which to study food preferences, farm animals clearly have the advantages of availability and docility. Also, information obtained may be of direct practical value to the farmer. One of the problems of learning about the taste responses of

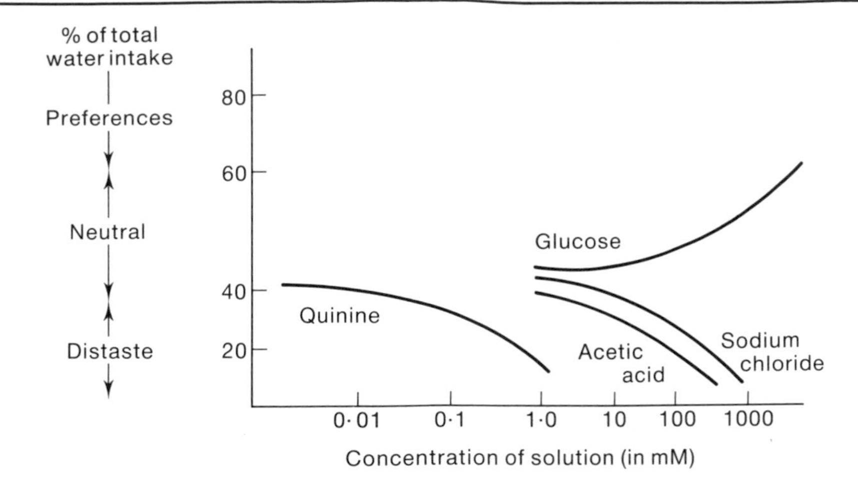

Figure 6.1 The taste responses of sheep to solutions of pure chemicals

such animals is, of course, their inability to express their preferences in human terms. It is an experimental disadvantage that cows cannot talk! However, the problem can be overcome to some extent by studying the response of cows and sheep to pure dilute solutions of chemicals, as a part of their normal daily water intake. Such studies do not reflect the natural situation where the animals have a wide choice of plant species in pastures, whose tastes and flavours are compounded of many different chemicals. Nevertheless, such work clearly indicates the importance of taste and smell to these animals in their choice of food.

Five basic substances in solution have been used to represent the main tastes in such experiments: sodium chloride (saltiness); sucrose or glucose (sweetness); acetic or citric acids (sourness); quinine (bitterness); and tannic acid (astringency). In general, the chemicals are detected at a fairly low threshold concentration and are accepted; increasing the concentration eventually leads to rejection. On a molar basis, bitterness is the first taste response to be rejected (Fig. 6.1). In general, the results clearly illustrate the fact that farm animals discriminate between chemicals and respond variously in degree to these major tastes (Arnold and Hill, 1972).

As may be expected, there are minor variations in response between individuals of the same species and also between different breeds of the same species. Larger differences in response occur when comparisons are made between cattle and goats, sheep and goats and cattle and sheep. In most cases, cattle are the most sensitive to chemical solutions, with goats being intermediate and sheep being the least sensitive. Only in the case of bitterness is this order reversed, goats being more sensitive than sheep with cattle showing the least sensitivity. To conclude, the results suggest at least that ruminants are not all that different from humans in their ability to recognize basic tastes and flavours in the plants they eat.

B. Responses to Chemicals Present in Plants

What is known of the sensory responses of ruminants to pure chemicals provides a guide to the way that these animals respond to stimuli from chemicals present in plant tissues. All sorts of complexities have to be taken into account in relating these feeding experiments to the situation in the field. Nutritional aspects are clearly important, but even here it is clear that animals cannot possibly respond to such commonly measured features of fodder plants as total nitrogen, crude fibre or ash content. Their nutritional response must be at the molecular level to specific chemical substances.

Although there is still much surmise, something can be said about the probable responses of most cattle, sheep and goats to several of the major classes of plant substance found in fodder plants. These data are discussed briefly in the following paragraphs.

Sugars. Sweetness is a clear preference in feeding, one shown by all ruminants that have been examined. There is also evidence that the form of the sugar can affect the response. Thus, calves show a greater response to sucrose solutions than to solutions of other sugars, while sheep appear to prefer glucose to sucrose at high concentrations. The importance of the response to sweetness in leaf can be gauged by the fact that cattle will eat normally unacceptable 'dung patch' herbage after it has been sprayed with sucrose solution. Objectionable stimuli can thus be swamped or suppressed by the presence of sugar as an attractant (Arnold and Hill, 1972).

Organic acids. There is some evidence that both cattle and sheep respond to the levels of particular organic acids present in temperate grasses used as fodder. Jones and Barnes (1967) have reported a positive correlation between the levels of citric and shikimic acids and the feeding preferences of ruminants. Acidity is, therefore, a recognizable component of the make-up of the plant tissues fed upon by farm creatures.

Tannins. There is now good evidence that high tannin levels are a significant deterrent to cattle. Thus, there was a drop in voluntary intake of 70% in cattle feeding on the legume herb *Lespedeza cuneata* when the tannin content rose from 4.8 to 12% dry weight (Arnold and Hill, 1972). Similarly, Cooper-Driver *et al.* (1977) noted that cattle and deer avoid eating bracken fronds during the months of August and September, when the tannin levels rise above 5%. Again Cooper and Owen-Smith (1985) reported that the palatability of 14 species of woody plants growing in the northern Transvaal to goats and two other ungulates was most clearly related to the levels of condensed tannin in the leaf. The effect was a threshold one, with all plants containing more than 5% of tannin being rejected as food during the wet season.

Avoidance of fodder with high tannin levels by domestic animals is undoubtedly due in part to the astringent taste but there are other factors which we are only just beginning to understand. The animals certainly experience adverse toxic effects during digestion. Kumar and Singh (1984) suggest that there is binding of tannin to proteins in the animal gut, leading to griping diarrhoea or constipation. Such effects may be partly relieved by other dietary components. Thus the simultaneous consumption of dietary tannin and saponins (in the right proportions) may promote chemical interactions that inhibit the absorption of the toxins from the intestinal tract (Freeland *et al.*, 1985).

Coumarins. The pleasant sweetness of new mown hay is due to coumarin. It has a bitter taste and is apparently objectionable to sheep, who reject clover containing 0.5 to 1.0% of this compound. However, the substance is volatile and rapidly lost by evaporation, so that its effects wear off within an hour or two. It therefore only reduces palatability for a short period. It is doubtful whether coumarin, released whenever grasses containing the necessary precursor are cut, is a significant feeding deterrent in the long run. The precursor of coumarin, a glucoside of *o*-hydroxycinnamic acid, occurring in *Melilotus*, is some danger to sheep and cattle. However, this is because it can be converted to dicoumarol, a substance which becomes an anti-clotting factor if it reaches the blood (Ramwell *et al.*, 1964).

Cyanogenic glycosides. Studies of feeding on bracken, in which cyanogenesis is a polymorphic character, indicate that herbivores select the acyanogenic forms in natural conditions (Cooper-Driver *et al.*, 1977). Cyanogenesis, and specifically the production of the poisonous HCN by this plant, is thus a feeding deterrent. Bracken fronds from areas where the acyanogenic form predominated (up to 98% of individuals in the population) were heavily grazed by both deer and sheep. By contrast, bracken protected by the presence of prunasin and the enzyme needed for the release of HCN was untouched by the same herbivores. These results are in line with earlier reports of cyanogenesis acting as a deterrent to rabbits and voles feeding on *Trifolium* and *Lotus* (Jones, 1972; see also p. 90).

Essential oils. Although many species of plant that are browsed on by ruminants contain essential oils, there is rather little evidence that the presence of these volatile terpenes has a significant effect on choice of plant. It is well known that in the case of cows, some of the essential oils of plants can appear in the milk and taint it. For example, the Australian umbellifer *Apium leptophyllum* has an oil reminiscent of wild carrot and the milk of cows feeding on it in Queensland pastures smells strongly of carrot (Park and Sutherland, 1969). This rather indicates that the cow itself is relatively insensitive to these plant odours.

Oh *et al.* (1967) have examined the effect of different monoterpenes on rumen activity in deer and sheep and have noted that while the hydrocarbons and esters have little activity, the monoterpene alcohols such as linalol and α-terpineol do have a pronounced inhibitory action on the digestive processes. It is not clear, however, whether these effects are related to palatability differences. It is an interesting question whether these ruminants are capable of learning to avoid feeding on plants rich in monoterpene alcohols, while accepting other essential oil-rich plants into their diet.

Isoflavones. As has already been mentioned (p. 108), *Trifolium* species are rich in isoflavones, compounds which are oestrogenic and which have a deleterious effect on the reproductive capacity of mammals feeding in quantity on such clovers. Although isoflavones thus can produce serious disturbance in animal reproduction, there is no evidence of selective preferences for clover lines deficient in isoflavone. Tests show that sheep cannot discriminate between a high isoflavone strain of *Trifolium subterraneum* and a strain essentially lacking these compounds. Presumably isoflavones are not sufficiently repellent in taste to deter feeding so that plants rich in these potentially dangerous compounds are not obviously avoided.

Alkaloids. There is much evidence that alkaloids are significant feeding deterrents to grazing animals, particularly when their presence, as in most cases, is associated with a bitter taste. Ragwort, *Senecio jacobaea*, which is rich in pyrrolizidine alkaloids (see p.

93), is notably avoided by cattle and sheep and is one of the few weeds, other than thistles and nettles, to survive uneaten in English meadows. In the genus *Lupinus,* there is a species *L. angustifolius* in which there are alkaloid-rich (up to 2.5% dry wt alkaloid) and alkaloid-free strains. These differ by a single gene. It has been shown that sheep, when presented with a choice of 'bitter' and 'sweet' strains, avoid feeding on the alkaloid-containing strain if at all possible but readily graze on the alkaloid-free variety.

Other alkaloidal plants in which feeding experiments have been attempted (Arnold and Hill, 1972) are *Phalaris tuberosa* and *P. arundinacea,* two grasses containing the tryptamine-based alkaloids, gramine and hordenine. Experiments with sheep indicate that gramine at low levels (0.01%) actually stimulates feeding while higher levels (up to 1%) lead to rejection.

C. Feeding Preferences

One can conclude from the limited data available, then, that ruminants respond to a range of chemical stimuli as a guide to feeding. These animals show a significant preference for sweetness, but also exhibit some requirement for a balanced taste involving the sourness of organic acids. They clearly avoid alkaloid-, cyanogen- and tannin-containing plants whenever they can. Domestic cattle, however, will not have learnt to avoid all poisonous or dangerous plants present in wild pastures and will obviously not be able to survive unscathed when exposed to unfamiliar grazing lands. Not surprisingly, cattle death due to plant toxins is a familiar agricultural hazard, especially where a wide range of potentially dangerous wild species are present in the native flora (as in the South African veldt or the North American prairies). Some of the symptoms of toxicity have already been mentioned earlier in Chapter 3.

Detailed knowledge of the response of ruminants to the chemical composition of their foods and especially of their response to nutritional factors is still largely unavailable. Much work is needed in the future to analyse with more refined techniques the behavioural responses of farm animals to the plants they ingest.

III. Wild Animals

What applies to sheep and cattle does not necessarily apply to wild animals and their feeding preferences may be determined by other factors than those mentioned so far. Wild animals in desert habitats, for example, may be guided in their feeding largely by an unending search for water in an otherwise arid environment and they will learn to feed on almost any succulent plant tissue that offers itself. This is true of the antelope-like oryx, which lives in southern Arabia on *Tamarix,* the root parasites *Cynomorium* and *Orobanche, Tribulus* and a sweet grass (*Aristida* sp.) (Shepherd, 1965).

Wild herbivores are probably better adapted to plant toxins than domestic animals and may be able to feed on many potentially toxic plants which are also nutritious. Evidence of such adaptation to cardiac glycoside poisoning is shown in the behaviour of the hyrax (*Procavia*) and of gazelles which feed on leaves of oleander, *Nerium oleander,* without tragic consequences. Other toxic plants eaten by the hyrax include *Phytolacca dodecandra* (rich in saponin), leaves of figs and various Euphorbiaceae (with toxic compounds in the latex) and members of the Solanaceae (an alkaloid-rich family) (Rothschild, 1972).

Even the hyrax, if hungry enough, may suffer the effects of plant toxins. Five animals were starved in captivity and then offered shoots of one of their available food plants, *Pituranthos triradiatus* (Umbelliferae), which contains between 0.6 and 1.7% dry wt of furanocoumarins. Four of the five died within 20 h, suffering from the symptoms of photosensitization (Ashkenazy *et al.*, 1985). In natural habitats, the hyrax will only eat the sprouts of this plant, where the furanocoumarin content is much lower. However, there is the clear presumption from this rather drastic experiment that most of the plant is well protected from herbivory by the furanocoumarins, even from a generalist browser.

With such wild fauna as the hyrax and the mountain goat, which is also unselective in its choice of food plant, it is apparent that survival is based on eating small amounts of many different plants. Presumably, they have an efficient detoxification system and can deal with most of the toxins that they are likely to encounter. Other browsers in more extreme climates may not be so fortunate in having a large number of plant species to feed on. For hares (*Lepus* spp.) living in northern climates during the winter, when the ground is covered with snow, the main sources of nourishment are the trees (e.g. alder, pine, spruce and birch). In such cases, it has been shown that the hares feed selectively on those tree tissues that are available to them; the more vital parts (bud, staminate, catkin, etc.) are protected chemically with resin and are therefore avoided (see Chapter 7).

Desert animals may be forced to feed on plants with high toxin content in the absence of others which are both nutritious and low in toxins. This applies to the wood rat, *Neotoma lepida,* which lives in the Mohave Desert in California and during the winter has only the creosote bush, *Larrea tridentata,* to provide it with food. It generally avoids the antinutritional effects of the phenolic resin in this plant by consistently feeding on low-resin plants. When the rat is provided with 12% resin in an artificial diet it continues to feed but it declines in body weight and may die. Here the herbivore is making the most of a relatively unpalatable plant by selecting out the less well defended members of the population. On the other hand, natural selection will be operating in the plant population to increase the levels of defensive phenols (Meyer and Karasov, 1989).

More useful information on feeding preferences of mammals can probably be obtained by considering species which have a wide choice of plants available to them in a habitat of high rainfall. Such an animal is the mountain gorilla of the African Congo and much is known about its feeding habits from the studies of Schaller (1963).

These gorillas are vegetarians, with an enormous daily capacity to eat leaves, stems and roots, topped up with fruits and seeds. Presented with a range of literally hundreds of different angiosperm species to feed on, the gorillas ignore most of them and concentrate on only some 29 species. What is remarkable about these 29 species, however, is the fact that a significant number of them contain what appear to be effective feeding deterrents and yet they are still consumed. Up to 30% of them are significantly bitter in taste, at least to man, and they are still eaten. One of them, namely *Laportea alatipes,* is covered with viciously stinging hairs. As Schaller writes: 'The virulence of (these) nettles was such that they readily burned through two layers of clothing . . . yet gorillas handled without hesitation and fed on stems and leaves that bristled with white hairs—the animals were apparently insensitive to them'.

The question may be asked: what determines the choice by the gorilla of these 29 plant species out of the several hundred it could choose from? If normal barriers to feeding such as bitterness and stinging hairs fail to deter gorilla appetites, what can be present in other plants of the Congo which restricts the gorilla to such a small selection of taxa? The answer to this question is one that can only be inferred; experimental proof is still required. However, as Bate-Smith (1972) has pointed out, the plants eaten by this animal, while they come from over a dozen families, have one thing in common—they are essentially lacking in condensed tannins. The only exceptions are two species of Rosaceae, where the bark, rather than the leaf or stem, is consumed. By contrast, the vast majority of plants avoided by the gorilla are probably woody angiosperms, a major feature of which is large quantities of tannin in the leaf tissues (Bate-Smith and Metcalfe, 1957).

One may hypothesize, then, that the gorilla makes its basic choice of plants simply on the level of tannin present, significant quantities being deterrent. This choice is determined by taste, by the astringency of the tannin-containing plants. It may be that tannin, after ingestion, has toxic consequences by binding to the wall of the stomach or gut, the gorilla not being adapted through evolution to deal with this (see Chapter 7). Its main choice then decided, the gorilla then eats almost all other plants open to it and, among these, develops a 'taste' for a number which have bitter constituents. In this case, as in human feeding responses, familiarity leads to acceptability and even liking, so that a basically repellent character becomes an attractive one. Finally, the gorilla's ability to cope with stinging hairs is proof that obvious physical barriers to feeding can be successfully circumvented.

The above hypothesis developed by Bate-Smith (1972) from observations by Schaller (1963) on the food plants of the mountain gorilla has been subsequently borne out by phytochemical analyses of the food plants of another herbivorous primate group, of colobus monkeys. An ecological study of the food plants selected by the black-and-white monkey *Colobus guereza* living in the Kibale forest reserve of western Uganda has been followed up by chemical determinations of their tannin and alkaloid contents (Table 6.2). These data confirm that tannin content is a key factor in determining the choice of plants and generally speaking plant parts with over 0.2% tannin (dry wt) are likely to be rejected by these monkeys (Oates *et al.*, 1977). Other factors are also important as feeding determinants. For example, mature leaves tend to be rejected because the tannin content usually increases significantly with leaf age. However, older leaves are also rejected by monkeys because of the increased toughness (higher lignin content) and the lower nutritional value.

A significant finding in the data of Table 6.2 is that the alkaloid content *per se* seems to have little effect on feeding choice. Presumably, these monkeys have a considerable capacity for alkaloid detoxification. Also, as with the mountain gorilla, it is possible that a certain degree of bitterness in the leaves may be a feeding attractant. Nevertheless, there may be some interaction between tannin and alkaloid content. For example, mature *Celtis durandii* leaves are fairly frequently eaten, although they have a higher than average

Table 6.2 Tannin and alkaloid contents of plants eaten or avoided by the black-and-white colobus monkey

Leaf age	Plant species	% Frequency on consumption	Tannin (mg/g)	Alkaloid (μg/g)
Young	*Celtis durandii*	35.0	0.30	1.58
Young	*Markhamia platycalyx*	8.9	0.02	8.1
Mature	*Celtis durandii*	5.2	1.12	3.48
Young	*Celtis africana*	2.5	0.20	6.5
Mature	*Olea welwitschii*	2.0	0.71	7.5
Young	*Olea welwitschii*	0.5	0.77	0.95
Mature	*Diospyros abyssinica*	0.5	2.85	11.3
Mature	*Celtis africana*	0.04	0.45	4.2
Mature	*Trema orientalis*	not eaten	81.5	5.02
Mature	*Markhamia platycalyx*	not eaten	1.0	9.7

Data from Oates *et al.* (1977). Chemical analyses refer to dry weight of leaf.

tannin content; the high tannin content might be counterbalanced by the fairly low alkaloid concentration in this species. Other classes of secondary constituent, not detected by Oates *et al.* (1977), might be of some importance in some of the species eaten by colobus monkeys. Mature leaves of *Celtis africana* are hardly eaten at all, although they have only moderate tannin and alkaloid levels (Table 6.2); possibly, some other chemical deterrent may be present here.

The over-riding influence of phenolic tannin content on monkey food choice is also apparent from related studies of the black colobus *Colobus satonas* living in the Duoala-Edea reserve of Cameroon (McKey *et al.*, 1978). Here it was found that mature leaves of the abundant trees contained twice the phenolic (and tannin) content of similar vegetation from the Kibale reserve, previously examined. This difference is apparently due to the much lower nutrient content of the soil in Cameroon, as compared to western Uganda. As a result of the higher concentrations of tannins and other phenols in the leaves, colobus monkeys living in the Douala-Edea reserve avoid leaves of nearly all the abundant tree species. Instead, they feed selectively on leaves of relatively rare deciduous trees and second growth vines in the area. This is not enough, however, to satisfy their hunger, so these monkeys, unlike other *Colobus* species, have to feed on plant seeds, which are fortunately low in phenolic content. Indeed, seeds make up about half their diet. Thus, as a direct consequence of the phenolic barriers present in the most abundant vegetation available, this monkey species has had to change its feeding strategy partly from leaves to seeds.

Even elephants appear to avoid grazing on plants if the phenolic and/or tannin content is too high. Other secondary metabolites, e.g. steroidal saponins, and a high lignin content also may deter feeding. Jachmann (1989) in a study of plants eaten by elephants noted that immature leaves of many species were regularly rejected. He suggests that the destructive feeding behaviour of elephants on trees represents an attempt to feed on the more palatable leaves higher up the trunk, which have a lower phenolic content.

IV. Birds

Birds are visually gifted and one might think that their choice of food plant, for those birds that are herbivorous, would be guided by colour rather than by taste. Nevertheless, birds do have a sense of smell and flavour and there are chemicals which deter them. We know from laboratory experiments (Chapter 3, Section IV) that the blue-jay rapidly learns to avoid eating monarch butterflies, because of the emetic, bitter-tasting cardenolides present in their tissues. From recent work, it appears that other plant secondary compounds affect the food choice of grazing birds. In particular, phenolic compounds of various types seem to be implicated as feeding deterrents.

It has been known for some time that sorghum varieties rich in tannins are bird-proof (e.g. McMillan *et al.*, 1972) and that varieties bred for low seed tannin content become liable to bird attack. More recently, Greig-Smith (1985), having found that bullfinches in Britain feeding on the developing flower buds of pear trees avoid certain varieties, looked for similar chemical deterrence. There was, however, no correlation between feeding preference and tannin content, as might be expected from the sorghum experience. Further investigation showed, more successfully, that the concentration of simple phenolic acids in the buds explained the differential feeding. When these phenols were tested individually on captive bullfinches, they indeed proved to be deterrent.

Phenolics have also been implicated as feeding deterrents to ptarmigans and grouse, during their search for food during winter feeding in the subarctic zones of North America and Europe. In this case, phenolic resins at the plant surface protect the few species of gymnosperm and angiosperm which are available to these birds. The phenolic resins are antimicrobial and are effective detences against the birds because they interfere with the digestion of the plant material by the microbial flora of the caecum. Birds feed on tissues low in the resin content and there is also some evidence of adaptation to dietary resin by these grazers (Bryant and Kuropat, 1980).

In a similar way, a phenolic of the quaking aspen, *Populus tremuloides*, namely coniferylbenzoate (Fig. 6.2), has been identified as a feeding deterrent to the ruffed grouse, *Bonasa umbellus*. This compound occurs in the staminate flower buds and catkins, a major food resource, and the grouse will only feed on the quaking aspen when the levels are relatively low. The substance produces a burning sensation in the human palate and presumably this could be perceived by the grouse when the concentration in its food is above a certain level (Jakubas and Gullion, 1990).

The most extensive investigation of a plant–bird interaction has been that of the Canada goose, *Branta canadensis*, which feeds on the coastal marshes of New England. Here again phenolics have been shown to guide the choice of plant species that are eaten (Buchsbaum *et al.*, 1984). As will be seen from Table 6.3, there is an inverse correlation between the

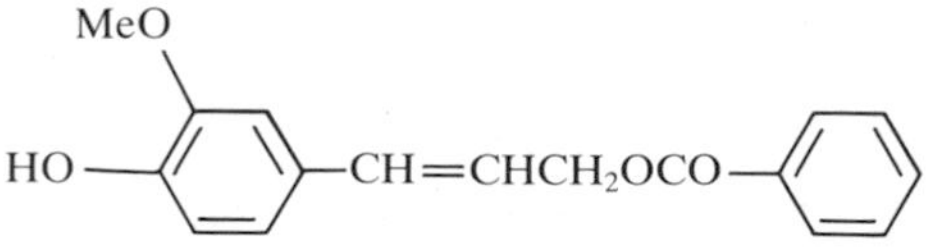

MeO

HO—C$_6$H$_3$—CH=CHCH$_2$OCO—C$_6$H$_5$

Figure 6.2. Structure of coniferylbenzoate, a feeding deterrent of the ruffed grouse

Table 6.3 Correlation between food plant choice by Canada geese and the soluble phenolic content of the leaves

Plant species	Frequency of feeding[a]	Soluble phenolic content[b]
Zostera marina	413	0.37
Poa pratensis	235	1.50
Enteromorpha sp.	142	0.07
Juncus gerardi	138	2.36
Spartina patens	131	1.64
Triglochin maritimum	56	3.19
Iva frutescens[c]	16	6.44
Phragmites australis[c]	1	1.57
Limonium carolinianum[c]	0	9.44

[a] Number of feeding observations per plant (similar results were obtained from both wild and captive geese) (Buchsbaum *et al.*, 1984).
[b] Values are % dry wt.
[c] These species are astringent in taste, and contain other chemical barriers.

frequency at which a plant species is devoured and the soluble phenolic content of that plant. One or two species do not fit in with this correlation (e.g. *Phragmites australis*) but in these cases there is evidence of some astringent chemical causing the deterrence. Nutritional factors are equally important and in low phenolic plant species, food choice is determined by carbohydrate and nitrogen levels; *Zostera marina* (Table 6.3) is probably the favourite food plant of the geese because it has the highest soluble carbohydrate content.

In summary, then, it can be seen that grazing birds, like other vertebrates, are sensitive to the secondary chemistry of the plants they feed on. The phenolic content is a significant barrier in many plants, although we know little at present of the detailed chemistry of the substances causing avoidance. This does not rule out the possibility that other classes of compound (e.g. terpenoids) may be involved. A co-evolutionary adaptation to plant chemistry in the birds is likely, although this has yet to be fully established by proper experimentation.

V. Man

A. The Choice of Plant Foods

Early man chose to eat from a wide range of plants. This is apparent from the plant remains found at the dwelling sites of Neolithic man. It is also apparent from the sort of plants eaten by primitive tribes today, e.g. the Bushman of Africa and the Veddhas of Sri Lanka. It is only in relatively recent times, as a result of civilization and urbanization, that the choice of food plants has become restricted. The limited basis of present-day diets is reflected in the small number of staple crops of agriculture; in the UK, for example, there are less than a dozen major crop plants grown for human consumption.

Two major refinements have had a profound effect on man's choice of food—the cultivation of plants and the cooking of foods. Human selection among wild species for favourable strains to cultivate has meant that many deleterious chemicals have disappeared or been reduced in quantity from plant tissues. Wild potato tubers, for example, are often bitter and toxic mainly because of the high steroidal alkaloid content. This characteristic was largely bred out by the South American Indians who first cultivated the crop and these alkaloids are almost completely absent from modern cultivars. It may be noted, however, that the Aymara Indians of highland Bolivia have continued to use varieties with potentially toxic levels. They have, indeed, a finely developed taste for bitterness in their diet and avoid the toxicity by eating the potatoes mixed with clay, which binds the alkaloid (Johns, 1990). By contrast to these deleterious chemicals, the favourable properties of plants have been enhanced by breeding and selection and cultivated fruits such as the apple, pear and strawberry have a much higher sugar content and are more attractively coloured than the wild relatives.

The effect of cooking on diet has also been important in changing the taste properties of plant tissues and also their acceptability. The potato, already mentioned above, would hardly have become such a staple item of the diet if it was eaten raw, since its starch tends to be indigestible before cooking. Again, cooking and other processing of plants has meant that many toxic components are destroyed or removed, so that a wider range of plants become available for eating. The trypsin inhibitors, an undesirable feature of raw soya beans, are destroyed by cooking. Fruits of quinoa, *Chenopodium quino*, contain a toxic saponin, but this can be leached out by steeping overnight in water, and the seed so treated can then be milled into flour for baking bread.

Given freedom of choice, then, man's selection of food among plants would be guided by a variety of qualities, such as colour, shape, smell, flavour, taste and texture. As with other animals, sweetness of taste and aroma is highly attractive to man's palate, while sharpness, bitterness and astringency are general repellents. However, some degree of acidity, bitterness or astringency in the presence of sugar increases the acceptability of otherwise insipid foods. A taste for bitterness can be developed—witness the increasing preference among habitual beer drinkers for bitter rather than mild beers. Bitterness here is due to the hop constituents, the hupulones and lupulones, which were originally added to beer only in order to improve the keeping qualities of the brew.

While much could be written on the biochemical basis of human response to flavours, there is a gap in our knowledge when it comes to the selection of plants as food, since we know so little about the detailed chemical make-up of plants in relation to palatability. Here, three areas of this subject are considered in more detail: the chemistry of flavour, the chemical basis of sweetness and the flavour potentiators.

B. The Chemistry of Flavours

Much progress has been made in recent years in the chemical analysis of food flavours. Some typical results are shown in Table 6.4. The identification of plant volatiles has been enormously aided by modern instrumentation, particularly gas liquid chromatography, which is ideal for the rapid and sensitive separation of components of smell. Nevertheless, there are still experimental problems when characterizing an active agent which is

Table 6.4 Chemical components of fruit and vegetable flavours

Plant	Components identified as flavour principles
FRUITS	
Apple	Ethyl 2-methylbutyrate
Banana	Amyl acetate, amyl propionate and eugenol
Coconut	α-Nonalactone
Grapefruit	(+)-Nootkatone, 1-*p*-menthene-8-thiol
Lemon	Citral
Mandarin orange	Methyl *N*-methyl anthranilate and thymol
Mango	Car-3-ene, dimethylstyrene
Quince	Ethyl 2-methyl-2-butanoate
Peach	Undecalactone
Pear	Ethyl *trans*-2,*cis*-4-decadienoate
Raspberry	1-(*p*-Hydroxyphenyl)-3-butanone
Vanilla	Vanillin
VEGETABLES	
Bell pepper	2-Isobutyl-3-methoxypyrazine
Celery	3-Butylphthalide, 3-butyltetrahydrophthalide, apiole, myristicin
Cucumber	$CH_3CH_2CH{=}CHCH_2CH_2CH{=}CHCHO$
Garlic	Di-2-propenyldisulphide
Onion[a]	Dipropyldisulphide, propanethiol
Mushroom	Lenthionine

[a] Raw onions also contain the lachrymatory principle, propanethial *S*-oxide, EtCH=S → O.

often only present in plants as a trace constituent. It was necessary, for example, to extract 5 tons of celery in order to identify its flavour volatiles.

It must also be remembered that the ultimate identification of a particular flavour principle depends on the subjective judgement of the human subject. Only the human nose can detect and the human brain record that a particular chemical or mixture of chemicals has the smell of raspberry, blackcurrant or peach. Much care has to be given to the selection of taste panels and to the use of organoleptic controls in flavour research. Special descriptive terms have to be adopted in order to standardize procedures. Several classifications of chemical odours have been attempted and a comprehensive survey of odour descriptions is available (Harper *et al.*, 1968b; see also Harper, 1975).

As can be seen from Table 6.4 and Fig. 6.3 many fruit flavours have now been identified (see also Nursten, 1970 and Kameoka, 1986). A few are probably given by single compounds, e.g. apple, peach, coconut and pear. Others are provided by several components; e.g. banana essence is composed of two aliphatic esters and the aromatic phenol, eugenol. Yet others are more complex; some ten monoterpenes are apparently active in the flavour principle of apricots.

Two fruits which have so far defied analysis are blackcurrants and strawberries. In both cases, over a hundred volatiles have been isolated from them but the aroma principles remain unknown. In the case of strawberry, a purely synthetic compound, ethyl-1-methyl-2-phenylglycidate, is available for use in artificial strawberry essence. Coffee,

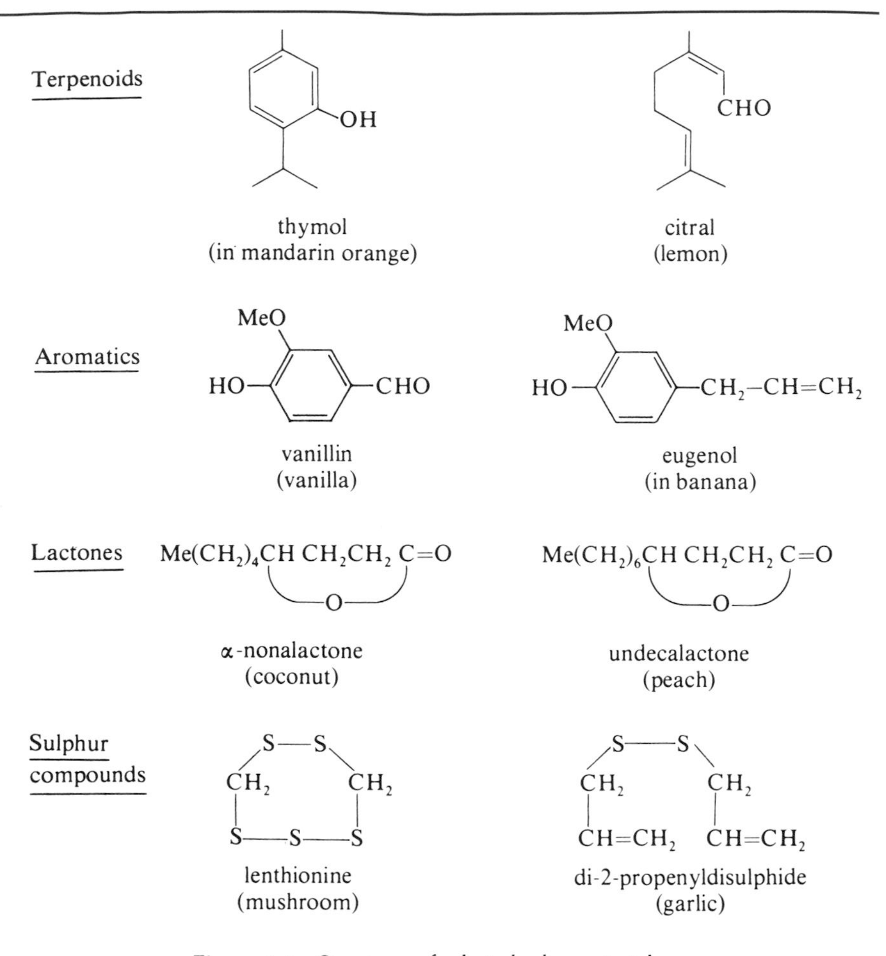

Figure 6.3 Structures of selected odour principles

chocolate and cocoa have variously yielded over 700 compounds on analysis, but it is still not entirely clear which components contribute most to the aromas. In coffee, alkylpyrazines appear to be of some importance, as well as sulphur compounds (e.g. furyl 2-methanethiol) and aliphatics (Dart and Nursten, 1985).

It is clear from this and other work that trace components are often important and indeed they may contribute more to the characteristic flavour than the major volatiles present. This is true of lemon, in which limonene makes up 70% of the oil; it is the content of citral (less than 5% of the oil) which gives the lemon flavour. The effectiveness of some volatile odour principles is exemplified by the case of cucumber smell, which is due to the aldehyde, nona-2,6-dienal (Table 6.4). This substance has an odour threshold of 0.0001 ppm!

The relationship between odour and chemical structure is hardly a straightforward one. Who would suspect that two such similar structures as undecalactone and α-nonalactone (Fig. 6.3) actually provide the odours of peach and coconut respectively? How is it that

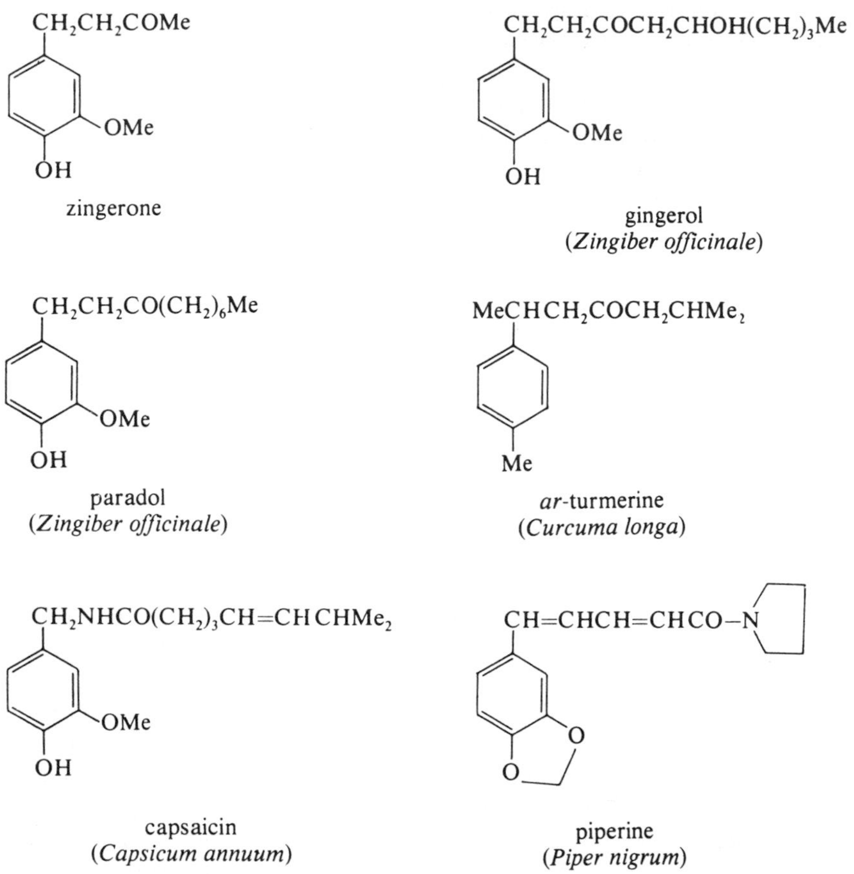

CH_2CH_2COMe
OMe
OH
zingerone

$CH_2CH_2COCH_2CHOH(CH_2)_3Me$
OMe
OH
gingerol
(*Zingiber officinale*)

$CH_2CH_2CO(CH_2)_6Me$
OMe
OH
paradol
(*Zingiber officinale*)

$MeCHCH_2COCH_2CHMe_2$
Me
ar-turmerine
(*Curcuma longa*)

$CH_2NHCO(CH_2)_3CH{=}CHCHMe_2$
OMe
OH
capsaicin
(*Capsicum annuum*)

$CH{=}CHCH{=}CHCO{-}N$
O
O
piperine
(*Piper nigrum*)

Figure 6.4 Structures of pungent principles in plants

the simple difference in the number of exocyclic methylene groups should change the odour so profoundly? Another remarkable relationship in odour character can be seen in the molecules responsible for the vanilla and ginger principles. Vanillin, the vanilla pod essence, has a pleasant sweet aromatic odour. Condensation of vanillin with acetone (and reduction of the double bond) gives zingerone, a compound which is now hot and pungent!

Analysis of the root of ginger, *Zingiber officinale*, has, in fact, shown that zingerone is an artifact derived from the true natural principles, gingerol and paradol, which are also hot and pungent in flavour. Related structures which provide pungency in *Curcuma longa* (also Zingiberaceae), in the pepper and in the capsicum are illustrated in Fig. 6.4. Some relationships between degrees of pungency and structure are apparent here. Pepper is less pungent than capsicum, which is in keeping with the hypothesis that a 4-hydroxy-3-methoxyphenyl group, present in zingerone and capsaicin, contributes to pungent properties. Replacement by a 3,4-methylenedioxyphenyl group, as in piperine, reduces pungency.

In any discussion of food flavours, the contribution of sulphur compounds to plant volatiles must be mentioned. The mustard oils, of acrid flavour, in mustard and other crucifers, have already been described as repellents in relation to insect feeding (p. 138). While attractive to man in trace amounts, as a garnish to meat, mustard is clearly a repellent when taken in any quantity. Cultivated mustard owes its properties to a blend of the pungent allyl isothiocyanate with the 'hot'-tasting *p*-hydroxybenzyl isothiocyanate. Undesirable trace components that can arise during manufacture and spoil a good mustard include butenyl isothiocyanate, which has a bad smell, and allyl nitrate, which gives it an onion flavour.

Other vegetables containing sulphur components are the onion and garlic, the principles of which are mainly aliphatic disulphides (Table 6.4; see also Johnson, *et al.*, 1971). Again, the flavour is two-edged, being attractive to some but objectionable to others. Finally, one remarkable sulphur-containing principle, which has no undesirable flavour qualities, is the compound lenthionine, the structure of which includes five sulphur atoms joined in a ring system with two methylene groups (Table 6.4). This substance is the delectable and highly prized aroma principle of the Japanese mushroom, *Lentinus edodes.*

C. The Chemistry of Sweetness

Sweetness in plant tissues is usually provided by a mixture, in varying proportions, of the three common sugars, glucose, fructose and sucrose. The disaccharide sucrose probably predominates in most plants; it is a major storage form of sugar and accumulates, sometimes in massive quantity as in the stems of sugar-cane and the roots of sugar-beet. As far as man is concerned, sucrose is used as the standard for sweetness, and all other sweet compounds are compared with it in solution on a molar basis. As indicated in Table 6.5, its two component sugars differ in sweetness from it, glucose being less sweet and fructose sweeter. Other naturally occurring monosaccharides and oligosaccharides are also usually sweet to taste, although this is not invariably so. While the disaccharides maltose, gentiobiose and lactose are all sweet, the trisaccharide raffinose is tasteless.

Sweetness, however, is not a unique property of plant sugars and, as is well known, some purely synthetic compounds have considerably greater sweetness than sucrose. Two of the most familiar—cyclamate and saccharin—are 30 and 500 times as sweet. Synthetic sweeteners have certain dietary disadvantages. Not only may they differ from sucrose in having an aftertaste (as does saccharin), but also they are 'suspect molecules' in that their intake in quantity over a period of years could conceivably cause cancer in humans. Although the evidence of suspected carcinogenicity is very weak, cyclamate has already been banned in some countries as a food additive and even the use of saccharin has fallen under some suspicion. Use of synthetic sweeteners in foods and drinks, of course, is artificial in the sense that it meets man's craving for sweetness without burdening him with the calorific content of sugar. The dangers of over-consumption of sugars in man's diet has been emphasized in recent years, notably by Yudkin (1988). There is also a more justifiable use of artificial sweeteners in the diet of diabetics.

For these various reasons, the search for natural plant sweeteners has been intensified in recent years and several such compounds are now being developed for commercial

Table 6.5 Relative sweetness of organic molecules

Compound	Sweetness, on a molar basis, relative to sucrose
Glucose	0.70
Sucrose	1.0
Fructose	1.3
Cyclamate	30
Glycyrrhizin	50
Aspartame	200
Acesulfame K	200
Stevioside	300
Saccharin	500
Naringenin dihydrochalcone	500
Sucrolose	650
Hernandulcin	800
Neohesperidin dihydrochalcone	1000
Serendip protein	3000
Thaumatin protein	5000

application. The fact that certain plant molecules other than carbohydrates have intense sweetness has been known for some time. One of the best known is stevioside, a diterpenoid glycoside which occurs in the leaves of *Stevia rebaudiana* (Compositae). It is used as a sweetening agent in Japan (Kinghorn and Soejarto, 1985). Another is glycyrrhizin, a triterpenoid glucuronide from liquorice root, which has the disadvantage of a liquorice-like aftertaste. A third is hernandulcin, which occurs in leaves and flowers of *Lippia dulcis* (Verbenaceae), a plant known to the Aztecs as 'sweet herb'. The structure of some of these and other sweet molecules are shown in Fig. 6.5 and their sweetness ratings are given in Table 6.5.

The most remarkable discovery of natural sweeteners has been of proteins occurring in the fruits of two West African plants: *Dioscoreophyllum cumminsii* (Menispermaceae) and *Thaumatococcus daniellii* (Marantaceae). The enormously intense sweetness of these fruits is reflected in their wide use in native cultures and their common names: serendipity berry and the miraculous fruit of the Sudan, respectively. Although first thought to be glycoproteins (see Inglett, 1975), the purified sweeteners are, in fact, simple proteins. They have been variously called monellin, serendip and thaumatins I and II. On hydrolysis, they give all the standard amino acids, with the notable exception of histidine. The simple dipeptide, aspartylphenylalanine methyl ester, is known to be sweet so that the sweetness of these proteins could reside within a small part of the total amino acid sequence.

The secondary and tertiary structure of the protein must also be involved in sweetness, since the taste is lost when monellin is denatured by heating in solution at 70–75°C. Indeed, sequence analysis has shown that monellin consists of two subunits of 50 and 42 amino acids, and their separation causes the sweetness to disappear (Bohak and Li, 1976). There is evidence that cysteine and methionine residues occur at adjacent sites in that part of the molecule responsible for sweetness.

Finally, one other type of natural sweetener has been discovered in an unexpected way,

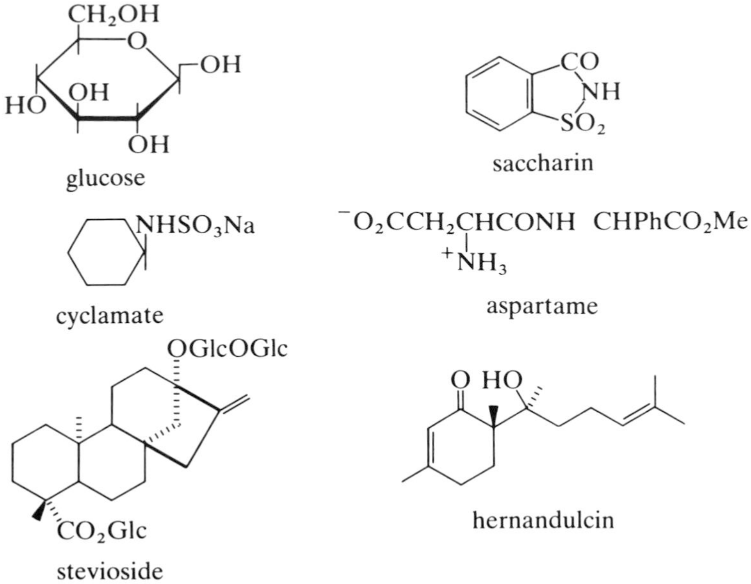

glucose

saccharin

cyclamate

$^{-}O_2CCH_2CHCONH\ CHPhCO_2Me$

$^{+}NH_3$

aspartame

stevioside

hernandulcin

Figure 6.5 Structures of some natural and artificial sweet compounds

initially from the study of bitterness in *Citrus* fruits (Horowitz, 1964). The water-soluble bitter compound was determined as naringin, the 7-neohesperidoside of naringenin, the formula of which is shown in Fig. 6.6. The structural requirements for bitterness were found to be highly specific, depending on the association of the flavanone nucleus with a disaccharide of glucose and rhamnose in which the inter-sugar linkage is precisely defined as $\alpha 1 \rightarrow 2$. A change in the mode of this linkage, as reflected in the structure of naringenin 7-rutinoside (rhamnose–glucose linkage $\alpha 1 \rightarrow 6$), another natural constituent of *Citrus*, destroys all bitterness and the substance is quite tasteless. The more important discovery came when the bitter naringin was chemically modified by opening of the central pyran ring and reduction of the isolated double bond. The molecule so produced—a dihydrochalcone—was found to be intensely sweet. Simple chemical manipulation can thus convert a molecule from being very bitter to great sweetness. As with bitterness, the structural requirement for sweet taste in this series is very specific and the special sugar moiety neohesperidose (Rha $\alpha 1 \rightarrow 2$ Glc) is essential for maximum sweetness. There are several naturally occurring dihydrochalcones known, but these only have glucose *or* rhamnose as the sugar moieties and they are only slightly sweet or bitter-sweet. By contrast, the dihydrochalcone derived from naringin is 500 times as sweet as sucrose, while that derived from neohesperidin (see Fig. 6.6) is a 1000 times as sweet on a molar basis. This latter compound is the sweetest dihydrochalcone known and it is being developed in the United States as a commercial sweetener.

The fact that it is possible to convert a bitter compound such as naringin, by simple rearrangement of the molecule, to a sweet compound does suggest that the sites of perception of sweetness and bitterness are related to each other. This is also apparent in the

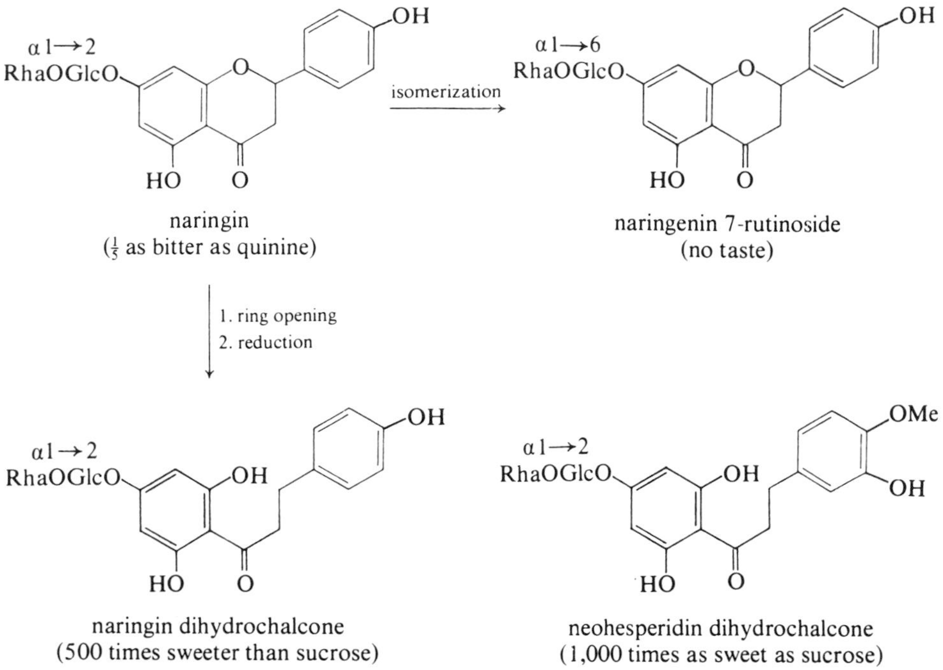

Figure 6.6 Reversal of bitterness to sweetness in the flavonoid series

facts that some sugars (e.g. mannose) are bitter-sweet and that certain modified sugars (e.g. penta-acetyl glucose) are actually bitter (see Birch and Lee, 1971). Tasting experiments with methyl α-D-mannoside and other model sugars have indicated that the sweet and bitter receptor sites on the tongue are very close to each other, indeed within 3 to 4 Å of each other (Birch and Mylvaganan, 1976). These authors suggest that sugar molecules are polarized on taste receptors and fit on as indicated in Fig. 6.7. Interaction at the two

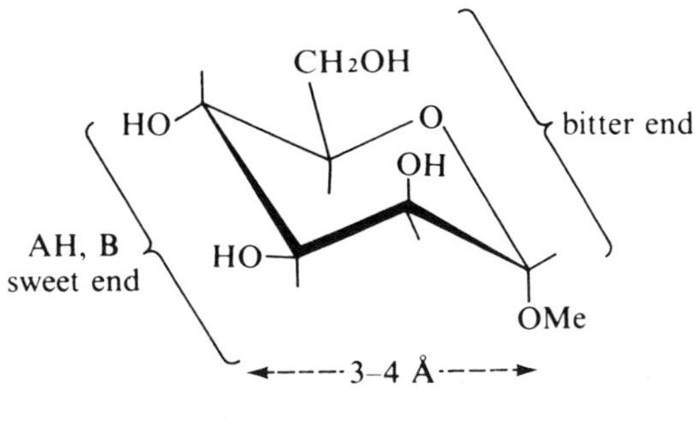

methyl α-D-mannoside

Figure 6.7 Structural requirements for sweet and bitter receptor sites in man sites was observed when attachment by presaturation with either sweet or bitter molecules lowered the subsequent response to either bitterness or sweetness respectively.

A third structural requirement for sweetness has been suggested: the presence of a hydrophobic site, X, within the molecule, but not all sweeteners have it. The latest discovery of an artificial sweetener is sucralose, a derivative of galactosyl sucrose in which the hydroxyl groups at the 4,1′- and 6′-positions are replaced by chlorine; this is 650 times as sweet as sucrose (Hough and Emsley, 1986).

D. Flavour Potentiators and Modifiers

Finally in this chapter, some mention is appropriate of flavour potentiators, since such compounds occur naturally and they could significantly influence the palatability of plant tissues to both insects and higher animals. Flavour potentiators are simply molecules which on their own have little effect but which with other taste molecules enhance the flavour. They are added to food to bring out the natural flavour or aroma. The most familiar potentiator is sodium chloride, a substance which has to be used judiciously in small amounts because of its own salty taste. In recent years, an amino acid monosodium glutamate and a purine base 6-hydroxypurine 5′-mononucleotide have been used in the food industry as excellent flavour enhancers. Both are naturally occurring, the latter compound being isolated in the first instance from bonito fish.

A more unusual flavour potentiator or modifier is the glycoprotein miracularin, of molecular weight 44,000, which occurs in miracle fruit *Synsepalum dulcificum* (Sapotaceae). This has the property of eliminating sourness or acidity, without disturbing the sweet response. Its most amazing effect can be seen when sour lemons full of citrate are eaten, following the chewing of the berries of this plant. The lemons taste as sweet as oranges! The effect is limited and wears off within an hour or so. In its action, it is assumed that the taste membrane is physically bound to this glycoprotein so that it is completely ineffective in responding to sourness.

Another equally remarkable modification in taste is due to a mixture of pentacyclic triterpenes, the gymnemins, which occur in the leaves of *Gymnema sylvestre* (Asclepiadaceae). Here, the effect is on the sweetness receptor; after these leaves are chewed, any food eaten lacks all sweetness and even bitterness is partly repressed. Thus it has the general property of dulling some of the most important taste sensations. The ability to remove sweetness, of course, which is a main attractant to eating, could be of value to plants in their battle against being eaten. Indeed, the effect of gymnemins in repelling herbivores of the plant in which they occur has been investigated by Granich *et al.* (1974). They found that gymnemic acids do act as feeding deterrents to the caterpillar of *Prodenia eridania*. Deterrence operates even with a sugar-free diet, so that it does not apparently modify the insect's taste responses. Presumably, the sweetness receptor sites in insects and mammals differ in some fundamental way.

One can conclude that the gymnemic acids have a dual action as defensive chemicals, acting as simple biting deterrents to insects but having a more profound effect on taste responses of mammalian feeders. These substances may only be the first of a series of such complex chemical agents to be implicated as deterrents in plant–animal interactions.

The above flavour modifiers exert gustatory effects not only in man, but also in the monkey *Cercophithecus aethiops*. However, gymnemic acid and miraculin have less effect in the dog or rabbit and apparently do not modify sweetness and sourness respectively when tested in the pig, rat and guinea-pig (Hellekant, 1976). The same author has also noted that the sweet proteins monellin and thaumatin are no longer sweet when tasted by the dog, hamster, pig and rabbit. Evidence is thus accumulating that the nature of the sweet receptor site varies significantly within the mammalian kingdom.

VI. Conclusion

Most of our information about the feeding preferences of vertebrates relates to man himself. We respond to a very wide range of chemical stimuli and our modern choice of food plants may be more closely linked to flavour, taste and odour than to the nutritional composition. The flavour principles of some of the common fruits and vegetables have been determined, but there are still many more awaiting full identification.

Something is known of the factors which guide gorillas and monkeys in the African continent in their search for food plants and it is apparent that tannins are a significant barrier to feeding. The best protected plants, however, may have more than one class of secondary compound to defend them. Birds may also be affected by the tannin content of food plants but simpler phenolics seem to be particularly distasteful to them.

The feeding behaviour of farm animals—sheep, goats, cattle—is reasonably well documented, but we know less about the response of most wild animals (other than primates) to their potential food plants. However, most of the chemical barriers which limit insect feeding also operate in restricting vertebrate feeding. The general co-evolutionary theory establishing a *raison d'etre* for secondary plant constituents was originally established from insect studies (Chapter 5) but it does seem to hold true for animals in general. This theory will be considered in more detail in the following chapter.

Bibliography

Books and Review Articles

Arnold, G. W. and Hill, J. L. (1972). Chemical factors affecting selection of food plants by ruminants. In: Harborne, J. B. (ed.), 'Phytochemical Ecology', pp. 72–102. Academic Press, London.

Bate-Smith, E. C. (1972). Attractants and repellents in higher animals. In: Harborne, J. B. (ed.), 'Phytochemical Ecology', pp. 45–56. Academic Press, London.

Birch, G. G. and Lee, C. K. (1971). The chemical basis of sweetness in model sugars. In: Birch, G. G., Green, L. F. and Coulson, C. B. (eds.), 'Sweetness and Sweeteners', pp. 95–111. Applied Science, London.

Bryant, J. and Kuropat, P. (1980). Feeding selection by subarctic browsing vertebrates. *A. Rev. Ecol. Syst.* **11**, 261–285.

Harper, R., Bate-Smith, E. C. and Land, D. G. (1968a). 'Odour Description and Odour Classification', 191 pp. J. & A. Churchill, Ltd., London.

Horowitz, R. M. (1964). Relations between the taste and structure of some phenolic glycosides. In: Harborne, J. B. (ed.), 'Biochemistry of Phenolic Compounds', pp. 545–572. Academic Press, London.

Inglett, G. E. (1975). Protein sweeteners. In: Harborne, J. B. and Van Sumere, C. F. (eds.), 'The Chemistry and Biochemistry of Plant Proteins' , pp. 265–280. Academic Press, London.

Johns, T. (1990). 'With Bitter Herbs They shall Eat It'. University of Arizona Press, Tucson.

Johnson, A. E., Nursten, H. E. and Williams, A. A. (1971). Vegetable volatiles: a survey of components identified. *Chem. Ind.* 556–565, 1212–1224.

Kameoka, H. (1986). In: Linskens, H. F. and Jackson, J. F. (eds.), 'Modern Methods of Plant Analysis', new series, Vol. 3, pp. 254–276. Springer, Berlin.

Moncrieff, R. W. (1967). 'The Chemical Senses', 3rd edn., 760 pp. Leonard Hill, London.

Nursten, H. E. (1970). Volatile compounds: the aroma of fruits. In: Hulme, A. C. (ed.), 'The Biochemistry of Fruits and their Products', Vol. 1, pp . 239–268. Academic Press, London.

Palo, R. T. and Robbins, C. T. (1991). 'Plant Defenses Against Mammalian Herbivory'. CRC Press, Boca Raton, USA.

Ramwell, P. W., Sherratt, H. S. A. and Leonard, B. E. (1964). The physiology and pharmacology of phenolic compounds in animals. In: Harborne, J. B. (ed.), 'Biochemistry of Phenolic Compounds', pp. 457–510. Academic Press, London.

Rohan, T. A. (1972). The chemistry of flavour. In: Harborne, J. B. (ed.), 'Phytochemical Ecology', pp. 57–71. Academic Press, London.

Rothschild, M. (1972). Some observations on the relationship between plants, toxic insects and birds. In: Harborne, J. B. (ed.), 'Phytochemical Ecology', pp. 1–12. Academic Press, London.

Schaller, G. B. (1963). 'The Mountain Gorilla'. University of Chicago Press, Chicago and London.

Shepherd, A. (1965). 'The Flight of the Unicorns'. Elek Books, London.

Yudkin, J. (1988). 'Pure, White and Deadly'. Penguin Books, London.

Literature References

Amoore, J. E. and Forrester, L. J. (1976). *J. Chem. Ecol.* **2**, 49–56.

Ashkenazy, D., Kashman, Y., Nyska, A. and Friedman, J. (1985). *J. Chem. Ecol.* **11**, 231–239.

Bate-Smith, E. C. and Metcalfe, C. R. (1957). *J. Linn. Soc. (Bot.)* **55**, 669–705.

Birch, G. G. and Mylvaganan, A. R. (1976). *Nature, Lond.* **260**, 632–634.

Bohak, Z. and Li, S. L. (1976). *Biochem. Biophys. Acta* **727**, 153–170.

Buchsbaum, R., Valiela, I. and Swain, T. (1984). *Oecologia* **63**, 343–349.

Cooper, S. M. and Owen-Smith, N. (1985). *Oecologia* **67**, 142–146.

Cooper-Driver, G., Finch, S., Swain, T. and Bernays, E. (1977). *Biochem. Syst. Ecol.* **5**, 177–183.

Dart, S. K. and Nursten, H. E. (1985). In: Clarke, R. S. and Macrea, R. (eds.), 'Coffee, Vol. 1, Chemistry', pp. 223–265. Elsevier, London.
Freeland, W. J., Calcott, P. H. and Anderson, L. R. (1985). *Biochem. Syst. Ecol.* **13**, 189–193.
Granich, M. S., Halpern, B. P. and Eisner, T. (1974). *J. Insect Physiol.* **20**, 435–439.
Greig-Smith, P. (1985). *New Scientist*, 21 November, pp. 38–40.
Harper, R. (1975). *Chemical Senses and Flavour* **1**, 353–357.
Harper, R., Bate-Smith, E. C., Land, D. G. and Griffiths, N. M. (1968b). *Perfumery and Essential Oil Record*, 1–16.
Hellekant, G. (1976). *Chem. Senses Flavor* **2**, 89–95, 97–106.
Hough, L. and Emsley, J. (1986). *New Scientist*, 19 June, pp. 48–54.
Jachmann, H. (1989). *Biochem. Syst. Ecol.* **17**, 15–24.
Jakubas, W. W. and Gullion, G. W. (1990). *J. Chem. Ecol.* **16**, 1077–1087.
Jones, D. A. (1972). *Genetica* **43**, 394–406.
Jones, E. C. and Barnes, R. J. (1967). *J. Sci. Fd Agric.* **18**, 321–324.
Kinghorn, A. D. and Soejarto, D. D. (1985). In: Wagner, H., Hikino, H. and Farnsworth, N. R. (eds.), 'Economic and Medicinal Plant Research', Vol. 1, pp. 2–52. Academic Press, London.
Kumar, R. and Singh, S. M. (1984). *J. Agric. Fd Chem.* **32**, 447.
McKey, D., Waterman, P. G., Mbi, C. N., Gartlan, J. S. and Struhsaker, T. T. (1978). *Science, N. Y.* **202**, 61–63.
McMillan, W. W., Wiseman, B. R., Burns, R. E., Harris, H. B. and Green, G. L. (1972). *Agron. J.* **64**, 821–822.
Meyer, M. W. and Karasov, W. H. (1989). *Ecology* **70**, 953–961.
Oates, J. F., Swain, T. and Zantovska, J. (1977). *Biochem. Syst. Ecol.* **5**, 317–321.
Oh, H. K., Sakai, T., Jones, M. B. and Langhurst, W. M. (1967). *Appl. Microbiol.* **15**, 777–784.
Park, R. J. and Sutherland, M. D. (1969). *Aust. J. Chem.* **22**, 495–496.
Swain, T. (1976). In: Bellairs, A. d'A. and Cox, C. B. (Eds.), 'Morphology and Biology of Reptiles', *Linn. Soc. Symp.* **3**, 107–122.

7 The Co-evolutionary Arms Race: Plant Defence and Animal Response

I. Introduction

Ehrlich and Raven (1964) were among the first to propose a new theory of biochemical co-evolution, in which the synthesis of secondary compounds as plant toxins is specifically related to patterns of host plant utilization by phytophagous insects (see Chapter 5, Section I). According to this theory an evolutionary scenario can be envisaged (Table 7.1) in which the production and accumulation of a particular toxin (e.g. an alkaloid) is followed by a reciprocal response in the insect to the toxin, e.g. adaptation by detoxification and excretion, so that it can then feed on that plant. Since only one or a few species of insect are thus adapted to feeding on that plant, there is only limited predation and a balance may be struck for a long period when plant and insect are in equilibrium. However, the possibility of increased grazing on that plant species due to adaptation by further insect species can be considered, whereby the plant responds by the synthesis and accumulation of a second toxin. This may synergize with the first toxin to protect the plant from grazing. A new generation of insect predators would then arise to become adapted to the doubly protected plant and so on.

This co-evolutionary theory was originally based on circumstantial evidence, chiefly the known feeding habits of phytophagous insects and the fact that the majority of

Table 7.1 A scenario of plant–animal biochemical co-evolution

Sequence of event	Response in plant	Response in animal
1	Synthesis and accumulation of toxin I	Avoidance by all species
2	Continued synthesis	Adaptation by few species, avoidance by most species
3	Survival with only limited predation	Toxin I becomes feeding attractant to adapted species
4	—	More species become adapted, causing herbivore pressure
5	Synthesis and accumulation of toxin II	Avoidance by all species
6[a]	Continued synthesis of toxins I and II	Adaptation by few species, avoidance by most species

[a] Further events might include the disappearance of toxin I from the plant and the synthesis of further more effective toxins III and IV, etc.

such insects are monophagous or oligophagous rather than polyphagous. The toxic effects of secondary compounds in insects have also been well documented and it is known that insects often feed on apparently toxic plants. The discovery that a small number of insect species have managed to capitalize on plant toxins and utilize them for their own purposes (Chapter 3, Section IV) also provided additional support for the theory. Studies of the feeding behaviour of animals other than insects have suggested that this theory applies to animals in general, including primates (Chapter 6).

Following the first proposal of this theory, deliberate attempts have been made to test it experimentally. At the same time, much new biochemical information has accrued on the interaction between plants and animals, which has a bearing on the theory. In particular, the fact that some plants can respond dynamically to insect predation by becoming unpalatable has become apparent in very recent years. New information has been obtained on the adaptive processes occurring in animals which deal with toxic plant constituents.

The purpose of this chapter is therefore to review these findings and to consider how far they support, or otherwise, the theory that there is a continuing co-evolutionary arms race between plant and animals for mutual survival. Can this theory explain the intricate patterns of plant utilization by animals that can be observed today? These and other questions provide the background for what follows.

II. Static Plant Defence

A. The Cost of Chemical Defence

The synthesis of secondary metabolites is costly to the plant, requiring as it does a steady flow of precursors from primary metabolism, together with enzymes and energy-rich co-factors (ATP, NADPH, etc.) to drive the biosynthetic pathways. Photosynthesis normally ensures a more than adequate supply of precursors for carbon compounds (e.g. terpenoids). By contrast, nitrogen uptake by the plant is limited and the synthesis of nitrogen compounds (e.g. alkaloids) can compete for precursor with protein synthesis. Indeed, the cost of synthesis of alkaloids has been estimated at 5 g of photosynthetic carbon dioxide per gram of toxin compared to a figure of 2.6 g for a phenolic (Gulmon and Mooney, 1986). This cost has to be balanced against the cost of new plant growth. Thus, all plants face a dilemma expressed in the words of Herms and Mattson (1992): to grow or to defend.

The competition for resources may be met in different ways and two related theories have been put forward in recent times to explain the phenotypic variation that occurs in secondary metabolism and hence in chemical defence. One is the carbon-nutrient balance (CNB) hypothesis (Bryant *et al.*, 1983) which predicts that carbon-based metabolites will be positively correlated with the carbon–nutrient balance and conversely that nitrogen-based metabolites will be negatively correlated with this balance. There is much evidence supporting this hypothesis (see Waterman and Mole, 1989) although not all data are consistent with it.

The growth–differentiation balance (GDB) hypothesis (see e.g. Tuomi *et al.*, 1990) has as its fundamental premise the existence of a physiological trade-off between growth and differentiation, the latter term including secondary metabolism synthesis. This hypothesis suggests that perennial plants might be divided into two groups. There will be growth-dominated plants, with rapid growth, poor chemical defence but with a highly inducible resistance system. Then, there will be differentiation-dominated plants with a slow growth rate, well defended with high levels of toxin but with poorly developed inducible resistance. Again, there is evidence supporting such a dichotomy in growth characteristics (Herms and Mattson, 1992).

These hypotheses help to explain the many differences that are apparent in the chemical armoury of the flowering plants (see Fig. 7.1 and Section IIB). They also indicate why the concentrations of secondary metabolites can increase in plants as a result of environmental stress. Growth on low-nitrogen soils or under drought stress can cause plants to stop producing new leaves, with the precursors and energy thus released flowing into secondary synthesis. Experiments with the composite plant *Heterotheca subaxillaris*, which is defended by monoterpenoids, illustrate the changes that can occur in growth and metabolism. The younger, softer leaves have higher terpene levels than the older, tougher leaves. Feeding leaves taken from plants growing in nutrient-rich soils to the generalist insect *Pseudoplusia includens* results in a survival rate of 78 and 98% according to leaf age (Mihalaik and Lincoln, 1989). Plants growing in soils with low nitrate levels immediately become more resistant to insect feeding and survival rates are reduced to 14% on young leaves and 38% on old leaves.

In this case, when nutrient levels in the soil are adequate for growth, the plant accepts a degree of insect feeding on the leaves and maintains monoterpene production at a moderate level, principally to protect the very young leaves. However, nitrate stress induces increases in terpene production so that insects can no longer feed successfully. This change can be seen as a dynamic response to herbivory when the plant cannot grow new leaves and is then more vulnerable to leaf loss.

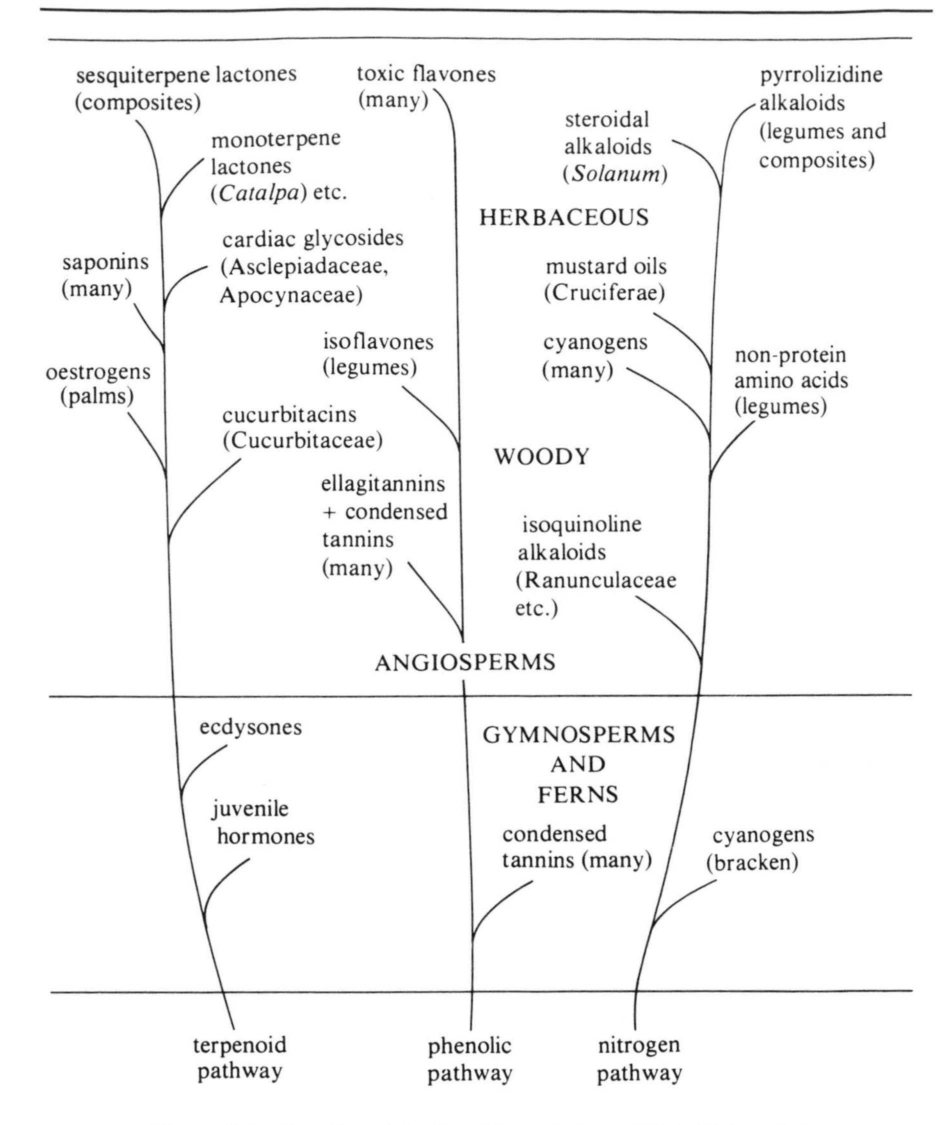

Figure 7.1 Evolution of feeding deterrents (*sensu latu*) in higher plants

B. Evolution of Feeding Deterrents

According to the theory of biochemical co-evolution, it should be possible to observe an evolving pattern of feeding deterrents within the plant kingdom. As the angiosperms have evolved, they should appear to have developed different modes of protection from animal feeding. This is, in fact, apparent, if one considers the various plant deterrents, which have been discussed earlier in Chapter 3, and their distribution in the different plant families. Some deterrents are highly sophisticated in their action, i.e. by affecting the hormonal balance within the animal. Other deterrents are highly toxic (e.g. cyanogens, alkaloids) and deter largely on account of their poisonous nature. Yet other deterrents essentially reduce palatability (e.g. cucurbitacins, tannins). It is possible to summarize the distribution of these various deterrents (or toxins) in the evolutionary scheme of Fig. 7.1.

Within the evolutionary series: ferns → gymnosperms → woody angiosperms → herbaceous angiosperms, there is a trend towards chemical complexity in deterrent structures. Any one of the three main biosynthetic pathways may be brought into operation: the terpenoid (mevalonate), the phenolic (shikimic) or nitrogen (amino acid) pathway. Other biosynthetic routes not included in Fig. 7.1 are occasionally important, e.g. the toxic polyacetylenes in umbellifer roots are derived from the fatty acid pathway. One plant family may concentrate on one type of deterrent molecule, e.g. the Cruciferae with their mustard oils. Within such a family, individual members may have developed further barriers to feeding; in the Cruciferae, a few species have cardenolides or cucurbitacins as well as mustard oils (see Chapter 5, Section IIIC). Other families may diversify their toxins, e.g. the Leguminosae with their non-protein amino acids, alkaloids, cyanogens and isoflavones.

All these many and varied compounds are produced by plants in the first instance as protective devices against insect feeding. However, in almost every instance, insects and other animals have evolved defence or detoxification mechanisms. Indeed, almost every type of toxin is utilized by a particular insect in a positive way as a feeding stimulant or attractant. This is true of the alkaloids (sparteine), the mustard oils (sinigrin), the cardiac glycosides, and the cyanogens. The only exception at present seems to be the condensed tannins, which have not been reported as yet as positive attractants to any animal.

The production and storage of secondary compounds by plants requires precursors and an energy supply derived from primary metabolism, and there is a significant 'metabolic cost' to the plant of producing toxins as herbivore defence (see Section IIA). Under these conditions, one might expect occasionally to find plant groups where chemical defence is minimal. Indeed, this seems to be true of the family Gramineae, where defensive compounds (e.g. tryptophan-derived alkaloids) are indeed recorded but, by and large, they are infrequent and do not often accumulate in quantity. This infrequency of toxins in the grasses seems to have been exploited particularly by members of one family of insects, the Acridoidea, most species of which are largely polyphagous. Indeed the graminivorous species, i.e. grasshoppers and locusts, exploit a wide range of grass species for food. Notably, they avoid feeding on broad-leaved plants, so that they are apparently not adapted to toxins. It is interesting that Bernays and Chapman (1978), in a survey of acridoid feeding behaviour, conclude that the Acridoidea are one of the few exceptional groups of insect where the diversity of speciation has not yet been affected by variations in host plant chemistry. This is partly related to their greater mobility as compared to other insect groups, such as the Lepidoptera and Coleoptera.

In avoiding the metabolic burden of toxin synthesis, it is possible that some grasses may have become so adapted to the pressures of herbivore grazing that their physiological processes are stimulated by the regular cropping of the leaves by insect or ungulate fauna. Indeed, Dyer and Bokhari (1976) suggest that the effects of grasshopper feeding on

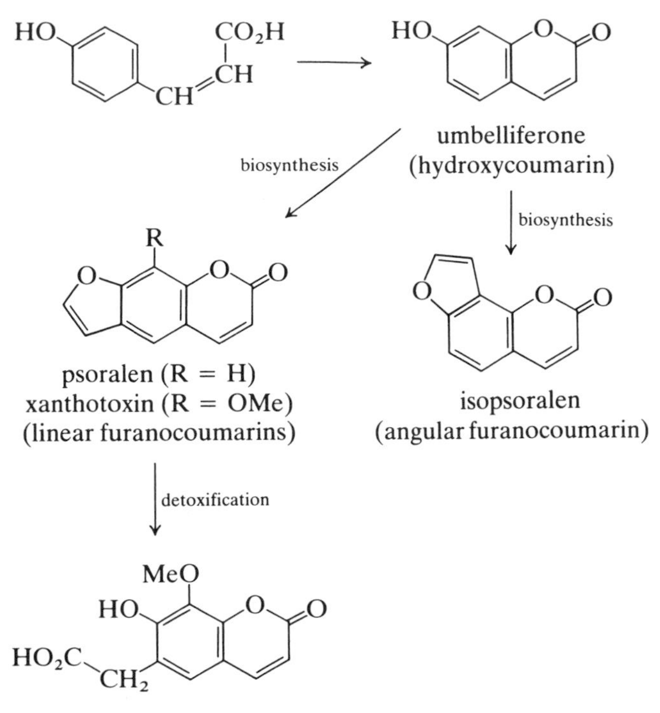

Figure 7.2 Biosynthetic and metabolic relationships of the three main classes of coumarin found in the Umbelliferae

blue gram grass can actually be favourable to continued growth. They indicate that animal feeding stimulates respiration and metabolism and also increases the rate of transport of metabolites to the root. While salivary factors may thus stimulate growth in some grasses, they decrease growth in others. Furthermore, there is no evidence of enhanced reproductive fitness in those species responding positively (Rhoades, 1985), so that even in these cases herbivory cannot ultimately be considered to be beneficial to the plant.

Returning to the great majority of plant families which rely on secondary constituents for protection from herbivory, one might speculate that within such a family, the more advanced members might be better protected than the less specialized species. In the Umbelliferae, there seems to be good evidence that this is so (Berenbaum, 1983). Here, plant defence is based mainly on coumarin synthesis and a series of simple hydroxycoumarins, linear furanocoumarins and angular furanocoumarins are widely present. These three types of coumarin are clearly related biosynthetically (Fig. 7.2) and toxicologically. Furanocoumarins are more toxic to insects than hydroxycoumarins because they are phototoxins and in the presence of UV light have the ability to bind to DNA. Likewise, angular furanocoumarins are more toxic than the linear compounds, since umbellifer-feeding insects can tolerate the latter but not the former class.

One measurement of evolutionary advancement in plants is increasing speciation. If one employs such a criterion in the Umbelliferae (Table 7.2) it is apparent that as a result

Table 7.2 Correlations between type of coumarin present and numbers of species per genus in the Umbelliferae

Generic groupings	Average number of species per genus
27 genera with only hydroxycoumarins	12
24 genera with linear furanocoumarins	17
11 genera with linear and angular furanocoumarins	67

of co-evolutionary pressures, these plants have developed increasing resistance to insect feeders by synthesizing more elaborate coumarins. This co-evolutionary model is supported by the facts that both behavioural and metabolic adaptations have been observed in umbellifer specialists. Some of these insects avoid the UV-induced phototoxicity by leaf-rolling (see Chapter 3, Section IIB). Others, such as *Papilio polyxenes*, have developed a very efficient detoxification system based on a cytochrome P-450 oxidase which results in opening of the furan ring. Detoxification of an oral dose of xanthotoxin at the rate of 5 $\mu g/g$ in this caterpillar was complete within 2 h (Bull *et al.*, 1984).

Non-adapted insects are poisoned by the furanocoumarins of umbellifers (see Berenbaum, 1983). There is thus evidence in the Umbelliferae of the synthesis of increasingly toxic coumarins and of counter-adaptation by specialized umbellifer feeders to these toxins. There is, additionally, a metabolic cost to these insects who have to cope with the furanocoumarins in their diet. Some umbellifer specialists (e.g. *Papilio* butterflies) have the opportunity to move onto other families with similar chemistry, but possibly less toxic components. Other specialists may not be so fortunate. The parsnip webworm, *Depressaria pastinacella*, is entirely restricted to the parsnip, but it does seem to have reached an equilibrium position *vis à vis* the toxic coumarins that are present (Berenbaum *et al.*, 1986).

The furanocoumarins of the Umbelliferae usually occur as mixtures of related structures rather than as single components. In the parsnip fruit, for example, there are six present. Why should this be? The answer seems to be that the plant can maintain its phototoxicity to insects in differing environments by varying the proportions of the six constituents. Indeed, feeding experiments showed that the mixture of the six were more damaging to larvae of *Heliothis zea* than an equimolecular amount of just one of them (Berenbaum *et al.*, 1991).

These various plant–insect interactions within the Umbelliferae and similar examples from other families are consistent with the idea of reciprocal evolutionary interactions, based on secondary chemistry. These interactions certainly help to explain the considerable diversity of both insect and plant species that we can observe to be present in nature today.

C. Localization of Toxins in the Plant

If secondary compounds do have a protective function against herbivory, then they are most likely to be located where they are most readily perceived by animals, namely at the plant surface. There is increasing evidence that this is so for many different kinds of plant; toxins are indeed concentrated at or near the plant surface. Secondary compounds have

Table 7.3 Trichome toxins and their effects on insects

Plant	Trichome toxins	Effects on insects
Tomato, *Lycopersicon esculentum*	Rutin, chlorogenic acid, 2-tridecanone	Reduce larval growth or kill
Solanum berthaultii	Polyphenol/phenolase system, *E*-β-farnesene	Aphid trapping
Stylosanthes spp.	α-pinene + sticky resin	Immobilize cattle ticks
Parthenium hysterophorus	Parthenin	Toxic at 0.01% concn
Pelargonium x *hortorum*	Anacardic acids, e.g. *o*-pentadecenylsalicylic	Toxic to two-spotted spider mite
Primula obconica	Primin	Antifeedant

been detected variously in glandular hairs (or trichomes), in leaf waxes, in leaf resins and in bud exudates. Even when they are present in the leaf vacuole, they are often found specifically in the epidermal cells of the upper surfaces (Wink, 1984). Again, in the case of laticiferous plants, toxins are found in the latex and are detected by animals as the latex exudes from the leaf following damage.

Localization at the surface particularly affects insect feeders and many of the adverse effects of surface toxins are discussed in two books on plant–insect interactions edited by Rodriguez *et al.* (1984) and Juniper and Southwood (1986). Some of the salient points of these studies will now be mentioned.

Many toxic terpenoids and phenolics have been identified in plant trichomes; a few examples are listed in Table 7.3, with the structures shown in Fig. 7.3. One of the most bizarre types of trichome defence has been observed in the wild potato species *Solanum berthaultii*. The defence is two-pronged, with different agents secreted in morphologically distinct trichomes A and B, and is aimed primarily at aphids. Type B trichomes contain a volatile signal, *E*-β-farnesene, which occurs with other sesquiterpenes. Farnesene is actually a well-known aphid alarm pheromone and its release by the plant causes a similar disturbance to aphid feeding. The second defence in the A-type trichome consists of a phenolic-containing exudate, linked to a phenolase/peroxidase enzyme system. When an aphid lands on the leaf, it disrupts the droplet,, the enzyme then reacts with the phenolic substrate and a brown, sticky residue is formed. This immobilizes the aphid, prevents it from feeding and causes its death from starvation (Gregory *et al.*, 1986). This defence must be effective since *S. berthaultii* is virtually free of insect attack; indeed, potato breeders have considered hybridizing this species with *S. tuberosum* in order to introduce aphid resistance into the domestic potato.

The toxicity of glandular trichomes in the genus *Stylosanthes*, a group of legumes found in tropical pastures, is also of practical value since the insects trapped by this plant are cattle ticks. In fact, the presence of these plants in cattle-grazing areas provides an effective means of getting rid of the cattle tick *Boophilus microplus*. The tick is immobilized by the sticky secretion of the trichome and is then killed by the volatiles present, which include α-pinene (Sutherst and Wilson, 1986). These trichome toxins in *Stylosanthes* are presumably just as effective against insects trying to feed on the plant.

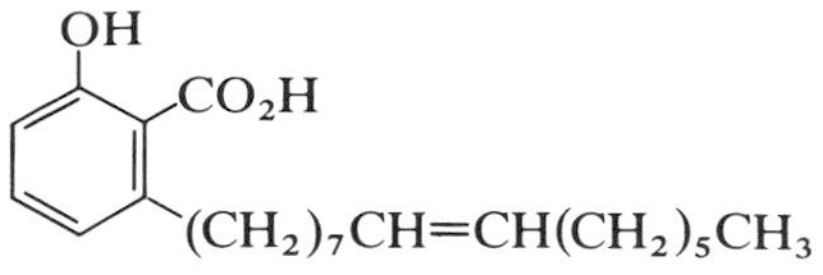

OH
CO_2H
$(CH_2)_7CH{=}CH(CH_2)_5CH_3$

o-pentadecenylsalicyclic acid
(*Pelargonium*)

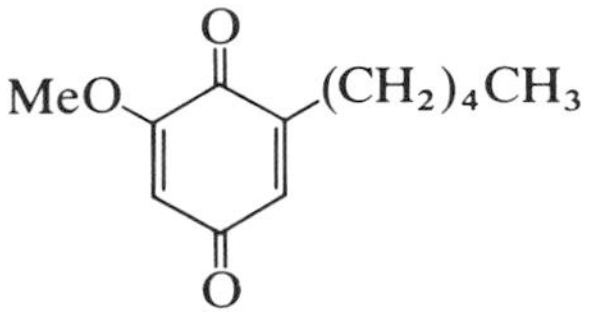

MeO
O
$(CH_2)_4CH_3$
O

primin
(*Primula*)

HO
O
O
O

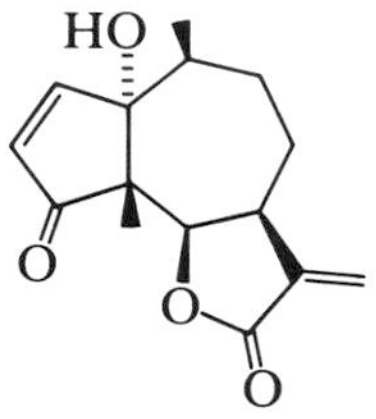

parthenin
(*Parthenium*)

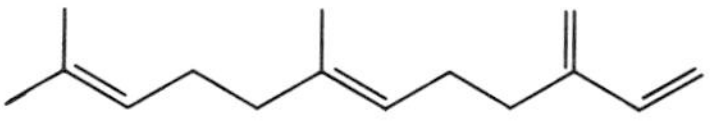

E-β-farnesene
(*Solanum*)

Figure 7.3 Some examples of trichome toxins

The range of substances found in trichomes is considerable and includes some water-soluble compounds, as in the case of the tomato. Here, two classes of toxin are present: the simple hydrocarbon 2-tridecanone and the phenolic compounds, rutin and chlorogenic acid. The effects of these substances on tomato pest insects have already been described in Chapter 5 (Section IV). Such toxins secreted in the leaf glands do not necessarily have to kill insect herbivores. All they need to do is to generally weaken them, retard their growth or delay their pupation. As a consequence, the insects will be more vulnerable to disease, to predation and to the environment (Stipanovic, 1983) and the plant will benefit from the reduced herbivory.

One final point needs to be mentioned about trichome constituents: they are almost inevitably multifunctional in their effects. Many of the same substances which are toxic to insects are antimicrobial (see Chapter 10) and also have deleterious effects on mammalian herbivores. The allergenic effects in man of the trichome constituents of *Parthenium hysterophorus* and *Primula obconica* are well known (Table 7.3). Many other allergenic plants have similarly yielded, on chemical examination, related toxins in the trichomes (Rodriguez *et al.*, 1984).

Leaf waxes provide a second line of defence in some plants, since they may be a barrier to certain insect feeding. More importantly, at least 50% of angiosperm species contain 'extra' secondary constituents (e.g. sterols, methylated flavonoids, etc.) at the leaf surface mixed in with the wax (Martin and Juniper, 1970). These substances almost certainly have a protective role, although less work has been done with them than with the trichome compounds. In the case of apple leaf wax, it is known that the dihydrochalcone phloridzin is present. Although this is used by the apple aphid as a signal to feed, it is known to be repellent to the pea aphid *Acyrthrosiphon pisi* (Klingauf, 1971). There is also evidence that the constituents of the leaf wax may, on occasion, be repellent to insects. Certain varieties of *Sorghum* are distasteful to *Locusta migratoria*, because of the presence of leaf alkanes of chain length C_{19}, C_{21} and C_{23} and of fatty acid esters of chain length C_{12}–C_{18}. These and other effects of surface wax constituents on insects are reviewed by Woodhead and Chapman (1986).

A third line of defence in plants from insect predation is latex production. Latex has been reported in over 12,000 plant species and one of its main functions would appear to be to protect plants from herbivory. The effectiveness of latex, a white viscous liquid consisting of a suspension of rubber particles, as a feeding deterrent is often reinforced by the presence of terpenoid toxins. One particularly unpleasant series of latex constituents are the phorbol esters of Euphorbiaceae. These were first isolated from croton oil, *Croton tiglium*, but they occur widely in the family. Although most notorious as skin irritants and co-carcinogens in mammals, these diterpene esters have recently been shown to be toxic to insects as well (Marshall *et al.*, 1985).

The effectiveness of a sticky latex as an insect feeding deterrent has rarely been tested but there is good presumptive evidence of such a role. For example microlepidopteran leaf miners are specifically known to avoid feeding on chicory, *Cichorium intybus*, and other members of the Cichorieae, the only tribe in the Compositae which is consistently laticiferous. They feed widely on members of other tribes in the same family and the latex appears to prevent them burrowing for food within the leaf.

The latex of chicory is rich in bitter-tasting sesquiterpene lactones (lactupicrin and 8-deoxylactucin, see Fig. 7.4) and feeding experiments against *Schistocerca gregaria* show that these are antifeedants at a concentration of 0.2% dry wt. Analyses of the concentrations of these lactones in roots, stems and leaves of chicory show that the levels are highest in the actively growing regions of the plant but they never drop below the threshold value required for deterrency. The chicory plant is remarkably free from insect pests and there is evidence here that the latex constituents comprise a major source of defence. The extreme bitterness of lactupicrin and 8-deoxylactucin to mammals undoubtedly provides some protection from other herbivores too (Rees and Harborne, 1985).

The most convincing evidence for the importance of the localization of secondary constituents at the plant surface comes from detailed investigations of the browsing behaviour during the winter months of the snowshoe hare, *Lepus americanus*, on the Alaska paper birch, *Betula resinifera*. and the green alder, *Alnus crispa*. Deterrent chemicals

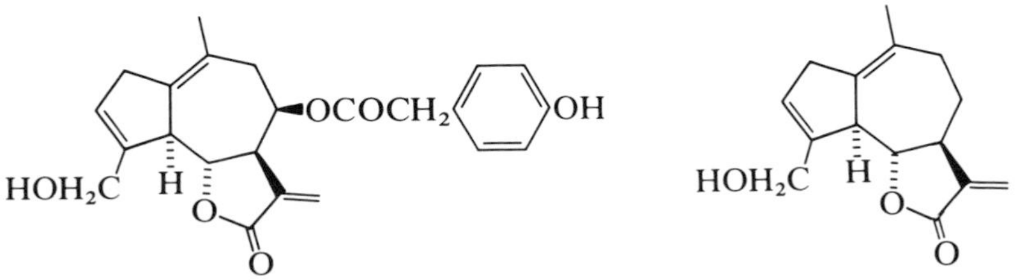

Figure 7.4 Feeding deterrents of chicory root, stem and leaf

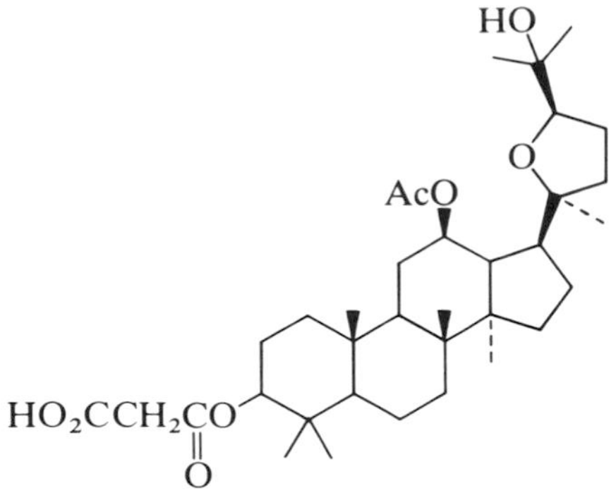

HO
AcO
O
HO_2CCH_2CO
O
papyriferic acid

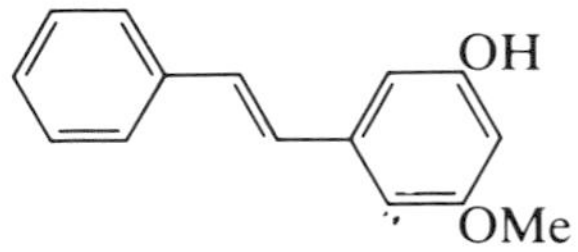

OH
OMe
pinosylvin methyl ether

Figure 7.5 Plant feeding deterrents to the snowshoe hare in Alaska

(Fig. 7.5) are specifically present in concentration in a resinous coating on the plant surface. Furthermore, they are only produced in quantity when they are needed, i.e. at particular stages in the life-cycle, at certain seasons, and in those tissues (e.g. reproductive organs) which require most protection. Otherwise, selective feeding takes place and the snowshoe hare exists in a state of equilibrium with its host trees.

The most dramatic example of chemical protection occurs in the paper birch where juvenile growth-phase internodes are made unpalatable to the hare by the enormous concentration (up to 30% of the dry weight) of the triterpenoid papyriferic acid. This is 25 times the level that is found in mature internodes. The unpalatibility of the juvenile tissues could not be explained by differences in the levels of any other chemicals (e.g. inorganic nutrients or phenolics). Furthermore, the triterpene was clearly identified as a feeding deterrent by feeding it to hares in oatmeal at 2% of the dry weight. The triterpene is deposited as a resinous solid on the surface of the juvenile twigs and, combined with volatiles in the resin, renders the juvenile tree highly deterrent to grazing by the hare (Reichardt *et al.*, 1984). There is provisional evidence that papyriferic acid protects the tree as well from browsing by moose and rodents. Rejection of plant tissue containing papyriferic acid by these herbivores is probably based on its potential toxicity and this is underlined by the fact that in laboratory tests, papyriferic acid at a concentration of 50 mg/kg will kill mice.

The snowshoe hare also feeds selectively on *Alnus crispa*, and here chemical protection is allocated to those organs which are dormant during the winter—i.e. the buds and the staminate catkins—in preference to the more expendable internodal tissues. The chemical agent in *Alnus crispa* is pinosylvin methyl ether, the concentrations that are encountered being $2.6 \pm 0.2\%$ in buds, $1.7 \pm 0.1\%$ in staminate catkins, and $0.05 \pm 0.0\%$ of the dry weight in internodes. Again, the stilbene is present in the resin, where it immediately deters feeding (Bryant *et al.*, 1983). The mountain hare (*Lepus timidus*) of northern Europe also grazes selectively on trees in the same way as the snowshoe hare. Species of *Populus* and of *Salix* are the main winter food and juvenile trees are protected by enhanced levels of phenolic glycoside (mainly salicin). These concentrations decline in mature tall willow species but remain high in low-growing willows, as an adaptation against mammals browsing from ground level (Tahvanainen *et al.*, 1985).

An extreme example of localization of toxin is the composite shrub *Baccharis megapotamica*, which contains the unusual trichothecins, macrocyclic toxins otherwise only known to be produced by soil fungi. The toxins were first thought to occur throughout the plant but detailed analysis of all the plant parts showed that they were exclusively concentrated in the seed coat. Here they protect the seed from insect attack and microbial infection (Kuti *et al.*, 1990).

D. Timing of Toxin Accumulation

One of the problems in establishing a defence function for secondary constituents is the bewildering fluctuations that can occur in the amounts that are present in the plant at any given time. It is true that some metabolites are present at an appreciable level throughout the life-cycle of the plant (e.g. the sesquiterpene lactones in chicory, see above), but others can vary considerably in their concentrations. This is particularly true of the plant alkaloids. And yet, these variations would be explicable if the pattern of accumulation correlated with a proposed defence strategy, i.e. the compounds were present in highest concentration when the plant is most vulnerable to predation. There is now rather good evidence in the case of the coffee plant, *Coffea arabica*, that the high levels of purine alkaloid present coincide precisely with those periods when the plant is most open to herbivory.

The alkaloid caffeine, which is responsible for the stimulating effects of the coffee we drink, is not restricted to the coffee bean but occurs throughout the plant. Its levels have been carefully monitored at critical points in the life-cycle. Thus, during germination, the coffee seeds are covered by a solid protective endocarp so that at first there is no large concentration of alkaloid in the developing shoot. However, as the endocarp decays, the concentration of caffeine in the young seedling more than doubles (Baumann and Gabriel, 1984).

Again, during leaf development, the concentration of alkaloid increases to a high level (4% of the dry wt) just at the time the young soft leaf is most vulnerable to grazing. However, as the leaf matures, the rate of biosynthesis decreases exponentially, from 17 to 0.016 mg/day/g of leaf tissue, and this low level is then maintained when the leaf is fully expanded (Frischknecht *et al.*, 1986). Eventually, the leaves at senescence are essentially alkaloid-free and it is likely that the caffeine is recycled within the plant so that the nitrogen can be incorporated into protein synthesis. Finally, during fruit development, when the pericarp is soft, there is a resurgence of caffeine biosynthesis, with accumulation of 2% dry wt. As the endocarp differentiates and hardens, the levels decrease and there is about 0.24% caffeine present at the end of the ripening process.

That caffeine is defensive in the coffee plant is apparent from feeding experiments with insects. It may either have a lethal effect, killing larvae of the tobacco hornworm, *Manduca sexta*, at a dietary concentration of 0.3% or it may cause sterility, as happens to the beetle *Caliosobruchus chinensis* at a concentration of 1.5% (Nathanson, 1984).

A related pattern of pyrrolizidine alkaloid accumulation has been observed in the groundsel, *Senecio vulgaris*, where, for example, the epidermal cells of the stem contain ten times the alkaloid levels of the remaining cells. During flowering, more than 85% of the total alkaloid is concentrated in the inflorescence to protect it from herbivory (Hartmann *et al.*, 1989). Alkaloid levels fluctuate considerably in other plants, e.g. quinolizidine alkaloids in lupins (Wink, 1984), and it is likely that these variations too may be linked to predator pressure.

Due to their nitrogen content, alkaloids could be considered as both primary and secondary metabolites. There is a clear advantage to the plant to reclaim the nitrogen locked up in the alkaloid molecule as soon as the alkaloid is no longer needed as a defensive agent.

E. Variability in Palatibility Within the Plant

Besides variation with time (Section IIC), variation between tissues is another common feature of secondary metabolite accumulation in plants. Variation between leaf and flower or root and stem is often manifest and may reflect the movement of secondary compounds within the plant (alkaloids may be synthesized in the root and transported to the leaf) or the fact that some parts of the plant are more expendable than others. Floral tissues, for example, may not be obviously protected from predation because they are generally so short-lived compared to leaves. Finally, there is the possibility of variation within the same tissues. This may be apparent especially in leaves of long-lived plants, i.e. trees, which are the most vulnerable of all plant tissues to herbivore attack. Deciduous trees produce leaves year after year, sometimes for centuries, so that insect populations can 'home in' on a readily available and predictable food resource. Since insect predation has the potential to remove such a tree, there must be some other factor which provides resistance in the plant. The most recent suggestions of ecologists are that this factor is variable defence (Denno and McClure, 1983; Whittam *et al.*, 1984).

The evidence in favour of chemical variation within leaf tissues is of two sorts: the patterns of insect grazing on trees and chemical analyses within tree canopies. A study of insect grazing on some trees, such as birch and hazel, shows that there is an overdispersion of grazing initiations and that a proportion of leaves receive a low level of grazing. Such a pattern would be consistent with between-leaf variations in palatability. This idea has been supported by chemical analyses of leaves within the canopy in *Betula lutea* and *Acer saccharum*, which indicate significant variations in chemical parameters from leaf to leaf on the same branch (Schultz, 1983). Palatability is principally defined by tannin levels, toughness (or degree of lignification), water content and nutrient content and all these measurements can change from leaf to leaf. Variation of this type may be due in part to genetic factors, the light regime, nutrient status of the tree, position of leaf on the tree and so on.

There may also be considerable variations in leaf chemistry in different individuals within a tree population. This occurs, for example, in the neotropical tree *Cecropia peltata* (Moraceae), where leaves of some individuals are rich in tannin and are low in herbivore damage and the leaves of others are low in tannin and suffer considerable insect feeding. The tannin levels in the leaves vary from 13 to 58 mg/g dry wt (Coley, 1986). Tests with the army-worm, *Spodoptera latifascia*, showed that there was reduced feeding on leaves of the high-tannin individual. There was, however, a significant cost to the tree in terms of tannin production and the high-tannin trees produced fewer leaves than the low-tannin trees.

For the insect seeking food on such trees, there are considerable risks, since having to spend more time moving around within the canopy must increase the chance of it being the victim of one of its predators. Certainly, within-leaf variation, where it occurs, must place a constraint on the insect grazer. As Schultz (1983) has put it: 'The situation can be said to resemble a 'shell game', in which a valuable resource (suitable leaves) is 'hidden' among many other similar-appearing but unsuitable resources. The insect must sample many tissues to identify a good one. The location of good tissues may be spatially unpredictable, and may even change with time. For a 'choosy' or discriminating insect, finding suitable food in an apparently uniform canopy could be highly complex.'

III. Induced Plant Defence

A. *De novo* Synthesis of Proteinase Inhibitors

One way that a plant might reduce the 'metabolic costs' of synthesizing and storing toxins is to only produce the defensive agent when it is actually needed, i.e. in direct response to feeding. Such a mechanism has already been mentioned in respect of aphids stimulating juvenile hormone analogue synthesis in fir trees (Chapter 4, Section V) and is well known in plant–fungal interactions as the phytoalexin response (Chapter 10, Section IIC). There is evidence from the work of Ryan and his co-workers (see Ryan, 1979) that plants may sometimes be able to respond rapidly to insect attack by the production of specific proteins which are proteinase inhibitors and are able ultimately to deter further feeding.

It has been shown by Ryan that a Colorado beetle feeding on potato or tomato leaves can cause the rapid accumulation of proteinase inhibitors, even in parts of the plant distant from the site of attack. The process is mediated by a proteinase inhibitor inducing factor, known as PIIF, which is released into the vascular system. Within 48 h of leaf damage, the leaves may contain up to 2% of the soluble protein as a mixture of two proteinase inhibitors. Subsequently, the presence of the proteinase inhibitors in the leaf is detected by the beetle, which avoids further feeding and moves onto another plant (see Fig. 7.6).

In theory, the inhibitors, if taken in the diet, will have an adverse effect on the insects' ability to digest and utilize the plant protein, since they inhibit the protein-hydrolysing enzymes, trypsin and chymotrypsin. In fact, proteinase inhibitors are well known as constitutive constituents of many plant seeds, where they have a similar protective role in deterring insect feeding.

PIIF induction is also brought about by mechanical wounding of plant tissue so that it is not yet entirely clear how specific this effect is to herbivore grazing. The nature of the chemical signal PIIF has been explored in tomato plants and it is a small protein, called systemin. This protein is 10,000 times more active than oligosaccharides, which also have the ability to trigger this defence system (Ryan, 1992). A volatile chemical, the fatty

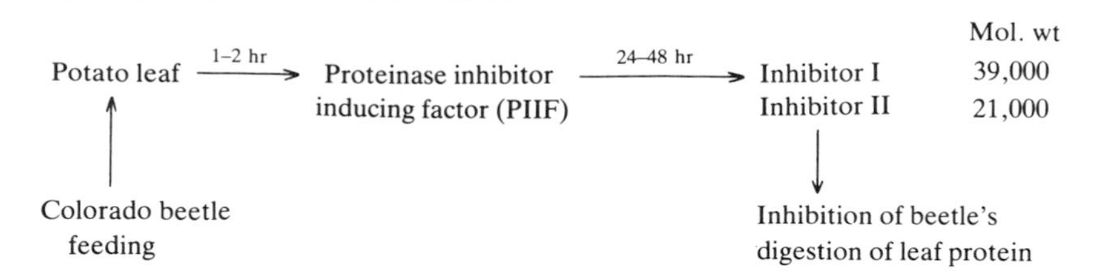

Figure 7.6 Mechanism of induction of proteinase inhibitors in plants in response to insect herbivory

acid metabolite methyl jasmonate, may also be involved in the signalling process (Farmer and Ryan, 1990).

The PIIF-like activities have been detected in extracts of 37 plant species representing 20 families (Ryan, 1979) so this mechanism may well be a general one. The general effectiveness of trypsin inhibitors in deterring herbivory has been elegantly demonstrated by Hilder *et al.* (1987) in genetic engineering experiments. A gene encoding a cowpea trypsin inhibitor was transferred to tobacco. As a result, transformed leaves were more resistant to budworm (*Heliothis virescens*) feeding than the original plants. Ecological experiments have also shown that PIIF induction in the tomato reduces the grazing by larvae of the army-worm *Spodoptera littoralis* within 48 h, with avoidance being most pronounced on the young leaves (Edwards *et al.*, 1992).

B. Increased Synthesis of Toxins

A related form of induced defence, apparently quite distinct from the PIIF system, has been observed in a variety of plants. The effect is relatively rapid and the leaves become unpalatable to animals within a matter of hours or a few days. It may be short term, disappearing after the insect has stopped feeding, or long term, extending over to the following season in trees. The chemical changes involve an increase in the concentration of existing toxins, sufficient to lead to herbivore avoidance. The effect is separate from a localized 'wound response', which only takes place immediately around the site of damage, since it can often be discerned throughout the plant.

Such increases in toxin synthesis have been observed in two alkaloid-containing plants. One is the wild tobacco species *Nicotiana sylvestris*, which contains nicotine and nornicotine as the major alkaloids. Larval feeding induced a 220% increase in alkaloid content throughout the plant over a period of 5–10 days. Mechanical damage which avoids cutting the secondary veins produced a smaller response (170%). In fact, the tobacco hornworm *Manduca sexta* when feeding on the tobacco leaf avoids cutting through the secondary veins. It thus avoids triggering off the fullest response in the leaf, which can be as much as 400% of the control if the simulated damage includes damaging the vein. The nicotine alkaloids are synthesized in the roots and transported up into the leaf and this was apparent in experiments in which pot-bound plants with confined roots failed to show any significant alkaloid increase after mechanical damage (Baldwin, 1988). Similar experiments with the tropane alkaloids in leaves of *Atropa acuminata* showed a maximum increase of 153–164% over the control 8 days after mechanical damage or slug feeding. Repeated mechanical damage at 11-day intervals initially increased the response to 186% of the control but this effect fell off with time. Further experiments showed that only 9% of the leaf area needed to be removed mechanically or by animal feeding to produce the maximum response (Khan and Harborne, 1991).

Another well-investigated example of induced chemical defence is the wild parsnip, *Pastinaca sativa*, which produces five furanocoumarins in the leaves. Artificial damage increased furanocoumarin synthesis to 162% of the control, while feeding by the generalist insect *Trichoplusia ni* increased it to 215%. Furthermore, larvae of *T. ni* grew very slowly on induced leaves, and larvae on artificial diet supplemented with furanocoumarins were similarly affected (Zangerl, 1990). The response of oil seed rape, *Brassica napus*, to insect infestation or leaf damage is quite distinctive and involves the massive accumulation of indole glucosinolates, which are barely detectable in the control. There is a corresponding reduction in the amounts of the aliphatic glucosinolates of the plant, but the total titre of glucosinolate does appear to increase under these treatments (Koritsas *et al.*, 1989).

Other examples where induced changes in protective chemistry have been recorded are given in Tallamy and Raupp (1991). For every plant that shows a positive response, there is another plant where no detectable change in palatability occurs (Edwards and Wratten, 1983). Rapid inducible resistance appears to be weak in or absent from the leaves of slow-growing plants (see Section IIA). Environmental factors also determine the magnitude of the response. Finally, the response may disappear as the plant grows older and concomitantly more resistant to grazing. For example, two-year-old trees of *Pinus contorta* respond to defoliation by increasing the concentration of both terpenes and tannins in the needles, whereas ten-year-old trees fail to show any increases (Watt *et al.*, 1991).

C. Release of Predator-attracting Volatiles

An even more interesting and remarkable plant–animal interaction involving induced chemical changes has been observed by Dicke *et al.* (1990). In response to herbivory, some plants have developed the means to release volatile chemicals, which are particularly attractive to parasitoids of their herbivores, which then visit the plant and destroy the herbivores. As Dicke puts it, plants may 'cry for help' when attacked by spider mites and predatory mites come to the rescue. Much research has been conducted on the spider mite *Tetranychus urticae*, the predatory mite *Phytoseiulus persimilis*, and the host plants. The chemicals released seem to be plant species-specific. Cucumber plants infested by the spider mite release β-ocimene and 4,8-dimethyl-1,3,7-nonatriene and are only moderately attractive to the predatory mites, while Lima beans release a cocktail of linalool, β-ocimene, the nonatriene and methyl salicylate which is highly attractive. A further advantage to the plant world is that the volatile released may alert uninfested neighbouring plants so that they become better protected from spider mite attack. Thus cotton seedlings, when infected by these mites, release volatile cues which both attract predatory mites and also alert neighbouring plants to withstand herbivore attack (Bruin *et al.*, 1992).

The systemic release of volatile chemicals which mediate in plant–herbivore–predator interactions has been observed in other plant systems. Turlings and Tumlinson (1992) have recorded that corn (*Zea mays*) seedlings respond to beet army-worm (*Spodoptera exigua*) attack by releasing volatiles, which attract parasitic wasps, *Cotesia marginiventris*, to attack the herbivore. The response occurs throughout the plant and not only at the site of damage. The chemicals involved include linalool, which is released at the rate of 1 ng/h before damage and at the rate of 110 ng/h six hours after army-worm attack.

A similar tritrophic system exists in the case of the soya bean plant, the soya bean looper *Pseudoplusia ineludens* and its parasitoid *Microplitis demolitor*. Here the volatiles again include linalool, but the more important attractants are guaiacol and 3-octanone. These latter two compounds do not appear to be released in appreciable amounts from the plant, but are released from the insect frass and are formed within the larvae from dietary sources (Ramachandran *et al.*, 1990). In other plants such as cotton and cow-pea, the release of green leaf volatiles (e.g. *E*-2-hexenal and *E*-2-hexen-1-ol) appears to be sufficient to attract parasitic wasps to attack leaf-feeding caterpillars (Whitman and Eller, 1990).

IV. Animal Response

A. Insects

The fact that insects can become adapted to plant toxins in their diet and have responded to the challenge of detoxifying such materials is a major tenet of the co-evolutionary theory. Insects that feed on toxic plants—and this includes representative species of all the major phytophagous groups—have adapted to the situation in a variety of ways, as indicated in Table 7.4. Most insect species are cryptic, since aposematic (or warning coloured) species and their mimics are relatively rare. The cryptic life-style would therefore seem to be the most successful in evolutionary terms and this usually involves metabolism and detoxification. Nonetheless, certain circumstances in the environment may favour a switch to a toxic host plant, so that a cryptic insect may be destined to become warning coloured (Rothschild, 1973). Such a switch may at first be thought to be a highly complicated process, yet some experiments of Rothschild *et al.* (1979) indicate that biochemically it may be quite simply achieved.

Table 7.4 Variations in life-style of insects feeding on toxic plants

Life-style	Fate of plant toxin	Comments
Cryptic	Metabolize/excrete	Also may avoid ingestion by selective feeding
Aposematic	Store unchanged or store metabolite	May also or instead synthesize its own toxin
Aposematic mimic	Metabolize/excrete	

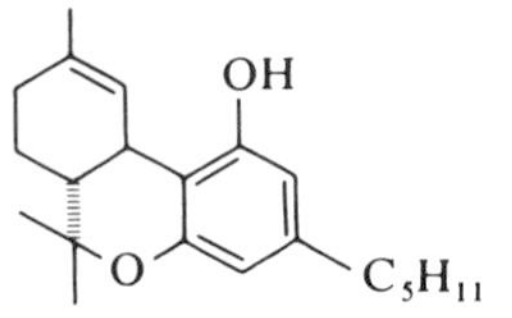

Δ^1-Tetrahydrocannabinol (THC), active principle of Mexican strain

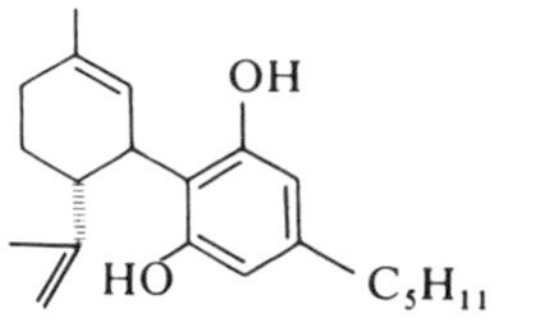

Cannabidiol (CBD), inactive principle of Turkish strain

Figure 7.7 Contrasting structures of active and inactive cannabinoids of the cannabis plant

The insect chosen for this investigation was the tobacco hornworm, *Manduca sexta* which has cryptic coloration and readily excretes toxins. When feeding on tobacco leaves, it excretes any ingested nicotine quite rapidly. Eggs of this insect were hatched and reared on a non-host solanaceous species, *Atropa belladonna*. Belladonna, instead of having the pyridine-based nicotine as major alkaloid, produces the tropane alkaloid atropine as major toxin. The pupae of the hornworm were later collected and fed to hens, which subsequently died within 1–7 days. Furthermore, analysis of the pupae showed the presence of atropine; thus, these insects are clearly able, when moved onto a strange plant, to store any new toxins that may be present. Acquisition of warning coloration could clearly follow this crucial switch in the insect's biochemistry.

A switch from one poisonous plant to another may occur with an aposematic insect and a further example records the versatility of such insects in dealing with strange toxins. Rothschild *et al.* (1977) allowed larvae of the tiger moth, *Arctia caja*, and the brightly coloured grasshopper, *Zonocerus elegans*, to feed on cannabis leaves, something they are not normally accustomed to. The plant *Cannabis sativa*, as is well known, contains a volatile psychoactive principle, Δ^1-tetrahydrocannabinol (THC) (see Fig. 7.7), which has been utilized by man as a source of apparently harmless pleasure for centuries. The phrase 'apparently harmless' is used because there is still controversy as to whether long-term exposure to THC is harmful to health. In terms of plant–animal interactions, THC must be regarded as a toxin and a potential threat to most insects feeding on the plant. Some tiger moth and *Zonocerus* larvae were indeed killed but others of both species survived this strange encounter with the cannabis plant. Also they readily stored the THC, i.e. they were preadapted to cope with a foreign plant toxin.

Populations of *C. sativa* vary in their THC content and strains can be identified with high THC content (Mexican strain) or with THC completely replaced by inactive cannabidiol (CBD) (Fig. 7.7) (Turkish strain). In another experiment, larvae of both insects were allowed to feed on this CBD-containing Turkish strain. Interestingly, both insects survived better in the presence of CBD, metabolizing and excreting the inactive material. Furthermore, when allowed a choice between leaves of the two cannabis strains, newly hatched larvae of the tiger moth showed a definite preference for the active but more toxic Mexican strain.

Thus, the larvae of this aposematic moth were clearly liable to become attracted to this

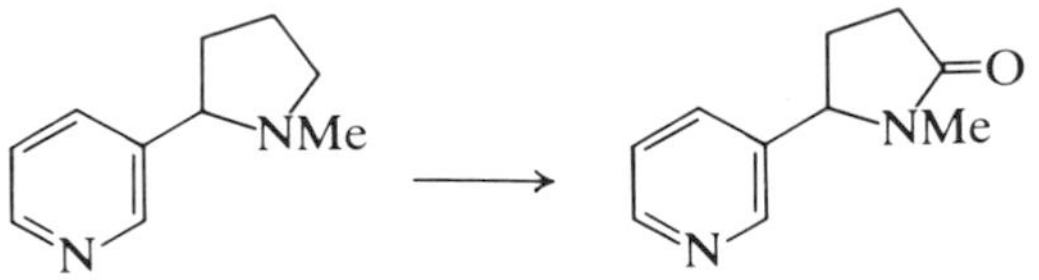

NMe → NMe =O

Figure 7.8 Detoxification of nicotine to cotinine in insects

material, just like cannabis smokers, so that it is possible that the plant, by producing THC in the leaves, is able to exert a subtle fascination on insect predators with possible lethal consequences. This example illustrates both the amazing achievement of insects to cope with plant poisons and the subtle dangers that lay in wait for insect species that may become hooked on a toxic plant.

Insect adaptation to toxic dietary components can take other forms besides sequestration (Blum, 1983). The tobacco hornworm, already mentioned above, feeds with impunity on tobacco, because it rapidly excretes (without metabolism) the dietary nicotine, which would otherwise be very poisonous. Trace amounts may be absorbed in the blood but it does have a second line of defence in that the membranes protecting the nerve cells are impermeable to nicotine. Other insects, such as the house fly, *Musca domestica,* which do not normally encounter dietary nicotine actually have a method of detoxifying it. They metabolize it to cotinine (Fig. 7.8), which is essentially non-poisonous.

Insect adaptation to toxicity may be behavioural. Thus the cucumber beetle feeding on squash avoids the induced chemical defence of the plant by cutting a circular trench around an area of the leaf, so that only a few veins and pieces of lower epidermis hold the encircled leaf in place. The beetle then feeds on the excised area quite safely (Carroll and Hoffman, 1980). For further studies on this phenomenon, see Tallamy and McCloud (1991).

We know little as yet of the many pathways that insects employ to detoxify plant toxins but there is increasing evidence of their success in accomplishing such a biochemistry. For example, it has been shown that the monoterpene carvone, when fed to the southern army-worm at a concentration of 100 μg/g body wt, induces an increase of 134% in the detoxifying enzyme, cytochrome P-450 oxidase (Brattsten, 1983). The main reaction is hydroxylation and the hydroxy derivative is then conjugated and excreted in the usual way (see Chapter 3, Section IIIC). We can also infer from studies of the effects of synthetic insecticides that insects have the capacity to metabolize them and hence develop resistance. Even the recently marketed synthetic pyrethroids are metabolized by a variety of pathways in pest insects (Soderlund *et al.*, 1983).

B. Kangaroos

The ability of some kangaroo populations in Australia to resist fluoracetate poisoning after feeding on legume pasture plants containing this toxin represents one of the best examples of a co-evolutionary response by an animal to a plant poison (Mead *et al.*, 1985a). Most mammals are fatally poisoned by fluoracetate when it is imbibed at a concentration as little as 1 mg/kg body wt. It blocks respiration, being converted to fluoro-

Figure 7.9 Distribution of fluoroacetate-containing plants in Australia (////) and likely evolutionary routes of the grey (A) and rat (B) kangaroos

citrate, which is a competitive inhibitor of aconitase hydratase, which then stops the Krebs tricarboxylic acid cycle at the citrate stage. Fluorocitrate also binds via thiolester links to two mitochondrial membrane-bound enzymes involved in citrate transport, and inhibition of these enzymes effectively prevents the influx and efflux of citrate.

In spite of its highly poisonous nature, fluoracetate is a natural plant product, being produced in some 34 species of *Gastrolobium* and *Oxylobium* which are native to Western Australia. The leaf concentration may reach levels of 2.65 g/kg fresh wt, which represents far more than is necessary to cause death to a potential herbivore. One explanation for these high levels is that the populations of grey and rat kangaroos in Western Australia have evolved resistance to these plants. Increased synthesis of fluoracetate would thus represent a further co-evolutionary response by the plant to develop protection against animal grazing.

It is noteworthy that kangaroo populations within Australia differ in their ability to tolerate fluoracetate. While the rat kangaroo (*Bettongia* spp.) of Western Australia is highly adapted, that of southeastern Australia, where fluoracetate-containing plants do not grow, shows no tolerance to the poison. These data suggest that the rat kangaroo probably originated in the east and radiated westward some thousands of years ago, developing resistance to fluoracetate in the new environment of Western Australia (Fig. 7.9). By contrast, all extant members of the grey kangaroo (*Macropus*) group in Australia are tolerant of fluoracetate. This would be expected if the genus first colonized the western areas and migrated eastward with time.

It is interesting that the eastern grey kangaroo still retains its ability to detoxify fluoracetate in spite of the fact that it is not normally challenged by a plant with this toxin present. However, even the western grey kangaroo, which is used to feeding on *Gastrolobium* species, will limit its intake of poison. If given a choice of two *Gastrolobium* species with low and high levels of fluoracetate, it will feed mainly on the former plant (Mead *et al.*, 1985a).

Glutathione + F $CH_2CO_2H \rightarrow F^- +$ *S*-carboxymethylglutathione
$\rightarrow$ glycine + glutamate + *S*-carboxymethylcysteine

Figure 7.10 Defluorination of fluoracetate via glutathione

The mechanism of detoxification of fluoracetate in tolerant kangaroos has not yet been fully worked out. *In vivo* defluorination certainly occurs and the tripeptide glutathione is involved in the process (Fig. 7.10). However, there is no direct relationship in different animals between tolerance to fluoracetate and the rate of defluorination (Mead *et al.*, 1985b). There must therefore be some further route by which detoxification takes place.

C. Rats and Man

As already described in detail in Chapter 6, plant tannins are a general feeding deterrent to mammals. The effect is quantitative, in that plants which have about 5% or more tannin per dry wt are the the ones that are rejected. Avoidance of tannin-rich plants is linked to the astringent taste imparted by the material. The anti-nutritional effects of dietary tannins in mammals are linked to: (1) inhibition of digestive enzymes; (2) formation of relatively less-digestible complexes with dietary protein; (3) depressed growth rate; and (4) inhibition of microbial flora. There may also be direct toxic effects, due to the binding of the tannins to the digestive tract.

The most impressive evidence confirming that tannins are a major feeding barrier is the discovery by Butler *et al.* (1986) that rats fed on a diet containing *Sorghum* tannins are able to adapt to their adverse effects. In these rats there is enormously increased synthesis of a series of unique proline-rich proteins (PRPs) in the parotid glands. These salivary proteins have a high affinity for condensed tannins and remove them by binding to them at an early stage in the digestive process. The high affinity of PRPs depends on the presence of up to 40% by weight of carbohydrate, the sugar units keeping the polypeptide chain in an open conformation so that it can bind strongly via hydrogen bonding to the tannin (Asquith *et al.*, 1987). Adaptation in rats takes place within 3 days of commencing the dietary tannin and there is a 12-fold increase in PRPs in the parotid glands. The adapted animal can subsequently thrive on the tannin-containing diet without any adverse effects.

Conversely, the harmful results of dietary tannin in a non-adapted animal are nicely demonstrated by the feeding of *Sorghum* tannins to the hamster. Weaning hamsters that are treated with a diet containing 4% dry wt of tannin suffer severe weight loss and then perish within 3–21 days. Analyses of their parotid glands show that the hamsters are unable to increase the synthesis and secretion of PRPs (Butler *et al.*, 1986).

Man himself is also able to respond to dietary tannin by maintaining the parotid glands in an induced state. About 70% of our parotid gland secretions are of the PRP type. This explains why high-tannin sorghum varieties are still grown as staple cereal in parts of Africa and India. It also explains our ability to acquire a taste for red wines, many of which are relatively rich in soluble tannin.

A survey of mammals (Mole *et al.*, 1990) showed that rabbits and hares, like rats, are well endowed with the ability to produce PRPs when fed a tannin diet. As expected, these proteins are lacking in the saliva of carnivores such as dogs and cats. The response is weak in ruminants and there is evidence in sheep that an alternative endocrine adaptation occurs, with an increase in the rate of glycerol release from adipose tissue when tannin is present in the diet (Barry *et al.*, 1986).

V. Conclusion

In this chapter, the evidence for a co-evolutionary arms race between plants and animals has been considered. It is suggested that plant defence is of two types: static or constitutive and dynamic or induced; and both types may operate in the same plant. Chemical defences are often concentrated in those parts of the plant, e.g. the leaf surface, where they are most readily perceived by animals. The most vulnerable plant tissues (young leaves, buds) are more likely to be better protected than old, mature tissues. Not all plants and not all plant parts need to be protected by toxic chemicals for predation. Variation in chemistry at many different levels—between plants, within plant parts, within leaves on the same branch—confounds the herbivore and makes it more difficult for it to seek its food and browse.

Animal responses to chemically defended plants likewise take many forms. Rhoades (1985) divides the offensive strategies of phytophagous animals into two groups: stealth and opportunism. Stealthy herbivores minimize plant damage and suppress induced defences, while opportunists may employ mass-attack to stress their food plant into submission.

In biochemical terms, one can observe that some insects can pass plant poisons through their systems without any ill effects. Others metabolize and detoxify and yet others sequester and safely store plant toxins. In mammals, elaborate detoxification systems are present and even the deadly poisonous fluoracetate ion can be dealt with by the kangaroo's system. Adaptation to toxicity by the synthesis of proline-rich proteins represents a more sophisticated response in mammals to the inherent dangers of high-tannin diets.

Whether these adaptations and counter-adaptations represent co-evolution in action or are simply pre-adaptations for survival is an arguable point. Nevertheless, the co-evolutionary hypothesis of Paul Ehrlich and Peter Raven has been a most useful catalyst for scientific investigation. It should continue to inspire further research in an area where our present knowledge is still quite fragmentary.

Bibliography

Books and Review Articles

Berenbaum, M. (1983). Coumarins and caterpillars: a case for coevolution. *Evolution* **37**, 163–179.

Bernays, E. and Chapman, R. G. (1978). Plant chemistry and acridoid feeding behaviour. In: Harborne, J. B. (ed.), 'Biochemical Aspects of Plant and Animal Coevolution', pp. 99–142. Academic Press, London.

Blum, M. S. (1983). Detoxification, deactivation and utilisation of plant compounds by insects. In: Hedin, P. A. (ed.), 'Plant Resistance to Insects', pp. 265–278. Amer. Chem. Soc., Washington, D.C.

Brattsten, L. B. (1983). Cytochrome P-450 involvement in the interactions between plant terpenes and insect herbivores. In: Hedin, P. A. (ed.), 'Plant Resistance to Insects', pp. 173–198. Amer. Chem. Soc., Washington, DC.

Carroll, C. R. and Hoffman, C. A. (1980). Chemical feeding deterrent mobilised in response to insect herbivores and counteradaptation by *Epilachna tredecimnota*. *Science, N. Y.* **209**, 414–416.

Denno, R. F. and McClure, M. S. (eds.) (1983). 'Variable Plants and Herbivores in Natural and Managed Systems'. Academic Press, New York.

Edwards, P. J. and Wratten, S. D. (1983). Wound induced defences in plants and their consequences for patterns of insect grazing. *Oecologia* **59**, 88–93.

Ehrlich, P. R. and Raven, P. H. (1964). Butterflies and plants: a study in co-evolution. *Evolution* **18**, 586–608.

Gulmon, S. L. and Mooney, H. A. (1986). In: Givnish, T. J. (ed.), 'On the Economy of Plant Form and Function', pp. 681–698, University Press, Cambridge.

Juniper, B. and Southwood, T. R. E. (eds.) (1986). 'Insects and the Plant Surface', 360 pp. Edward Arnold, London.

Martin, J. T. and Juniper, B. E. (1970). 'The Cuticles of Plants'. Edward Arnold, London.

Rhoades, D. F. (1985). Offensive–defensive interactions between herbivores and plants. *Amer. Nat.* **125**, 205–238.

Rodriguez, E., Healey, P. L. and Mehta, I. (eds.) (1984). 'Biology and Chemistry of Plant Trichomes', 255 pp. Plenum, New York.

Rothschild, M. (1973). Secondary plant substances and warning colouration in insects. In: van Emden, H. (ed.), 'Insect–Plant Interactions', pp. 59–83. Oxford Univ. Press.

Ryan, C. A. (1979). Proteinase inhibitors. In: Rosenthal, G. A. and Janzen, D. H. (eds.), 'Herbivores: their Interaction with Secondary Plant Metabolites', pp. 599–618. Academic Press, New York.

Schultz, J. C. (1983). Impact of variable plant defensive chemistry on susceptibility of insects to natural enemies. In: Hedin, P. A. (ed.) 'Plant Resistance to Insects', pp. 37–54. Amer. Chem. Soc., Washington, DC.

Stipanovic, R. D. (1983). Function and chemistry of plant trichomes and glands in insect resistance. In: Hedin, P. A. (ed.), 'Plant Resistance to Insects', pp. 69–102. Amer. Chem. Soc., Washington, DC.

Tallamy, D. W. and Raupp, M. J. (eds.) (1991). 'Phytochemical Induction by Herbivores'. John Wiley and Sons, New York.

Whittam, T. G., Williams, A. G. and Robinson, A. M. (1984). The variation principle: individual plants as temporal and spatial mosaics of resistance to rapidly evolving pests. In: Price, P. W., Slobodchikoff, C. N. and Gaud, W. S. (eds.), 'A New Ecology', pp. 16–51. Wiley, Chichester.

Literature References

Asquith, T. N., Uhlig, J., Mehansho, H., Putnam, L., Carlson, D. M. and Butler, L. (1987). *J. Agric. Fd Chem.* **35**, 331–334.

Baldwin, I. T. (1988). *J. Chem. Ecol.* **14**, 1113–1120.

Barry, T. N., Allsop, T. F. and Redekopp, C. (1986). *Br. J. Nutr.* **56**, 607–614.

Baumann, T. W. and Gabriel, H. (1984). *Plant Cell Physiol.* **25**, 1431–1437.

Berenbaum, M., Zangerl, A. R. and Nitao, J. K. (1986). *Evolution* **40**, 1215–1228.

Berenbaum, M., Nitao, J. K. and Zangerl, A. R. (1991). *J. Chem. Ecol.* **17**, 207–215.

Bruin, J. Dicke, M. and Sabelis, M. W. (1992). *Experientia* **48**, 525–529.

Bryant, J. P., Chapin, F. S. and Klein, D. R. (1983). *Oikos* **40**, 357–368.

Bryant, J. P. Wieland, G. D., Reichardt, P. B., Lewis, V. E. and McCarthy, M. C. (1983). *Science, N.Y.* **222**, 1023–1025.

Bull, D. L., Ivie, G. W., Beier, R. C., Pryor, N. W. and Oertli, E. H. (1984). *J. Chem. Ecol.* **10**, 893–912.

Butler, L. G., Rogler, J. C., Mehansho, H. and Carlson, D. M. (1986). In: Cody, V., Middleton, E. and Harborne, J. B. (eds.), 'Plant Flavonoids in Biology and Medicine', pp. 141–158. Liss, New York.

Coley, P. D. (1986). *Oecologia* **70**, 238–241.

Dicke, M., Sabelis, M. W. and Takabayashi, J. (1990). *Symp. Biol. Hung.* **39**, 127–134.

Dyer, M. I. and Bokhari, V. G. (1976). *Ecology* **57**, 762–772.

Edwards, P. J., Wratten, S. D. and Parker, E. A. (1992). *Oecologia* **91**, 266–272.

Farmer, E. E. and Ryan, C. A. (1990). *Proc. Natl. Acad. Sci. USA* **87**, 7713-7716.

Frischknecht, P. M., Ulmer-Dufek, J. and Baumann, T. W. (1986). *Phytochemistry* **25**, 613–616.

Gregory, P., Ave, D. A., Bouthyette, P. J. and Tingey, W. M. (1986) In: Juniper, B. and Southwood, T. R. E. (eds.), 'Insects and the Plant Surface', pp. 173–184. Edward Arnold, London.

Hartmann, T., Ehmke, A., Sonder, H., Borstel, K. V., Adolph, R. and Toppel, G. (1989). *Planta Medica* **55**, 218–219.

Herms, D. A. and Mattson, W. J. (1992). *Q. Rev. Biol.* **67**, 283–335.

Hilder, V. A. Gatehouse, A. M. R., Sheerman, S. E., Barker, R. F. and Boulter, D. (1987). *Nature* **300**, 160–163.

Khan, M. B. and Harborne, J. B. (1991). *Biochem. Syst. Ecol.* **19**, 529–534.

Klingauf, F. (1971). *Z. angew. Entomol.* **68**, 41–55.

Koritsas, V. M., Lewis, J. A. and Fenwick, G. R. (1989). *Experientia* **49**, 493–495.

Kuti, S. O., Jarvis, B, B., Rejali, N. M. and Bean, G. A. (1990). *J. Chem. Ecol.* **16**, 344.

Marshall, G. T., Klocke, J. A., Lin, L. J. and Kinghorn, A. D. (1985). *J. Chem. Ecol.* **11**, 191–206.

Mead, R. J., Oliver, A. J., King, D. R. and Hubach, P. H. (1985a). *Oikos* **44**, 55–60

Mead, R. J., Moudden, D. L. and Twigg, L. E. (1985b). *Aust. J. Biol. Sci.* **38**, 139–149.

Mihalaik, C. A. and Lincoln, D. E. (1989). *J. Chem. Ecol.* **15**, 1579–1588.

Mole, S., Butler, L. G. and Iason, G. (1990). *Biochem. Syst. Ecol.* **18**, 287–293.

Nathanson, J. A. (1984). *Science, N.Y.* **226**, 184.

Ramachandran, R., Norris, D. M., Phillips, J. K. and Phillips, T. W. (1990). *J. Agr. Fd Chem.* **39**, 2310–2317.

Rees, S. B. and Harborne, J.P. (1985). *Phytochemistry* **24**, 2225–2231.

Reichardt, P. B., Bryant, J. P., Clausen, T. P. and Wieland, G. D. (1984). *Oecologia* **65**, 58–69.

Rothschild, M., Rowan, M. G. and Fairbairn, J. W. (1977). *Nature, Lond.* **266**, 650–651.

Rothschild, M., Aplin, R., Baker, J. and Marsh, N. (1979). *Nature, Lond.* **280**, 487–488.

Ryan, C. A. (1992). *Plant Molecular Biology* **19**, 123–133.

Soderlund, D. M., Sanborn, J. R. and Lee, P. W. (1983). In: Hutson, D. H. and Roberts, T. R. (eds.), 'Progress in Pesticide Biochemistry and Toxicology', Vol. 3, pp. 401–435. Wiley, Chichester.

Sutherst, R. W. and Wilson, L. J. (1986). In: Juniper, B. and Southwood, T. R. E. (eds.), 'Insects and the Plant Surface', pp. 185–194. Edward Arnold, London.

Tahvanainen, J., Helle, E., Julkunen-Tiitto, R. and Lavola, A. (1985). *Oecologia* **65**, 319–323.

Tallamy, D. W. and McCloud, E. S. (1991). In: Tallamy, D. W. and Raupp, M. J. (eds.), 'Phytochemical Induction by Herbivores', pp. 155–181. John Wiley and Sons, New York.

Tuomi, J. P., Niemela, P. and Swen, S. (1990). *Oikos* **59**, 399–410.

Turlings, T. C. J. and Tumlinson, J. H. (1992). *Proc. Natl. Acad. Sci. USA* **89**, 8399–8404.

Waterman, P. G. and Mole, S. (1989). In: Bernays, E. A. (ed.), 'Insect–Plant Interactions', Vol. 1, pp. 107–134. CRC Press, Boca Raton.

Watt, A. D., Leather, S. R. and Forrest, G. I. (1991). *Oecologia* **86**, 31–35.

Whitman, D. W. and Eller, F. J. (1990). *Chemoecology* **1**, 69–75.

Wink, M. (1984). *Z. Naturforsch.* **39c**, 553–558.

Woodhead, S. and Chapman, R.F. (1986). In: Juniper, B. and Southwood, T. R. E. (eds.), 'Insects and the Plant Surface', pp. 123–136. Edward Arnold, London.

Zangerl, A. R. (1990). *Ecology* **71**, 1926–1932.

8 Animal Pheromones and Defence Substances

I. Introduction

The paramount importance of chemical communication in biological systems is now widely appreciated. Together with aural and visual modes of communication, olfactory signals play a vital role in most groups of animals. Examples have already been given in previous chapters where insect behaviour in pollination and feeding is controlled by such chemical signals. It is the purpose in this chapter to consider these phenomena in more detail, with especial emphasis on the chemical structures involved.

Chemical signals in their widest sense are a universal attribute of life. In some form, they exist within cells and within and between all organisms. In lower plants, for example, there is clear evidence of such interaction. In the slime mould *Dictyostelium discoideum*, for example, cyclic AMP acts as an aggregation pheromone while a simple chlorinated phloroglucinol derivative induces differentiation (Morris *et al.*, 1987). In higher plants, too, volatile chemicals are involved in interactions between organisms; for example, detrimental effects involving monoterpenoids and other compounds occur between one higher plant and another, a phenomenon called allelopathy (see Chapter 9). It is, however, only in the animal kingdom that chemical signals of an olfactory type are generally present, to serve an enormous variety of different purposes. Such signals are used in relation to an animal's need for food, reproduction and protection from predation. They are also important, in the case of social animals, for communication between individuals of the same species. Volatile chemicals used for communication *within* species are termed 'pheromones', while chemicals used *between* different species are called 'allomones'. The distinction between the two classes of pheromone is sometimes blurred, since the same compound may occasionally serve both purposes.

Although pheromones are known throughout the animal kingdom, most of our information on these substances is derived from work on insects. This is partly because with insect pheromones it is relatively easy to monitor their activities; it is more difficult to do this with mammalian pheromones. It is also partly because there has been a practical incentive for the study of insect pheromones, since their identification at once provides a means of monitoring populations and thus, in the case of agricultural pests, of pest control.

The existence of pheromonal interactions in mammals, is, however, also now well documented. Obvious examples occur in the sexual life of animals living in groups, e.g. caged mice and rats. An odorant in the urine of male mice, for example, induces and accelerates the oestrus cycle in the female. This effect is more pronounced in those females whose cycles have been suppressed by grouping them together in the absence of a male, another pheromonal response. In humans, this can be seen in the effect on the menstrual cycle in female students living together in university halls of residence; their cycles eventually become synchronized in such a way that most members of the group eventually menstruate at the same time. Although such pheromonal interactions in mammals are now well accepted, the chemicals concerned are often unknown and further discussion must await the time when the molecular basis of these signals is better understood.

In insects, pheromones are secreted in exocrine glands and are transmitted to other members of the species in vapour form. The effectiveness of some insect sex pheromones is proverbial. Only a few molecules are apparently needed to produce a response and the same few molecules can be effective over considerable distances. The great signalling power of sex pheromones is reflected in the fact that the release of less than 1 $\mu g/s$ by the female can attract the male gypsy or silkworm moth; the male begins to react when the molecular concentration is as low as 100 molecules/ml of air. A single female moth releasing its pheromone downwind from a particular site will produce what Wilson (1972) has called 'an active air space' several kilometres long and over a hundred metres in diameter. Any male entering this active space will then turn upwind and fly towards the female. The size of the active space will vary with the wind velocity, an increase in velocity decreasing its volume. For a chemical to be active in such a system, it must be highly volatile and of a relatively low molecular weight. Indeed, most sex pheromones fall into this category in being hydrocarbon derivatives of carbon number between C_5 and C_{20} and of molecular weight between 80 and 300.

By contrast to such airborne pheromones, those used by aquatic organisms clearly have to be less volatile, of a higher molecular weight and presumably water soluble. Their effectiveness must depend on their rate of diffusion in water, this being speeded up if they are placed in natural or artificially produced currents. In keeping with these ideas, it should be noted that some waterborne pheromones are protein in nature. This is true of the female substance produced in *Volvox* (Chlorophyta) (Starr, 1968) and of the substances controlling the attraction and settling of larvae of barnacles *Balanus balanoides* (Crisp and Meadows, 1962). Other waterborne signals are steroidal in nature. Thus 17α,20β-dihydroxy-4-pregnen-3-one, a sex pheromone of *Carassius duratus*, synchronizes male–female spawning readiness in this goldfish (Dulka *et al.*, 1987).

To the biochemist, one of the most intriguing aspects of animal pheromones is their biosynthetic origin. In the case of insects, it is probable that many are synthesized *de novo* within the animal body from simple starting materials. Others could, however, be obtained from plant sources and used directly or modified biochemically before use. In the case of animal defence compounds (allomones), there are many examples where arthropods have taken over plant chemicals and used them for protection. The very fact that animals make use of plant toxins in this way argues in favour of a function for these substances in the plants themselves. Animals are hardly going to borrow or adapt chemical defences from plants unless they have already proved of value as a defensive barrier in the plant to herbivore attack.

One particularly interesting group of plant substances used by animals are the alkaloids, compounds well known for their physiological activities in animals. There is good evidence that plant alkaloids are used by some animals for defence, e.g. the *Senecio*-feeding cinnabar moth protects itself from avian predators by accumulating alkaloids at the larval stage in its tissues (see p. 93). Another example discussed in this chapter concerns aristolochic acid, which is used by the butterfly *Pachlioptera aristolochiae* for the same purpose. Even more remarkable is the fact that certain animals appear to mimic plants in synthesizing their own alkaloids as defensive agents. This is true of millipedes, ants, ladybirds, water-beetles and frogs.

One other group of defensive substances which may be both of dietary origin and the subject of direct synthesis are terpenoids. The larva of the saw-fly, for example, has the disconcerting habit of discharging at its predators an oily effluent of dietary-derived monoterpenes, previously stored in a pouch of the foregut. By contrast, the meloid beetle *Lytta vesicatoria* in order to discourage its attackers exudes a terpenoid toxin, synthesized *de novo*, by bleeding at its knee joints. One final group of defence compounds, discussed later in this chapter, are quinones which are, like the toxin of the meloid beetle, synthesized by the animals. These substances are produced as 'hot secretions' by the bombardier beetle in special reactor glands, as the need arises.

Information on pheromones and defence substances has accumulated rapidly, so much so that it is impossible in the space available to present a completely comprehensive account. It is only possible to illustrate the situation here with selected examples. More detailed information on insect pheromones is available in the books of Ritter (1979), Birch and Haynes (1982) and Bell and Carde (1984). Mammalian pheromones are discussed by Albone (1984) and Stoddart (1980a,b). Defence substances are reviewed by Schildknecht (1971), Eisner (1980), Blum (1981), Pasteels *et al.* (1983) and Prestwich (1983, 1986). Insect substances are also dealt with extensively in Rockstein (1978).

II. Insect Pheromones

A. Sex Pheromones

The term sex pheromone refers to a compound liberated by a female, with the dual purpose of both attracting the male from a distance and also of inciting it to copulation when

Table 8.1 Structures of some typical aliphatic insect sex pheromones

Structure and name[a]	Sex	Organism
$CH_3(CH_2)_3CO_2H$ Valeric acid	♀	Sugar-beet wireworm (*Limonius californicus*)
$CH_3CO(CH_2)_5CH{=}CHCO_2H$ (*E*)-9-keto-2-decenoic acid	♀	Honey-bee (*Apis mellifera*)
$CH_3(CH_2)_2CH{=}CH(CH_2)_7OAc$ (*Z*)-8-dodecenyl acetate	♀	Oriental fruit-fly (*Grapholitha molesta*)
$CH_3CH_2CH{=}CH(CH_2)_{10}OAc$ (*Z*)-11-tetradecenyl acetate (*E*)-11-tetradecenyl acetate	♀	Oak leaf roller-moth (*Archips semiferanus*)
$CH_3(CH_2)_{15}OAc$ Hexadecanyl acetate; $CH_3(CH_2)_4CH{=}CH(CH_2)_{10}OAc$ (*Z*)-11-octadecenyl acetate	♂	Butterfly (*Lycorea ceres ceres*)
$CH_3(CH_2)_9CO(CH_2)_3CH{=}CH(CH_2)_4CH_3$ (*Z*)-6-heneicosen-11-one	♀	Douglas fir tussock-moth (*Orygia pseudotsugata*)

[a] In the older literature, *cis-* and *trans-* are used instead of (*Z*) and (*E*) to indicate differences in stereochemistry around the double bond.

at close quarters. The same term also applies to substances produced by the males to excite females. Such compounds are sometimes referred to as aphrodisiacs, although this term strictly speaking applies to drugs which excite venereal desires in man. Sex pheromones are probably the most widely studied group of insect allelochemics and they have now been recognized and characterized in many different species. In the case of the Lepidoptera, pheromones have been identified in over 500 species and new ones are being reported almost daily.

In terms of chemical structure, the simplest sex attractant is valeric acid, the female pheromone from the sugar-beet wireworm. The majority, however, are long chain unsaturated alcohols, acetates or carboxylic acids (see Table 8.1); 77% of lepidopteran species have this kind of pheromone. One of the best known is undoubtedly 9-keto-2-decenoic acid, or the queen bee substance, which attracts the male drones to mate with the queen bee. It is, however, only one of 32 compounds of similar structure present in the head of the queen bee. The related 9-hydroxy-2-decenoic acid, for example, is also an active compound, causing clustering and stabilization of the worker swarms.

Cyclic structures occasionally act as pheromones, some typical examples being illustrated in Fig. 8.1. The cyclic pheromones of the pine bark beetles, which have already been described earlier in Chapter 4 (Section VI), are further examples. The male swift moth, *Hepialus necta*, releases the pyran shown in Fig. 8.1, together with two more complex structures, to attract the female after sunset, at the same time displaying its big hind-leg scale-brushes; this male attractant is reported to smell of wild strawberry or pineapple (Sinnwell *et al.*, 1985).

Another cyclic structure, *R*-mellein, is reported to be the male wing-gland pheromone of the bumble bee wax moth, *Aphomia sociella* (Kunesch *et al.*, 1987). In this case, the pheromone appears to be of fungal origin, since a fungus known to produce *R*-mellein,

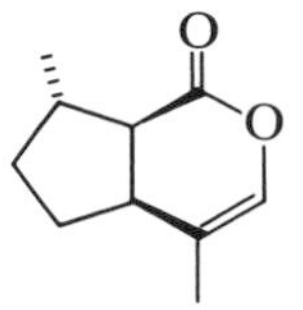

nepetalactone
(aphid: *Myoura viciae*)

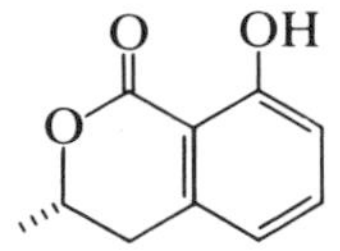

R-mellein
(wax moth: *Aphomia sociella*)

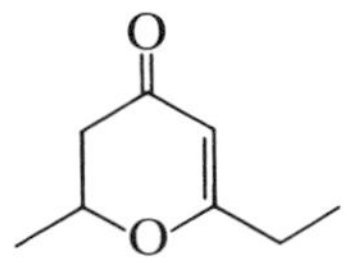

6-ethyl-2-methyl-3,4-dehydro-2H-pyran
(swift moth: *Hepialus necta*)

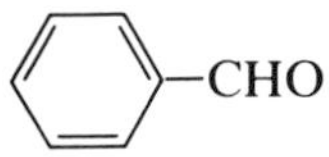

benzaldehyde
(moth: *Leucania impuris*)

Figure 8.1 Structures of some sex pheromones which are cyclic

namely *Aspergillus ochraceous*, has been detected in the intestines of the last instar larvae and also in the bumble bee nest on which the larvae feed.

Perhaps the most unexpected insect pheromone is nepetalactone (Fig. 8.1), which has been identified together with the corresponding lactol in the hindleg secretion of the female vetch aphid, *Myoura viciae* (Dawson *et al.*, 1987). Nepetalactone was first found in plants, in *Nepeta cataria*, and is notable because it is a powerful attractant for cats (see Section IVB). This is a clear example of the same molecule having quite unrelated functions in the plant, insect and mammalian kingdoms.

Pheromones occur in very low concentration in insects and many specimens may be needed when isolating and identifying them. Each female Douglas fir tussock-moth contains about 40 ng of pheromone in the abdominal tip and 6000 insects were required in order to isolate enough material for characterization. In the case of the pink bollworm moth, nearly a million virgin female moths were extracted to yield 1.5 mg of its pheromone.

In any one insect, the chemical structure of the main pheromone is usually very specific and small changes in the molecule normally destroy or diminish the activity. In the case of most hydrocarbon pheromones, there is a single isolated double bond in the structure: its position and stereochemistry (Z- or E-) is vital to activity. This has been demonstrated in the case of the female pheromone of the cabbage looper, *Trichoplusia ni*, where a range of synthetic analogues have been tested; none has as much activity as the natural compound and most are completely inactive (Jacobson *et al.*, 1970). At first, it appeared that (Z)-7-dodecenyl acetate was the only pheromone in this insect, but more detailed examination of the female gland extract revealed five other minor components (Fig. 8.2). These five trace compounds seem to be mainly concerned with eliciting wing pencil display in the male, with the major component still acting as the primary attractant (Bjostad *et al.*, 1984).

In general, each insect species has its own special pheromone blend (see Fig. 8.2), but occasionally those of related species can be rather similar. This is true of the two tortricid

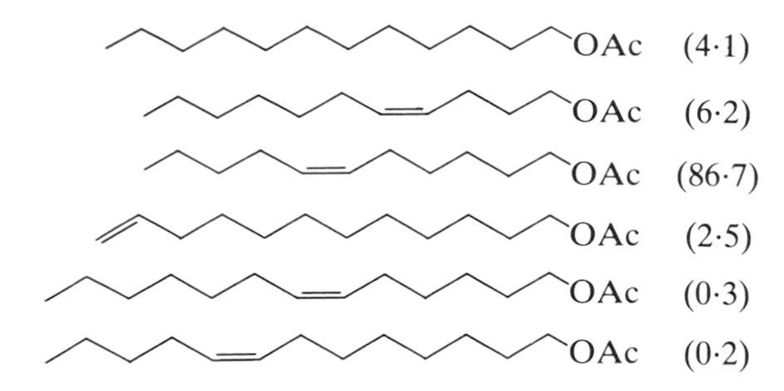

Figure 8.2 Pheromone blend of the female cabbage looper moth (with percentages present in parentheses)

moths (*Clepis spectrana* and *Adoxophyes orana*), the females of which both use (Z)-9- and (Z)-11-tetradecenyl acetates. Reproductive isolation is achieved by the fact that males respond to different ratios of the two components; the females of *C. spectrana* use a 1:3 blend, while the females of *A. orana* release a 3:1 blend. Again, in the case of eight European ermine moths, different blends of eight pheromones are employed variously by the females to attract the males for mating (Lofstedt *et al.*, 1986).

Although the structural requirements for sex pheromonal activity are relatively rigid, it is possible that completely unrelated compounds can produce the same signal. Evidence of this has been obtained with the male of the American cockroach, *Periplaneta americana*, which is sexually excited by compounds present in plant extracts as well as by the natural female pheromone (Bowers and Bodenstein, 1971). One such active substance, present in gymnosperms, was identified as D-bornyl acetate (for structure, see Fig. 8.3), which is active at a concentration of 0.07 mg/cm^2. It is noteworthy that the L-derivative has only a hundredth of the activity of the D-isomer, so that stereochemistry is in this case very important. Angiosperm species also give off volatile compounds with pheromonal effects on this cockroach—a survey of 100 such species showed eight with unknown active constituents. Purely synthetic organic substances can also sometimes mimic the activity of the natural

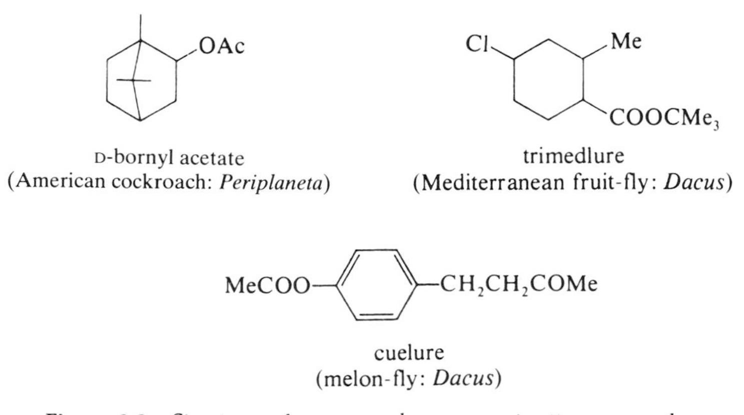

Figure 8.3 Structures of some sex pheromone mimetic compounds

pheromones and two which have been used commercially are trimedlure, which is active in the Mediterranean fruit-fly, and cuelure, which attracts males of the melon-fly, *Dacus cucurbitae*. The structures of these two latter molecules are shown in Fig. 8.3.

A more specific role for the plants on which insects live and breed has been discovered in the chemistry of sex attraction in some instances. Thus the production of female sex pheromones in moths may sometimes be triggered by a chemical signal from the host plant. This happens in the corn budworm, *Heliothis zea*, where the female delays reproduction until a suitable host plant is reached on which to deposit its eggs. Perception of volatile signals from the corn silk sets off the synthesis of the female sex pheromone and its subsequent release leads to mating. The triggering signal appears to be partly the presence of the plant ripening hormone, ethylene, together with some more specific corn volatiles (Raina *et al.*, 1992). A similar situation is probably present in the alfalfa blotch leaf miner, *Agromyza frontella*, the female pheromone of which is 3,7-dimethylnonadecane. Mating behaviour is determined by this pheromone but the presence of the host plant is also needed (Carrière *et al.*, 1988).

The presence in flies of anti-aphrodisiac pheromones (called abstinones) which are deposited on females by males during mating and turn off other males from later mating is a somewhat controversial subject, since the physical masking of active sex pheromones by inactive compounds has also been proposed. Carlson and Schlein (1991), however, have shown that in the tsetse fly, *Glossina morsitans morsitans*, such an abstinone is present. The compound turns out to be an unusual long chain alkene 19,23-dimethyltritriacont-1-ene. It is present at a concentration of 1–2 μg per male fly and is partially transferred to the female during mating. A dose-dependent anti-aphrodisiac effect was seen in exposed male flies, with 2–4 μg causing 80% loss of copulatory attempts.

Almost all sex pheromones are synthesized within the insect. The aliphatic compounds, which are fatty acid derived (Table 8.1), are formed in the usual way from acetyl CoA and malonyl CoA, followed by suitable modification. For example, the cabbage looper moth pheromones are produced from palmitic or stearic acids, via desaturation, chain-shortening, reduction and acetylation (Bjostad *et al.*, 1984). Only in a few exceptional cases, e.g. the pyrrolizidine-derived molecules of male danaid butterflies (see Chapter 3) and the terpenoids of bark beetles (see Chapter 4), are sex pheromones known to be of dietary origin. In these insects, sequestered plant material is either used directly or after appropriate chemical modification.

Sex pheromones are exploited by man as a means of trapping pest insects (see below) but interestingly he is not the only animal to use this means of catching insects. Bolas spiders are notable for their minimal type of web and the fact that they only capture one type of prey, male moths. Chemical analysis of the spiders has shown that this spider draws the moths to their death by mimicking the odour of the female moth, i.e. by releasing female pheromones. Three compounds, (Z)-9-tetradecenyl acetate, (Z)-9-tetradecenal and (Z)-11-hexadecenal have been identified in spider secretions (Stowe *et al.*, 1987) and the same compounds are pheromones in four moth species they catch. Since each moth tends to have its own species-specific pheromone blend, some variation in the lure of individual spiders might be expected and this seems to be so. A range of different moths are indeed trapped in this sophisticated way by these remarkable predators.

The practical application of sex pheromones to the problems of pest control has been developed widely, especially in the United States. Broadly, there are two approaches: one is to set up traps for the males of the species by releasing quantities of the female pheromone in areas where pest control is needed. This effectively prevents the males orientating themselves to the natural pheromone released by females and thus arrests mating. The quantities of pheromone needed in traps may be quite small. Field trials show that for the cabbage looper, 17-mg samples placed in 100 positions within a 27-m^3 plot prevent males from orientating themselves to living females. An alternative approach is to release more unspecific volatile chemicals which mask the effect of the pheromone and interfere with the signal.

While theoretically some insect pests could be completely eliminated by trapping procedures, in practice there are many factors limiting the effectiveness of sexual lures. The most widespread use of synthetic attractants to date has been in trapping insect samples for detecting the build-up of infestations and the size of a given insect population. This information can then be used to determine what control measures are to be taken, such as spraying the area with conventional pesticides.

In at least two instances, field trials have confirmed the feasibility of using pheromones directly for pest control. Thus Gaston *et al.* (1977) have successfully controlled the pink bollworm, *Pectinophora gossypiella*, the larvae of which attack the cotton crop, by applying the synthetic gossyplure, which disrupts normal pheromonal communication between adult moths. The cost of the operation, which involved the controlled release of a pheromone mixture diluted with hexane from special fibre containers, was comparable to that of applying conventional insecticide to the crop. Another synthetic material, multilure, has been incorporated into sticky traps and utilized effectively to control the spread of the European elm bark beetle, *Scolytus multistriatus*, in the eastern United States (Lanier, 1979). Mass trapping of the beetle, however, must be accompanied by the treatment of diseased trees so that they do not provide feeding sites for the small fraction of the beetle population which escapes the traps. The destructive effect of the beetle on the elm tree is due to the fact that it is the principal vector for a pathogenic fungus; the toxins produced by this fungus and their effects on the tree are described more fully in Chapter 10 (see p. 286).

B. Trail Pheromones

Trail pheromones, as the name implies, are used by social insects to lay down an odour trail which other members of the species can follow to guide them from the nest to a food source and back again. They are characteristically employed by ants, bees and termites, which produce them in a variety of special glandular tissues.

Chemically, trail pheromones are of a variety of structures (Fig. 8.4). That produced by the leaf-cutting ant, *Atta texana*, is the highly active substance, methyl-4-methylpyrrole 2-carboxylate (Tumlinson *et al.*, 1971). This compound is detected by ants at a concentration of 0.08 pg/cm of trail which is equivalent to 3.48×10^8 molecules/cm of trail. On this basis, it is possible to calculate that 0.33 mg of the substance would be enough to draw a detectable trail completely around the world! The biosynthetic origin of this

methyl 4-methylpyrrole 2-carboxylate (leaf-cutting ants: *Atta*)

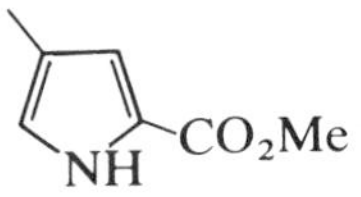

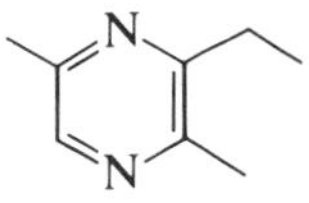

3-ethyl-2,5-dimethylpyrazine (red ants: *Myrmica*)

$$CH_3(CH_2)_2(CH{=}CH)_2CH_2CH{=}CH(CH_2)_2OH$$

(*Z*,*Z*,*E*)-3,6,8-dodecatrien-1-ol
(termites: *Reticulitermes*)

$$CH_3CH_2CH(CH_3)\overset{Z}{=}CH(CH_2)_2CH(CH_3)CH_2CH\overset{E}{=}CH(CHCH_3)_2CH_2CHO$$

faranal
(Pharaoh's ant: *Monomorium*)

Figure 8.4 Insect trail pheromones

very active pheromone is not yet clear but it could conceivably be formed by bacterial action in the gut on dietary tryptophan. The same pyrrole has been identified as a trail pheromone in a second species, *Atta cephalotes*, and it may well be present in other ants of the tribe Attini. Thus all but one of 12 such ant species tested (Robinson *et al.*, 1974) followed a trail of this substance, while non-attine ants completely ignored it.

The biochemistry of leaf-cutting ants is also of great interest from other points of view (see Martin, 1970). Besides producing these interesting trail pheromones, ants also synthesize three hormonal substances which they use to control the growth of the fungal colony which they supply with plant material and on which they ultimately feed. One chemical, indoleacetic acid, the auxin of higher plants, is supplied by the ant to encourage the growth of the fungus. A second substance, myrmicacin, $CH_3(CH_2)_6CHOHCH_2CO_2H$, is used to prevent the growth of undesirable (i.e. foreign) fungal spores. A third chemical, phenylacetic acid, $PhCH_2CO_2H$, is employed to keep the plot free from bacteria. All three control substances are synthesized in the metathoracic glands and are sprayed by the ant continuously onto the fungus and distributed over the whole nest. In this remarkable symbiotic association between the insect and the fungus, the ant uses chemical herbicides to control the growth of other undesirable micro-organisms on its fungal colony.

A trail pheromone from the venom gland of eight species of red ant (*Myrmica*) has been identified as 3-ethyl-2,5-dimethylpyrazine (see Fig. 8.4). All these different species thus utilize the same chemical signal, and the same compound has also been found in a single species of leaf-cutting ant, *Atta sexdens* (Evershed *et al.*, 1982). The pheromone of Pharaoh's ant, *Monomorium pharaonis*, has also been chemically investigated because this tropical insect has become something of a pest in bakeries and hospitals. Three pyrrolidine-based alkaloids have been identified in the ant's secretion but a fourth compound, a long chain aldehyde faranal (Fig. 8.4), seems to be the most active ingredient.

Some Lepidopteran larvae, such as the eastern tent caterpillar, *Malacosoma americanum*, live in colonies and lay trails to suitable food sources. The pheromone in this case has been characterized as 5β-cholestan-3,24-dione, a steroid presumably readily synthesized from the animal's cholesterol. This compound is just as active as ant pheromones, the caterpillar larvae having a threshold sensitivity to it at a concentration of 10^{-11} g/mm of trail (Crump *et al.*, 1987).

Termites also use odour trails (Prestwich, 1983) and a pheromone of *Reticulitermes virginicus* has been characterized as 3,6,8-dodecatrienol (Fig. 8.4). This alcohol occurs in the fungus-infected wood that the termites feed on and it is known to be synthesized by the fungus. However, present evidence suggests that in spite of the fact that the termite could obtain the alcohol from dietary sources, it does synthesize its own pheromone.

One insect which undoubtedly obtains its trail pheromone directly from plants is the honey bee, *Apis mellifera*, which uses the monoterpene geraniol as a trail substance. Geraniol is collected from flower scents, concentrated within the bee's body and then exuded when required as a food guide. Some of this geraniol is converted to a second pheromone, (Z-)-citral, in the bee's gland (Pickett *et al.*, 1981). Another compound, possibly of plant origin, is similarly used by bees of the genus *Trigona* and is benzaldehyde, which may be derived from the cyanogenic glycoside prunasin. This aldehyde, with its familiar almond-like odour, is an almost ideal pheromone for food trails, since it loses it potency after a time due to oxidation to benzoic acid, which is inactive. Unless reinforced, benzaldehyde trails can be timed to decay in potency and fade just as the food source for which they are a guide is used up by the insect.

C. Alarm Pheromones

Most insect alarm pheromones are produced and delivered from the mandibular or anal glands or from the sting apparatus. Production is often related to that of defence substances. In combat among social insects, the contents of the mandibular glands are discharged through the mandibles onto the enemy which is thus 'tagged' as an aggressor. Alarm is communicated to other members of the society by diffusion of the pheromonal vapours in the air.

The sting apparatus of bees and wasps contains several glands that produce alarm pheromones. The poison gland itself often produces alarm chemicals which are discharged with the venom. Wasps of the genus *Vespa* spray venom containing an alarm substance while honey bees leave traces of isopentenyl acetate and (Z)-11-eicosen-1-ol at the sting site, which induces other bees to sting at the same location. Comparison of European and Africanized honey bees show that the more active defensive behaviour (measured in number of stings) of the latter so-called 'killer bees' is correlated with the release of larger amounts of alarm pheromone. By contrast, the major sting pheromone isopentenyl acetate does not differ in amount in the two races of bee (Collins *et al.*, 1989). In other insects, the same substance can function both for alarm and in defence. This is true of the formic acid produced by ants of the genus *Formica*.

Most alarm pheromones identified in insects are of a relatively simple structure (Fig. 8.5). In certain ants, they consist of simple hydrocarbons, such as undecane, tridecane

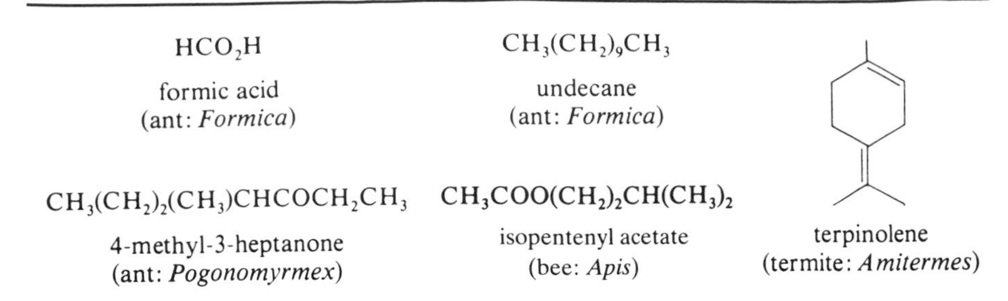

Figure 8.5 Structures of insect alarm pheromones

and pentadecane. In other species, the same hydrocarbons occur with aldehyde or ketonic functions. Essential oil components, including citronellol, citral, α-pinene, terpinolene and limonene, have been implicated as alarm odours in species of the Formicinae, Hymenoptera, Isoptera and Myrmicinae. More complex terpenoids, such as the unsaturated monocyclic sesquiterpenes germacrene A and (*E*)-β-farnesene, have been implicated as alarm pheromones in aphids. These sesquiterpenes are very labile; this has the advantage that they break down soon after the predator has moved on, so that aphids can reinfest the feeding site almost immediately.

Alarm pheromones appear to be the least specific of the various volatile hormones in insects. Thus different species of the same genus or even species in different genera may employ the same alarm signal. Nevertheless, ants have a high degree of olfactory acuity and display considerable sensitivity to their own alarm pheromones, closely related chemicals having little effect. The myrmicine ant *Pogonomyrmex barbatus*, for example, is 10,000 times less sensitive to 2-methyl-3-heptanone, an isomer of the natural pheromone 4-methyl-3-heptanone, than to the natural compound itself.

The amount of information conveyed by the alarm pheromones in ants can be quite considerable. For example, the weaver ant *Oecophylla longinoda*, when attacked by a predator, releases a pheromone bouquet containing four oxygenated hydrocarbons of differing volatilities. Other ants approaching the scene of the attack are first alerted by the most volatile component 1-hexanal, $CH_3(CH_2)_4CHO$, and then attracted towards the scene by the related alcohol, 1-hexanol. Finally, as they move in to join the first ant, they are provoked to attack the predator by a 'biting marker' 2-butyl-2-octenal; a fourth component, 3-undecanone, acts at the same time as a short-range orientation signal. Thus attack by predators leads to a well-orchestrated sequence of alarm behaviour which provides an effective defence of the ant colony (Bradshaw *et al.*, 1979).

While ants may be predated upon by other animals, they themselves have evolved well-defined predator–prey relationships with termites. In these interactions, chemical signals are of utmost importance for the purposes of recruitment and alarm. Indeed, recruitment pheromones may be released by ants in order to co-ordinate a foraging attack on a termite colony and the same signals are picked up by the termites, giving them advance warning of the attack. In this situation, in certain ant species, e.g. *Decamorium uelense*. the chemical recruitment signals have evolved in such a way that the termite prey are quite insensitive to the volatiles produced. A small change in the chemical structure of the pheromone, e.g. from an aliphatic aldehyde to the related alcohol, is sufficient to do this. This phenomenon has been aptly described as 'chemical crypsis' in analogy to the more frequently encountered visual crypsis (Baker and Evans, 1980).

A related chemical mimicry has been observed in the staphylinid beetle *Trichopsenius frosti*, which, like the above ants, preys on termite colonies. In this case, the beetle produces a complex series of cuticular hydrocarbon components which are identical in all respects to those found in the cuticle of the termite host (Howard *et al.*, 1980). This predatory beetle, by adopting the same chemical coating as its prey, has become a proverbial wolf in sheep's clothing.

III. Mammalian Pheromones

A classic response to danger is shown by the skunk *Mephitis mephitis*, which, when frightened, releases a secretion of scent from the anal glands while in a hand-stand posture. The resulting stink not only acts as a warning of danger to other skunks in the neighbourhood, but is also an important antipredator mechanism. It is, for example, highly effective in driving away human beings. The active principles of the revolting skunk odour include three sulphur compounds, crotyl and isopentyl mercaptan and methylcrotyl-sulphide. The release of scent during stress situations is only one of many examples where, in the animal kingdom, chemical communication occurs via olfactory means. The sulphur components of the skunk gland are both pheromones and allomones, and are similar in many ways to the odours produced in insects, already discussed in the previous section. Unlike the situation in insects, however, our knowledge of the chemistry of mammalian pheromones is still relatively primitive and in many cases we can only guess at the nature of the chemicals involved.

Other mammals besides the skunk produce strong odours from scent glands when under stress, e.g. the striped hyaena *Hyaena hyaena*, the house shrew *Suncus murinus* and the black-tailed deer *Odocoileus hemionus*. When threatened with danger, urine and faeces are often excreted, sometimes as an automatic response but occasionally as a controlled response. Chinchillas and guinea-pigs, for example, deliberately squirt urine at human handlers when disturbed. Practically all chemical signals in animals in fact originate either from the urine or faeces or else from anal glandular exudates. Special glands may be employed for the manufacture of the volatile pheromone, as in the case of the skunk, or the odours are produced in sex accessory glands or in the glands of the skin.

The range of chemical signals used in mammalian communication is very considerable and new examples are regularly being discovered (for review, see Albone, 1984). One of the advantages of an olfactory signal over an auditory or visual one is that the odour persists for some time after the sender has moved on. This is valuable in relation both to warning signals and also in scent marking of territorial rights. The persistence of olfactory signals is probably also advantageous in sexual arousal and in the preparation of both partners for mating.

Some of the complexities of mammalian odour signals used to transmit information concerning sex, age, identity and mood have been learnt from studies of the black-tailed deer. In this animal, six specialized gland areas are involved, the secretions being spread

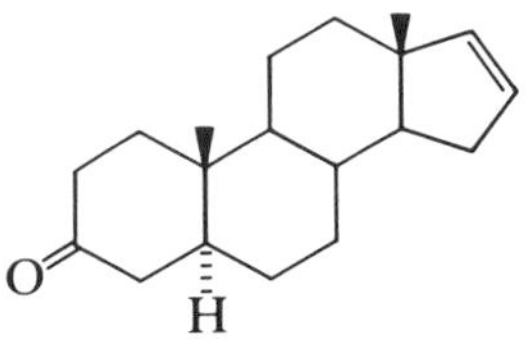

5α-androst-16-en-3-one
(boar: *Sus*)

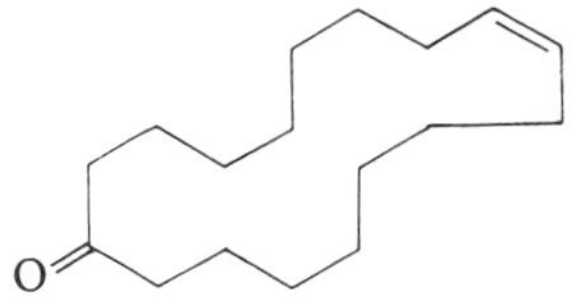

civetone
(civet cat: *Zibeth*)

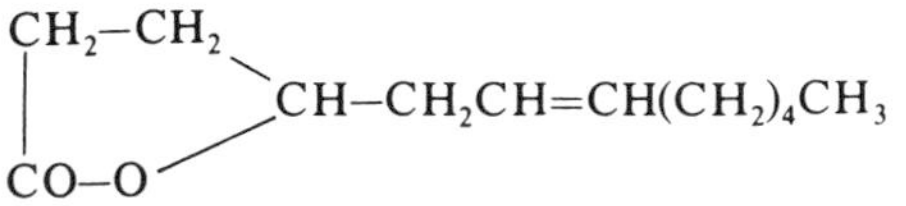

$\overset{CH_2-CH_2}{\underset{CO-O}{|}} \!\!> CH-CH_2CH=CH(CH_2)_4CH_3$

(*Z*)-6-dodecen-4-olide
(black-tailed deer: *Odocoileus*)

$CH_3COCOCH_2CH_2SCH_3$

5-thiomethylpentan-2,3-dione
(hyaena: *Hyaena*)

$CH_3CH=CHCH_2SH$
$CH_3CH=CHCH_2SSCH_3$
$(CH_3)_2CHCH_2CH_2SH$

(skunk: *Mephitis*)

$(CH_3)_2CH(CH_2)_{12}OH$
$(C_2H_5)(CH_3)CH(CH_2)_{11}OH$
$(CH_3)_2CHCH_2CO_2H$
$CH_3(CH_2)_2CH(CH_3)CO_2H$

(pronghorn: *Antilocapra*)

Figure 8.6 Structures of some mammalian odours

by the sender on another part of its own body or the substances being deposited on specific loci within its chosen environment. The secretions on the body of the sender diffuse into the air and are received by other deer of the same herd. The tarsal gland secretion of this animal has been characterized as a γ-lactone (see Fig. 8.6) (Brownlee *et al.*, 1969) which arises from the urine, but in general the chemistry that provides the basis of these various signals still remains to be elucidated.

Most mammals use a similar range of odour signals in their social and sexual life. In sheep, the ram can detect when ewes are ready for mounting by changes in the odour of their urine. This is presumably related to the excretion of increasing amounts of oestrogen at the time of oestrus. In the pronghorn deer *Antilocapra americana*, the male secretes a series of long chain alcohols and short chain organic acids (e.g. isovaleric) in the subauricular glands and uses them for marking vegetation within its territory (Muller-Schwarze *et al.*, 1974). A similar scent marking pheromone in the male Mongolian gerbil *Meriones unguiculatus* has been identified as phenylacetic acid $PhCH_2CO_2H$ (Thiessen *et al.*, 1974). The same compound incidentally has been detected in the exocrine secretion of the stinkpot turtle *Sternotherus odoratus* where its powerful malodorous odour is employed to warn off feeding predators (Eisner *et al.*, 1977).

One of the few cases where chemical signals of sexual arousal have been fully identified is that of the boar. The nature of boar odours was determined some years ago as a mixture of 5α-androst-16-en-3α-ol and the related 3-ketone (for formula, see Fig. 8.6). These two compounds, which are closely related in structure to the male sex hormones androsterone and testosterone, have a strong musky odour. Presumably there is a parallel here with the glandular secretions of the civet cat, civetone, and of the musk ox, muscone, both of which are also musky to smell. Indeed, the structural resemblance is clearly apparent if the civetone molecule is drawn on the same steroid template as that of the boar odour compounds (Fig. 8.6).

Quite inexplicably, the boar odour ketone has recently been detected in trace amounts (8 ng/g fresh wt) in two vegetables, parsnip roots and celery stalks (Claus and Hoppen, 1979). Its identification was based both on radioimmunoassay and combined gas chromatography–mass spectrometry. In keeping with its role as a sexual signal, its presence in these two vegetables was first noticed not by the two male researchers in this investigation but by one of their wives (see below). Celery has some popularity as a 'libido-supporting' vegetable but there is at present no clear connection between this supposed property and the presence of the boar odour compound.

The effectiveness of boar odour in arousing sows can be gauged from experiments where the chemical odour remaining in a pen after removal of the boar was found to be sufficient to induce 81% of females in oestrus to assume the mating stance. The boar odour occasionally taints the meat produced from them and it is significant that this musky taint is much more apparent to women eating the meat than to men. Indeed, it is possible that in human sexual contacts, there are pheromonal interactions based on the excretion of the appropriate male or female steroid hormones from under the armpit or the sexual organs. It is known, at least, that human males and females have recognizably different body odours (Stoddart, 1990).

The presence of 5α-androst-16-en-3-one, part of the boar odour, has been monitored in humans to determine if it has a pheromonal function (Bird and Gower, 1983). The ketone (measured by radioimmunoassay) was detected in the saliva of six out of nine men, at a concentration of between 0.8 and 1.8 nmol/l, whereas it could only be detected in one of four women, at a concentration of 0.83 nmol/l. It was also present, at variable concentrations, in the odour of human armpits (Bird *et al.*, 1985). Male armpits generated from 5 to 1019 pmol over a period of 24 h. In women, it was present but the concentrations were lower, being between 1 and 17 pmol per day. The ketone has a urine-like odour and was described by most women as unpleasant or repellent, so that it seems an unlikely candidate for a sex attractant. The differences in concentration between the sexes could simply reflect differences in the bacterial populations of armpits and the fact that women probably wash under the armpits more frequently than men!

Another interesting group of mammalian pheromones are the warning odours of predators, which can have a frightening effect on their prey. The secretions of the anal sac of seven species of mustelid have recently been compared (Crump and Moors, 1985). A variety of thietanes and dithiacyclopentanes were characteristically present, some of the sulphur compounds being species-specific. In the wolf, sulphur derivatives are not commonly present, the major volatiles being odoriferous alcohols, aldehydes, and ketones (Raymer *et al.*, 1985). The fox, by contrast, has a range of volatile nitrogen- and sulphur-containing compounds in its anal secretion and in its urine (see also below). Rats that have been reared in laboratories retain a hereditary stress response to the predatory odours of the red fox, and provide a means of testing the effectiveness

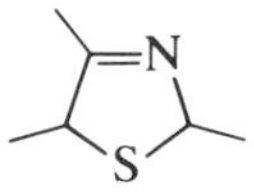

2,5-dihydro-2,4,5-trimethylthiazole

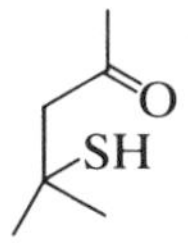

4-mercaptopentan-2-one

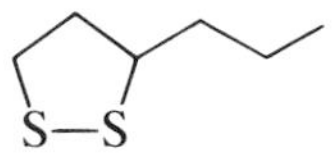

3-propyl-1,2-dithiolane

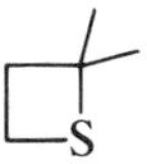

2,2-dimethylthietane

Figure 8.7 Stress-inducing odorants of mammalian predators

of individual components. Nine of the fox compounds—a dihydrothioazole, two cyclic polysulphides, five mercaptoketones, and a mercaptan—were found to induce such a stress response. Two of the most active compounds were a thiazole and a mercaptoketone (see Fig. 8.7), and it has been proposed that these should be used as area repellents to control rats. A similar method for control of snowshoe hares, *Lepus americanus*, has been proposed, using the volatile from the anal sac of the stoat, 3-propyl-1,2-dithiolane, or of the mink, 2,2-dimethylthietane. In bioassay, both compounds were more effective than a range of other warning odour principles from species of *Mustela* (Sullivan *et al.*, 1985).

In mice, chemical communication is important in reproductive physiology and behaviour, and a variety of pheromonal interactions have already been studied extensively (Bronson, 1979). Nevertheless, there still seem to be new pheromones to be found. A series of novel compounds were detected in the urine of female mice which had been brought into oestrus by hormone implantation. Enhanced levels of *n*-pentyl acetate, (*Z*)-pent-2-en-1-yl acetate, *p*-toluidine, heptan-2-one, (*E*)-hept-5-en-2-one, (*E*)-hept-4-en-2-one and (*Z*)-hept-3-en-2-one were observed (Schwende *et al.*, 1984). These may be attractants to male mice, but this has not yet been established. There are two points of general interest. The first is the occurrence in mammalian urine of volatiles that were previously only known to occur in insect systems; e.g. heptan-2-one is an alarm pheromone in ants. The second is that *n*-pentyl acetate and heptan-2-one may be by-products of the metabolism of fatty acids in the liver, and they may accumulate in the urine because of the effect of oestrogens on this metabolism. Thus male mice might recognize females that are in oestrus through the metabolic changes in primary metabolism that occur rather than through the production of specific pheromones.

One last group of odoriferous substances which may be of significance in mammalian interactions are various amines. For example, trimethylamine, NMe_3, which has a strong fishy odour, has been identified as a component of both human menstrual blood and the anal gland secretion of the red fox, *Vulpes vulpes* (see Amoore and Forrester, 1976). Curiously, the same compound occurs to the extent of 400 ppm (Cromwell and Richardon, 1956) in the aptly named plant stinking goosefoot, *Chenopodium vulvaria*. Indeed Linnaeus (1756), who named the plant, clearly recognized the odour and also recorded the fact that dogs become very excited when they approach this plant. In the case of the anal sac secretions of the red fox, two diamines, putrescine and cadaverine, are also present, together with volatile fatty acids (Albone and Perry, 1976). It is possible that all these compounds are formed in the fox from non-volatile precursors by microbial action within the gland. Their value as pheromones in foxes has not yet been fully clarified but they may be sex attractants or factors in group 'recognition'.

IV. Defence Substances

A. Distribution

Chemical defence is well known to operate as a form of protection against predation in many animals. A variety of defence mechanisms have been recognized and many chemical substances have been implicated in such interactions (Table 8.2). In the case of arthropods, which have been most widely studied in recent years, the chemicals may be either synthesized *de novo* by the animal or adapted from dietary sources. Some of the toxins are elaborated in special exocrine glands and others are contained in the blood or gut. Some glandular secretions are ejected with some force, others are sprayed onto the enemy and yet others simply ooze out of the creature. Most of the toxins have a broad spectrum of activity against many different kinds of predator.

From the viewpoint of the phytochemist, the most interesting aspects of arthropod defensive secretions is that, with few exceptions, the compounds present are of the same type known to occur in plants as secondary metabolites. Some of the substances used, such as (*E*)-2-hexenal, benzaldehyde, salicylaldehyde, citral and citronella are in fact widely distributed in plants. Even the mechanism of release may be the same. In plants, HCN is generated by hydrolysis of cyanohydrin glycosides (see p. 84); in larvae of certain chrysomelid beetles, the cyanogenic secretion contains both benzaldehyde and

Table 8.2 Chemical defence substances in animals

Class	Examples	Typical toxins
Fish	Puffer fish	Alkaloids
Amphibia	Frogs, toads, salamanders	Cardiac toxins, peptides, neurotoxins, alkaloids
Reptilia	Snakes	Peptide venom
Arthropoda		
Diplopoda	Millipedes	Alkaloids, quinones, cyanogens
Chilopoda	Centipedes	Alkaloids, quinones, cyanogens
Arachnida	Whipscorpions	Acetic acid, peptides
Insecta	Cockroaches	Aliphatic aldehydes
	Termites	Terpenes, quinones
	Beetles	Steroids, quinones
	Moths, butterflies	Cardiac glycosides, alkaloids
	Ants	Formic acid, terpenes
	Fireflies	Bufadienolides

Table 8.3 Some defence substances synthesized by both animals and plants

Toxin	Animal source	Plant source
Alkaloid: Anabaseine	Venom alkaloid of *Aphaenogaster* ant	One of several alkaloids of tobacco leaf *Nicotiana*
Cyanogenic glycosides: Linamarin and lotaustralin	Cyanide defence of *Zygaena* moths and *Heliconius* butterflies	Toxins of birdsfoot trefoil, clover and other plants
Phenol: Hydroquinone	Defence secretion of waterbeetle *Dytiscus*	Toxin in burs of *Xanthium canadense*
Terpenoid: β-Selinene	In lepidopteran larva *Battus polydamus*	In celery leaves *Apium graveolens*
Amine: 5-Hydroxytryptamine	Barbs of the tiger moth *Arctia caja*	Stinging hairs of nettle *Urtica dioica*

glucose, so that a similar mechanism probably operates (Moore, 1967). Some examples where the same defence substance has been found in both an animal and a plant source are shown in Table 8.3.

Defence secretions usually contain mixtures of chemicals and the toxins present may act synergistically in repelling predators. Volatiles, for example, are more effective when combined with lipophilic components, so that they can be spread over the cuticle of the predator. *n*-Nonyl acetate occurs admixed with formic acid in the defence spray of the carabid beetle *Helluomorphoides*; this greatly increases the burning sensation of the formic acid on the human skin.

Occasionally, chemical defences may have pheromonal properties and it is possible to talk of 'defensive pheromones'. For example, strong odorous substances are transferred by males to females during mating in the butterfly *Heliconius erato*. These are considered to be anti-aphrodisiac pheromones, since they prevent other males from mating subsequently with the same female (Gilbert, 1976). But the same odour could have a more general defensive function, warning off predators from the female. Similarly, male monarch butterflies sprinkle the females with 'love dust' containing pyrrolizidine alkaloids before mating (see Chapter 3). This release of a defensive secretion during mating may protect the insects during the copulation process, when they are often most vulnerable to attack.

Chemical defence has been most widely studied in arthropods and much information is available (see especially Blum, 1981). Something is known of the chemical ecology in the marine world (Naylor, 1984) and alkaloids have been identified in many marine organisms (Fenical, 1986). Here only a small selection of typical examples will be considered under four chemical headings: terpenoids, alkaloids, phenols and quinones.

B. Terpenoids

The lower terpenoids are relatively non-specific toxicants produced in defensive secretions of many insects (Fig. 8.8). Because of their volatility and powerful smell, their

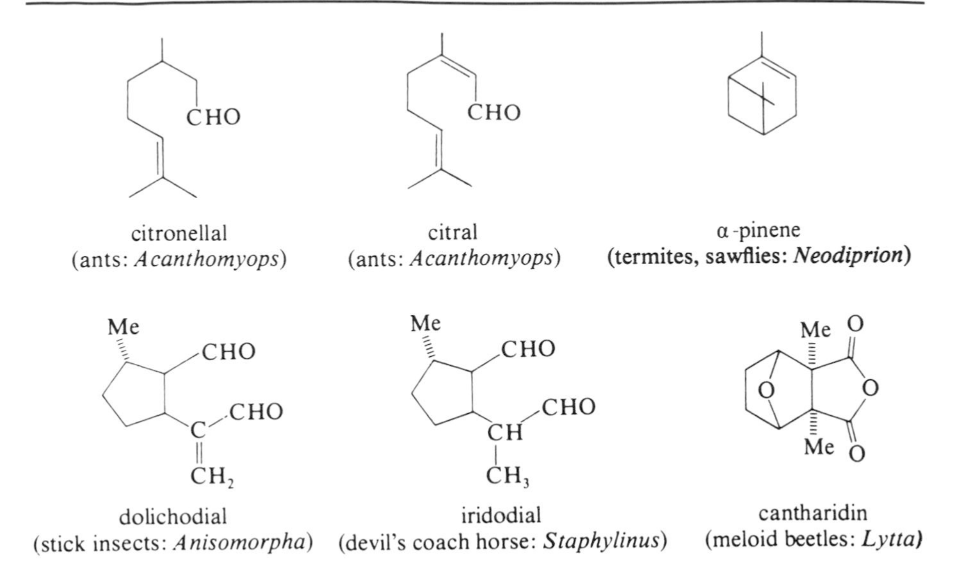

Figure 8.8 Lower terpenes as defence compounds in arthropods

odours may be sufficient to deter the attacker. The vapours may have an irritating effect and the oil on the predator's skin may cause burning and itching. A good example of the use of simple terpenoids in defence is the case of the larvae of the sawfly *Neodiprion sertifer* (Hymenoptera) (Eisner *et al.*, 1974). When disturbed, this insect discharges an oily effluent identical chemically in every way with the terpenoid resin of its host plant, *Pinus sylvestris*. What happens is that the larvae sequester the resin constituents during feeding and store them in two compressible diverticular pouches of the foregut. When approached by a predator, the fluid is discharged and is effective in repelling the majority of such attackers.

Analysis has shown that the same mixture of mono- and diterpenes occur both in the pine resin and in the insect secretion. Compounds present include α- and β-pinene, pinifolic acid, pimaric acid, palustric acid, dehydroabietic acid, abietic acid, neoabietic acid and (−)-pimaric acid. In this mixture, α- and β-pinene are probably the major deterrents, since these compounds are known to be highly obnoxious to most arthropods. The various diterpene acids present in the insect secretion presumably act mainly as a fixative for the two more volatile components. The pine resin is an important defensive secretion of the tree and this sawfly has clearly 'crashed' through the chemical defence of the host plant, at the same time appropriating the very same materials for its own purposes.

However, the food plant can vary in its resin content from 1.5 to 5.26% dry wt (Bjorkman and Larsson, 1991). As a result caterpillars on a pine with low resin acid content are more susceptible to ant attack than those that feed on high-resin acid trees. Furthermore, feeding on a high-concentration pine has a side-effect of reducing the growth rate. Therefore the larvae have to choose between being well defended and growing slowly and lacking defence and growing more quickly.

The sawfly, by adapting the plant toxins directly for its use in this way, might be

Table 8.4 Chemical defence systems of termite soldiers

Termite genus[a]	Chemical classes present
1. BITING AND INJECTING	
Macrotermes	Alkanes and alkenes
Cubitermes	Diterpene hydrocarbons
Armitermes	Macrocyclic lactones
2. POISON-BRUSHING	
Prorhinotermes	Nitroalkene and farnesene
Schedorhinotermes	Hydrocarbon ketones
Rhinotermes	β-Ketoaldehydes
3. GLUE-SQUIRTING	
Nasutitermes	Monoterpenes and cyclic diterpene alcohols

[a] Other termite genera tend to fall within these three classes. For further details, see Prestwich (1983, 1986).

regarded as an evolutionary advanced insect, since it is clearly economical not to have to synthesize defensive toxins *de novo*. Most other insects utilizing mono- and sesquiterpenes for defence appear to make their own toxins from simple starting materials. Thus, it has been shown by radioactive precursor feeding experiments that the walking stick insect *Anisomorpha buprestoides* and the ant *Acanthomyops claviger* produce their terpenes from acetate through mevalonate, according to the usual biosynthetic pathway. The former makes dolichodial, while the latter uses citronellal and citral as defence agents. These three compounds (Fig. 8.8) are typical plant terpenes and, although today they are made by the insects, they are substances which could have been obtained from dietary sources at some earlier stage in the evolutionary history of these insects.

In termites, chemical defence is provided by sterile soldiers, which make up between 10 and 30% of the colony. Defence is their only function and up to 8% of the fresh weight of the insect may consist of defensive chemistry. At least three main types of defensive strategy are adopted (Table 8.4) and a range of chemicals are used, the majority being terpenoid based. Several unique series of cyclic diterpenes, such as the trinervitenes (Fig. 8.9), have been identified in *Nasutitermes* and related genera (Prestwich, 1983, 1986). The second group of termite soldiers (Table 8.4) contain electrophilic lipids (Fig. 8.9) which are contact poisons, and which are liable to be released on conspecific workers as well as

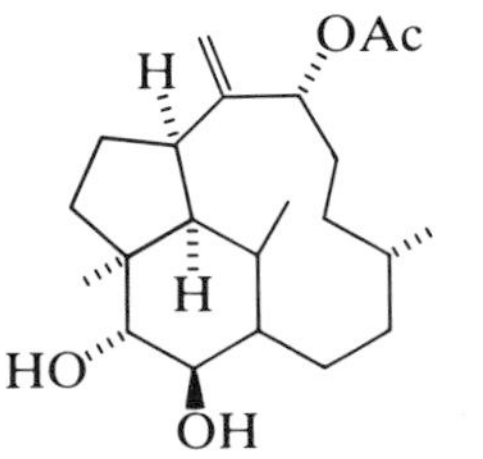

OAc
H
H
HO
OH

(trinervitene: *Nasutitermes*)

$CH_3(CH_2)_{11}CH{=}CHNO_2$
(nitroalkene: *Prorhinotermes*)
$CH_3(CH_2)_9COCH{=}CH_2$
(ketohydrocarbon: *Schedorhinotermes*)
$CH_3(CH)_9COCH{=}CHOH$
(β-ketoaldehyde: *Rhinotermes*)

Figure 8.9 Chemical variation in defensive secretions of termite soldiers

predators. These workers, however, are protected from the poisons, since they can detoxify them via substrate-specific reductases present in their tissues.

The termite defence diterpenes have been shown to be biosynthesized from acetate and mevalonate, by injecting labelled precursors into the abdomens of soldiers (Prestwich, 1986), and another insect terpenoid, cantharidin, is likewise known to be synthesized *de novo* in the beetle which contains it. Cantharidin occurs in the blood of the meloid beetle *Lytta vesicatoria*, but curiously, although present in both adult sexes, is only synthesized by the adult male. The adult female thus must depend on manufacturing it during the larval stage and storing it for subsequent use during adult life.

Cantharidin is a highly irritating material and is, in fact, the basis of the well-known 'aphrodisiac' Spanish fly. Its effect on human sexual performance is entirely due to its vesicant properties, causing marked irritation to the urogenital tract during its excretion. To use it is quite hazardous, since it is significantly poisonous in man, the lethal dose being about 0.5 mg/kg body wt.

In the beetle, cantharidin is released by reflex bleeding from the knee joints and it appears to act as a feeding deterrent to predaceous insects, largely because of its unpleasant taste. The amount present in meloid beetles (0.2 to 2.3% body wt), however, is enough to cause toxic effects when swallowed by vertebrates. Pederin, a second terpenoid more complex in structure than cantharidin, occurs in the blood of staphylinid beetles of the genus *Paederus*. Besides being a vesicant like cantharidin, pederin is also a cytotoxin, being effective in concentrations of 1.5 ng/ml (Blum, 1981).

Finally, it should be pointed out that some arthropods may use mixtures of compounds of different biosynthetic origin in their defence secretions. The devil's coach horse beetle *Staphylinus olens* secretes the terpenoid iridodial, together with 4-methylhexan-3-one, a ketone of fatty acid origin (Fish and Pattenden, 1975). This funereal-coloured beetle defends itself chemically by exuding the foetid-smelling mixture of the above two compounds from glands near the anus; at the same time, it holds its mandibles apart and snaps vigorously at any passing object. As with most arthropods, it combines chemical with physical means of defence.

Irododial of the devil's coach horse and dolichodial, the defence substance of stick insects, both belong to a group of cyclopentanoid monoterpenes which have their analogues among plant constituents. One such plant compound is nepetalactone, a constituent of the mint *Nepeta cataria*, well known for its peculiar ability to excite cats and other felids (Hill *et al.*, 1976). Clearly the function of nepetalactone in *Nepeta* cannot lie alone in its ability to attract cats to the plant. On the other hand, from its close structural resemblance to the above two insect defence compounds, its *raison d'être* could well be its ability to repel insects attacking the plant. This argument has been tested by Eisner (1964), who indeed was able to show that a majority of insects tested (17 out of 24) were repelled by a pure solution of nepetalactone. More work is obviously needed to prove that it has this function in the living plant.

A mixture of cyclopentanoid monoterpenes are also employed for defensive purposes by larvae of the willow leaf beetle *Plagiodera versicolora*. While primarily produced to deter bird predators, these terpenoids are also repugnant to conspecific adults and to the larvae of a second willow leaf beetle *Nymphalis antiopa*. Hence this first willow beetle stakes out

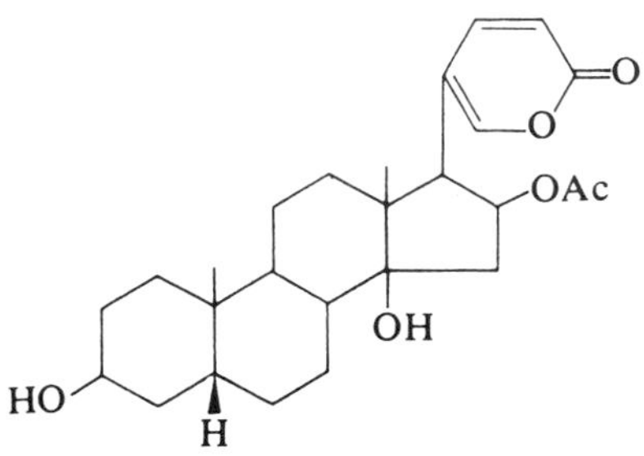

bufotalin
(common toad: *Bufo*)

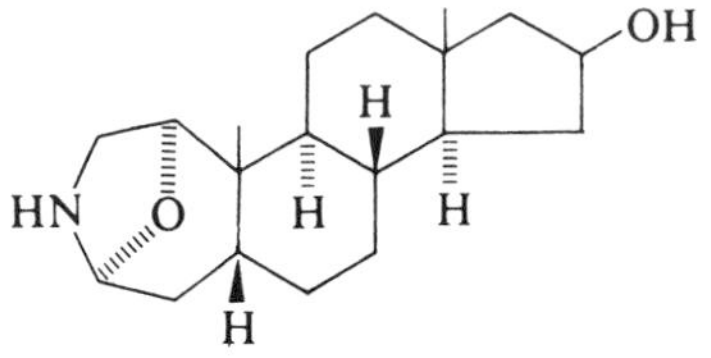

samandarin
(salamander toxin: *Salamandra*)

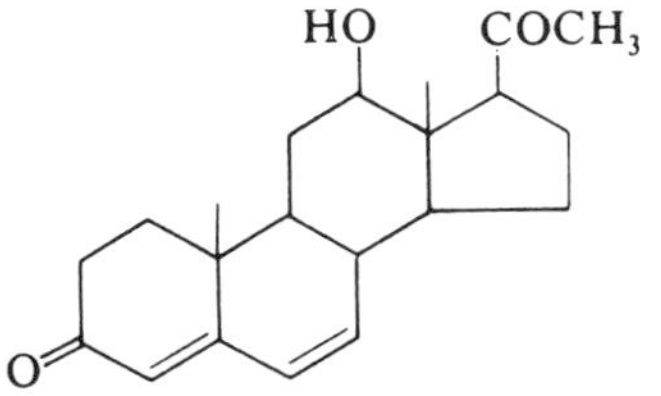

12-hydroxy-4,6-pregnadien-3,20-dione
(Mexican water-beetle: *Cybister*)

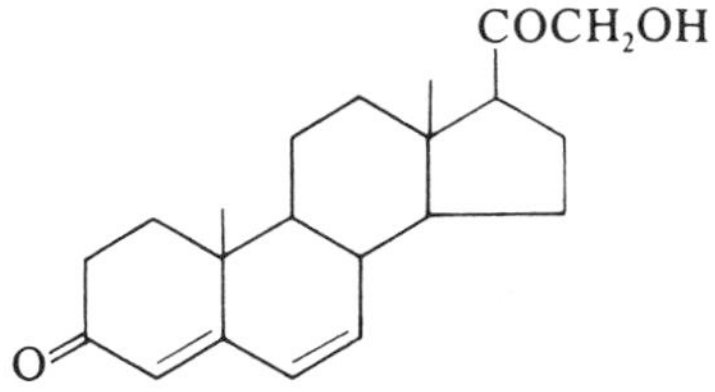

cortexone
(*Cybister*)

Figure 8.10 Steroids as defence compounds in animals

its claim to a particular food plant leaf, the defensive secretion protects it from competition and other herbivores maintain their distance (Raupp *et al.*, 1986).

The defensive role of higher terpenoids in insect–plant interactions has already been discussed under other headings in earlier chapters. There are, for example, the phytoecdysones, insect moulting hormones of plant origin, which can interfere with insect metamorphosis and which are thus potentially dangerous to insects (see Chapter 4). More directly relevant here are the cardiac glycosides which again are of plant origin but are used by insects, especially by monarch butterflies, to protect themselves from bird predation (Chapter 3). A range of other insects, it should be emphasized, use the same toxins of dietary origin in their defence. The grasshopper *Poekilocerus bufonius*, like the monarch butterfly, feeds on milkweed and accumulates the cardenolides. Unlike the monarch butterfly which makes only passive use of the toxins, this grasshopper, when attacked by a bird, ejects the cardenolide material as a noxious foam from special dorsally situated poison glands.

It is worth noting here that a number of animal toxins are closely related in structure to the plant cardenolides. There are the bufogenins, steroidal toxins which also act on the hearts of vertebrates and which are used by frogs and toads as defence agents. Bufotalin (Fig. 8.10), for example, is the bufogenin of the common toad *Bufo vulgaris* and is quite similar in structure to the cardenolide aglycone of milkweed (see Chapter 3). A not unrelated structure, samandarin, is in the defensive secretions in the skin of the salamander. However, unlike the bufogenins and the cardiac glycosides, it acts on the nerves and not on the heart. Another neurotoxin of steroidal structure is holothurin, a compound synthesized by sea cucumbers to deter their predators, especially fish.

Somewhat astonishingly, defensive steroids closely related in structure to the bufogenins of frogs and toads have been detected in the blood of several fireflies of the genus *Photinus* (Eisner *et al.*, 1978). Five have been characterized as esters of 12-oxo-2,5,11-trihydroxybufalin. These steroids, called lucibufagins, protect the fireflies from predation by thrushes, lizards and several mammals. These fireflies, however, do have an enemy in the form of a second group of fireflies, of the genus *Photuris*. The females of *Photuris*, appropriately named as 'femmes fatales', feed on male *Photinus*, luring them at night by imitating the flash signal of the *Photinus* female. These predatory females at the same time acquire the lucibufagins—which they are unable to synthesize themselves—and are thus protected from predation by jumping spiders and thrushes (Eisner, 1980). Hence in the world of fireflies, *Photuris* predates on *Photinus*, borrowing their powerful defensive agents in the same process.

Finally, returning to the arthropods, remarkably rich sources of steroidal defence compounds have been encountered in water-beetles of the subfamilies Colymbetinae and Dytiscinae (Schildknecht, 1971). These particular beetles store up their toxins as a poisonous milk in the prothoracic glands. The activity of the toxin is not appreciated by the predator until it has actually swallowed a beetle. Within a few minutes, it sickens and disgorges its prey. Predatory fish, feeding on the same beetles, fall into a narcotic state and learn from this to avoid future feeding on water-beetles.

The water-beetle toxins are pregnane derivatives; at least 15 structures have been identified variously in a similar number of beetle species. Each Mexican water-beetle, *Cybister tripunctatus*, contains as much as 1 mg of 12-hydroxy-4,6-pregnadien-3,20-dione poison, while each *C. limbatus* beetle contains the same amount of cortexone (Fig. 8.10). In addition to having these pregnane derivatives, one further genus of water-beetle, *Ilybius*, is remarkable in secreting mammalian sex hormones, namely testosterone, dehydrotestosterone, oestradiol and oestrone. Whether these serve a purpose as deterrents in upsetting hormonal balance in mammalian predators is not yet known.

C. Alkaloids

For many years alkaloids were considered to be exclusively plant products, part of the richness and variety of secondary metabolism in the plant kingdom. However, the identification of alkaloids in marine organisms (Fenical, 1986), in a number of arthropods (Jones and Blum, 1983) and in many brightly coloured neotropical frogs (Daley and Spande, 1986) shows quite clearly that the capacity for alkaloid synthesis is not confined to plants. In considering the contribution of alkaloids to animal defence mechanisms, however, it is probably true that plant alkaloids, accumulated following dietary origin, make a major contribution to protection in insects, especially in Lepidoptera.

The best known case is that of the cinnabar and tiger moths which feed on *Senecio*, accumulate pyrrolizidine alkaloids (e.g. senecionine) and are hence highly toxic to their predators; the same is true of certain species of danaid butterflies (see p. 95). At least four other insects feeding on alkaloid-containing plants have been shown to accumulate

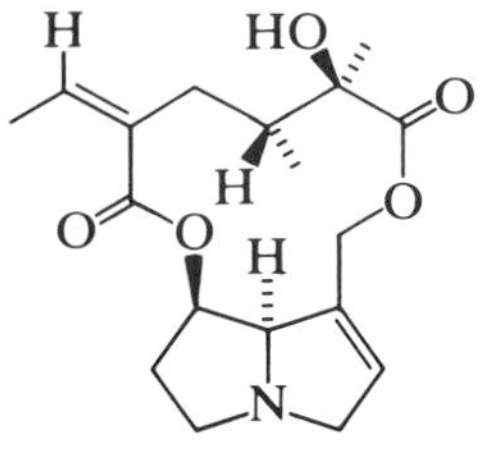

senecionine
(cinnabar moth: *Tyria*)

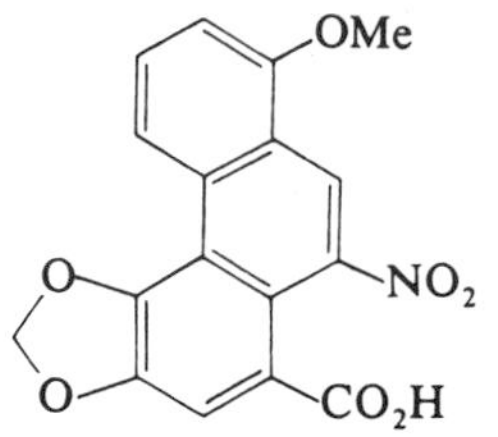

aristolochic acid
(butterfly: *Pachlioptera*)

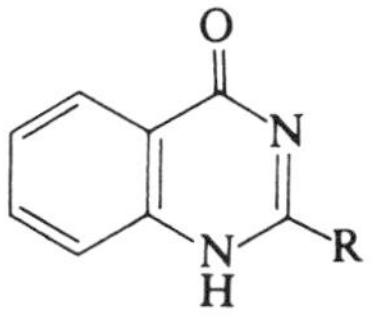

glomerin, R = Me
homoglomerin, R = Et
(millipede: *Glomeris*)

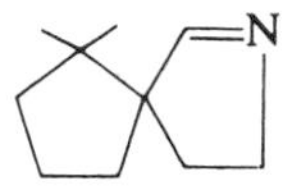

polyzonimine
(millipede: *Polyzonium*)

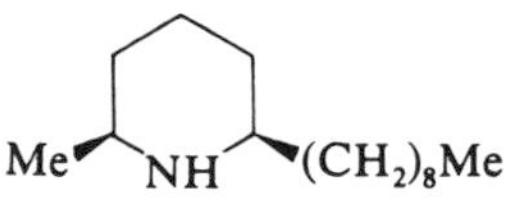

2-methyl-6-nonyl
piperidine
(ant: *Solenopsis*)

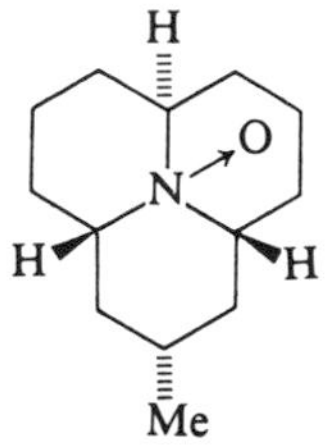

coccinelline
(ladybird: *Coccinella*)

Figure 8.11 Alkaloid defences of arthropods

alkaloids (Rothschild, 1973). Another related example is that of insects feeding on *Aristolochia clamatis* or *A. rotundo* which accumulate the nitro compound, aristolochic acid. The butterfly *Pachlioptera aristolochiae* has been particularly investigated but six other species feeding on Aristolochiaceae do the same thing. Indeed, there is evidence from the work of Rothschild (1973) that ingestion of plant alkaloids or cardiac glycosides is a widespread protective device among Lepidoptera, Hemiptera, Coleoptera and Orthoptera.

Turning now to alkaloids apparently of animal origin, i.e. bases synthesized *de novo* by the insect, these have been reported variously in millipedes, fire-ants, ladybirds and water-beetles (Tursch *et al.*, 1976). Some of the structures of these animal alkaloids are illustrated in Fig. 8.11. Two produced by the European millipede, *Glomeris marginata* are the quinazolinones glomerin and homoglomerin. Their effectiveness as toxins is shown by the fact that ingestion of the millipede causes death in mice and paralysis in spiders. There is good evidence that these two compounds are synthesized by the millipede, since feeding radioactive anthranilic acid to the insect produced labelled alkaloids. Another millipede with an alkaloid as defence is the species *Polyzonium rosalbum*. This substance, polyzonimine, acts as a topical irritant to predating insects, causing cockroaches to scratch themselves.

The potency of the red fire-ant is due to its venom which has haemolytic, insecticidal and antibiotic properties. The major principles of the venom have been characterized as a series of 2,6-dialkylpiperidines, of which the simplest is the 2-methyl-6-nonyl derivative (Fig. 8.11). Such compounds have been identified in the venom of all seven fire-ant species in the genus *Solenopsis* so far studied. A structural relationship of these ant compounds with the highly poisonous plant alkaloid conine (2-propylpiperidine) of the hemlock *Conium maculatum* is clear. The substances are unique in the animal kingdom as the first examples of venom constituents which are not peptides.

One-upmanship is not uncommon in the insect world and this can be attributed to various species of thief ant, which belong to the same genus as the above fire-ants. Thief ants secrete slightly different alkaloids, which have a five-membered instead of a six-membered ring; in *Solenopsis fugax*, the compound is (*E*)-2-butyl-5-heptylpyrrolidine. These substances are dual purpose; they are defensive (as in the fire-ants) and also act as active repellents to other ant species. Thief ants are thus enabled to steal, and subsequently eat, the larvae from proximate nests of other ant colonies, the poison gland secretions preventing the worker ants of these other species from defending their broods (Blum *et al.*, 1980). 2,5-Dialkylpyrrolidines are not entirely restricted to *Solenopsis* species, since they have been reported in the venomous secretions of Pharaoh's ant *Monomorium pharaonis*, an insect which also steals broods from other ants.

Ladybirds (Coccinellidae), when molested, emit haemolymph droplets at their joints (like the cantharidin-producing beetles, p. 230), a mechanism which is an efficient protection against their predators. That the droplets have a bitter taste has been known since the eighteenth century; the fact that alkaloids are present was not firmly established until 1971 (see Tursch *et al.*, 1976). A major component was identified as the alkaloid *N*-oxide coccinelline and a number of related structures were subsequently found in similar secretions.

Coccinelline (Fig. 8.11) is a representative of a new class of alkaloid and a type not known in plants. Biosynthesis has been shown to be endogenous. By feeding the ladybird with labelled 1-^{14}C-acetate and 2-^{14}C-acetate, labelled coccinelline was produced. In laboratory tests, this alkaloid was shown to be effective protection against attack by ants and quails. Many ladybirds are brightly coloured and the occurrence of these alkaloids was found to be correlated with the distribution of aposematic (warning) coloration in these insects.

Defensive alkaloids have been detected in many marine plants and animals (Fenical, 1986) and just one example of the complex utilizations that can occur is that involving the four bipyrroles, tambjamines A–D (Fig. 8.12). These have been isolated from the nudibranchs *Roboastra tigris*, *Tambje eliora*, and *T. abdere* and from the green bryozoan *Sessibugula translucens*. The bipyrroles occur to the extent of 0.45% of its dry weight in the bryozoan and are obtained through predation by the two *Tambje* species, which accumulate levels of 2.15 and 3.42% of their dry weight, respectively. These two animals are then further predated upon by the large carnivorous *R.tigris*, which tracks down the

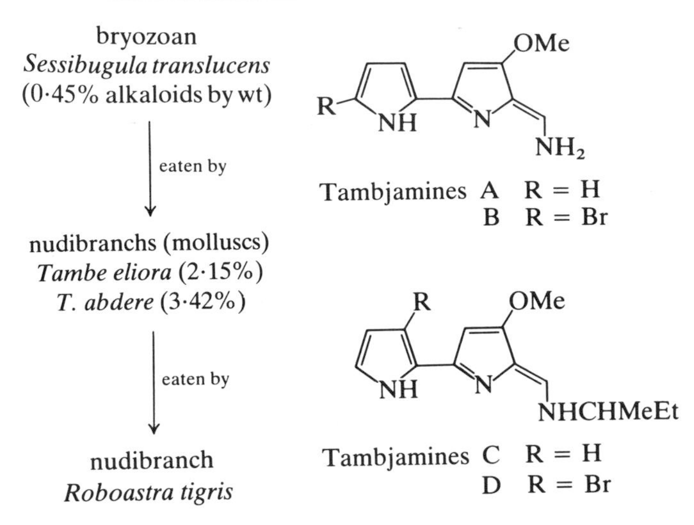

Figure 8.12 Four defensive alkaloids pass through a food chain in marine animals

Tambje prey through a slime trail that they leave and which contains the alkaloids. When attacked, *T. eliora* protects itself by producing a yellow mucus from goblet cells in the skin, but the other species, *T. abdere*, has no such defence and is the preferred diet of the predator (Carté and Faulkner, 1983). The toxic alkaloids are thus utilized by each animal in turn in the food chain to protect them from other predators or from unwanted microbial infections. The structures of the bipyrroles are reminiscent of the red bacterial pigment prodigiosin, and it is possible that they are produced within the bryozoan by some associated bacterial symbiont.

D. Phenols and Quinones

The most remarkable use of phenols in the chemistry of animal defence is that exhibited by the bombardier beetle, *Brachynus*, a creature commonly found on chalkland in Europe. This animal, when endangered, discharges a hot explosive cloud of toxin in the direction of the attacker. This unique system of defence, involving the production of secretions as hot as 100°C, is due to the beetle initiating at the moment of discharge a reaction between a phenol substrate, hydroquinone, H_2O_2 and the enzyme catalase (Fig. 8.13). A highly exothermic reaction occurs, with the phenol hydroquinone being oxidized to benzoquinone, a major product of defence. The reaction can occur with explosive force, the noise produced having been likened to the report of a pistol. Benzoquinone has a highly irritating vapour, producing damaging effects on eye tissue, so it is a simple but effective weapon.

The use of 'hot' quinones in this way is not restricted only to the bombardier beetle and certain other carabid beetles not related to *Brachynus* have glands for producing these explosive discharges (Eisner, 1980). Quinones, produced by more conventional means, also occur quite widely in defence secretions of arthropods and through their persistent

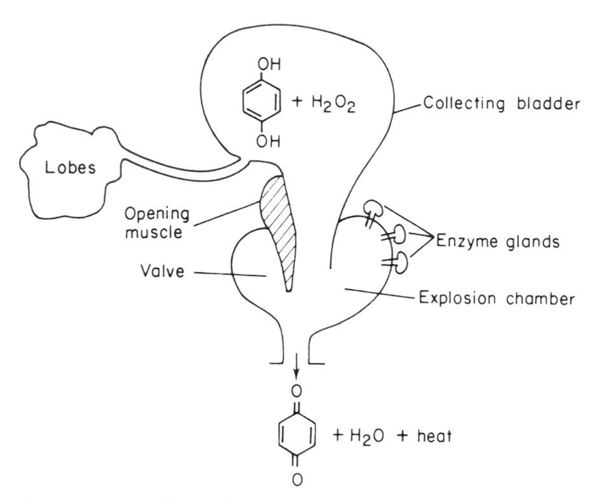

Figure 8.13 The defensive organ of the bombardier beetle

and disagreeable odours provide repellency to attackers. In the case of black beetles (Tenebrionidae), quinones are invariably present. In a survey of 147 species of beetles from 55 genera, Tschinkel (1975) reported that all contained 2-methyl and 2-ethylbenzoquinone, while rather few contained quinone itself, the major weapon of the bombardier beetle.

Quinones also occur variously in arachnids, millipedes, earwigs and termites. Beside quinone itself, several other simple substituted derivatives are present, including 2,3-dimethyl and 2,3,5-trimethyl benzoquinones. In the case of the termites, quinones when present occur in the frontal glands of a special caste within the community, the soldier termite, whose task it is to defend the colony. Although simple quinones are reported to occur in several genera of termites, a range of terpenoid defence substances are more characteristic of this insect order (see Table 8.4).

Higher quinones can also on occasion be toxic to insects. It has been known for many years that the cochineal insects *Dactylopius* spp. which feed on cacti produce large quantities of a red dyestuff—cochineal or carmine—the major component of which is the anthraquinone *C*-glucoside, carminic acid. Although cochineal is widely used in colouring food and is familiar to scientists because of its use in the cytological stain acetocarmine, no one has questioned until recently its *raison d'être* in the insect. It now appears from the experiments of Eisner *et al.* (1980) that it is a potent feeding deterrent to ants which would otherwise predate on these scale insects. The effectiveness of carmine as a defence is underlined by its utilization by an opportunist feeder on the cochineal insect. Thus the carnivorous caterpillar of a pyralid moth *Laetilia coccidivora* has overcome the toxic effects of the dyestuff and eats these insects. It sequesters the toxin and then utilizes it when attacked by ants by regurgitating the red dye at the predator.

Phenols, from which quinones can be produced by enzymic oxidation, are themselves relatively toxic molecules and undoubtedly provide a degree of protection to arthropods from predators. Indeed, simple phenols such as *m*- and *p*-cresol and salicylaldehyde have

p-cresol
(ground-beetle: *Calosoma*)

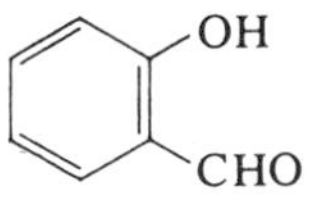

salicylaldehyde
(water-boatman: *Notonecta*)

hydroquinone, R = OH
p-hydroxybenzoic acid, R = CO_2H
(water-beetle: *Dytiscus*)

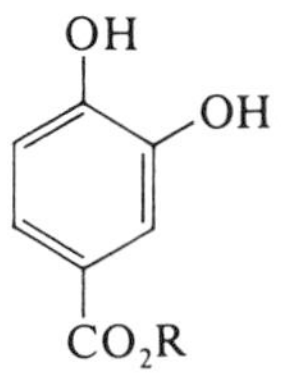

protocatechuic acid methyl
(R = Me) and ethyl (R = Et) esters
(water-beetle: *Dytiscus*)

Figure 8.14 Simple phenols as defensive agents in arthropods

been detected in some defence secretions (Fig. 8.14). While phenol and guaiacol (2-methoxyphenol) are known to be formed in millipedes and hemiptera from tyrosine (Duffey and Blum, 1977), other defence phenols secreted by insects could be of dietary origin, since phenolic derivatives are widespread in plants. It is indicative that 2.5-dichlorophenol, a compound clearly derived from ingested herbicide, was found in the defence secretion of a grasshopper (Eisner *et al.*, 1971).

Phenols are also highly toxic to micro-organisms and one of the more specialized uses of phenols as defensive agents relates to their production in the pygidial defence bladders of water-beetles of the Dytiscinae and Colymbetinae, creatures whose defence mechanisms have already been commented on earlier (p. 232). In these beetles, bodily hygiene is essential to life since although they are at home in water, they have to project their posteriors above the water surface from time to time to replace their air (Schildknecht, 1971). This is only possible as long as their chitin covering does not become wet. If algae, fungi or bacteria become attached to the beetle, water no longer flows off and the subsequent change in surface tension prevents the beetle suspending itself by its steering legs. The air cavity under the wings, essential for respiration, fills up with water and the luckless beetle then suffocates.

The need for an effective means of keeping the body clean is therefore self-evident and fortunately these beetles are excellently equipped for this cleaning operation. Using their rear legs as brushes, they distribute over their rear ends droplets of a secretion from the pygidial glands which are located on either side of the terminal intestine. Injurious micro-organisms are killed by the phenolic compounds (Fig. 8.14) present in the secretion; at the same time they are embedded in a glycoprotein network which is formed as cysteine, also present in the secretion, polymerizes on contact with air. When the beetle springs back into the water, the solidified secretion crumbles away; the beetle is clear of any debris and it can breathe freely once more.

V. Conclusion

Pheromones have now been investigated in many insects (Section II) and the chemical basis is well established (Table 8.5). Simple volatiles are mainly involved, but there are considerable complexities in the blends of related hydrocarbon derivatives that can occur in individual species, especially in the case of sex attractants. While sex pheromones have a high degree of species-specificity, other pheromones may be common to several species or genera. The same pheromone can sometimes serve more than one purpose. This is especially apparent in social insects, such as honey bees and pine bark beetles.

Our knowledge of mammalian pheromones is more limited, because of the experimental difficulties involved. Chemical communication systems can be very complex in foxes, deer, apes and man himself. Pheromonal signals may be released from the skin, scent glands, saliva, faeces and urine and multicomponent mixtures of volatiles have been detected from such sources (Albone, 1984). Chemically, there are straight chain and isoprenoid-based hydrocarbons, small and large ring lactones, mercaptans, dithiacyclopentanes and steroids. In humans, pheromonal interactions are clearly present but their chemical basis is still uncertain.

The greatest variation in chemical defence substances has been observed in arthropods and in this chapter (Section IV) attention has concentrated on these invertebrates rather than on other animals. As a result of much experimentation (Pasteels *et al.*, 1983), it is

Table 8.5 Chemical basis of insect pheromones

Pheromone class	Chemical types
Sex	Unsaturated hydrocarbon blends, cyclic derivatives, terpenoids, alkaloids
Trail	Pyrroles, pyrazines, long chain alcohols
Alarm	Organic acids, hydrocarbons, monoterpenes
Aggregation	Monoterpenes, bicyclic ethers

Table 8.6 Some principles of chemical defence in arthropods

1. Variable natural distribution; more common (?) in large, conspicuous, long-lived animals.
2. Defensive secretions are usually effective against many enemies.
3. Chemical toxins are broadly similar to those found in plants.
4. Chemicals may be sticky, irritant or poisonous in their effects on predators.
5. Mixtures of related chemicals are common and may be synergistic.
6. There may be chemical differences between the sexes, and between larvae and adults.
7. Defensive secretions may have pheromonal properties.

possible to see some general principles emerging (Table 8.6). An astonishing chemical armoury is present in arthropods, and no more so than in the termite soldiers. These insects are walking weapons, capable of biting, snapping, hole-plugging, squirting, oozing, daubing, defecating and exploding at their prey (Prestwich, 1986). The chemicals identified in termites include alkanes, alkenes, diterpenes, β-ketoaldehydes, macrocyclic lactones, nitroalkenes and quinones. This, however, is only a small part of the chemical arms race that is under way in the insect world, waged between one insect and another or between insect and predator. As Blum (1981) has pointed out, the majority of arthropods have yet to be examined for their defensive secretions, so that there is yet much to be learnt about this area of ecological chemistry.

Bibliography

Books and Review Articles

Albone, E. S. (1984). 'Mammalian Semiochemistry: the Investigation of Chemical Signals Between Mammals', p. 360. Wiley, Chichester.

Baker, R. and Evans, D. A. (1980). Chemical mediation of insect behaviour. *Chem. Brit.* **16**, 412–415.

Bell, W. J. and Carde, R. T. (1984). 'The Chemical Ecology of Insects', p. 524. Sinauer Associates, Sunderland, Mass.

Birch, M. C. and Haynes, K. F. (1982). 'Insect Pheromones', 60 pp., Studies in Biology, no. 147. Edward Arnold, London.

Blum, M. S. (1981). 'Chemical Defenses of Arthropods', 562 pp. Academic Press, New York.

Daley, J. W. and Spande, T. F. (1986). Amphibian alkaloids: chemistry, pharmacology and biology. In: Pelletier. S. W. (ed.), 'Alkaloids: Chemical and Biological Perspectives', Vol. 4, pp. 4–274. Wiley, New York.

Eisner, T. (1980). Chemistry, defense and survival: case studies and selected topics. In: Locke, M. and Smith, D. S. (eds.), 'Insect Biology in the Future', pp. 847–878. Academic Press, New York.

Fenical, W. (1986). Marine alkaloids and related compounds. In Pelletier, S. W. (ed.), 'Alkaloids: Chemical and Biological Perspectives', Vol. 4, pp. 276–330. Wiley, New York.

Jacobson, M., Green, N., Warthen, D., Harding, C. and Toba, H. H. (1970). Sex pheromones of the Lepidoptera. Structure–activity relationships. In: Beroza, M. (ed.), 'Chemicals Controlling Insect Behaviour', pp. 3–20. Academic Press, New York.

Jones, T. H. and Blum, M. S. (1983). Arthropod alkaloids: distributions, functions and chemistry. In: Pelletier, S. W. (ed.), 'Alkaloids: Chemical and Biological Perspectives', Vol. 1, pp. 33–84. Wiley, New York.

Martin, M. M. (1970). Biochemical basis of the fungus–attine ant symbiosis. *Science, N.Y.* **169**, 16–19.

Naylor, S. (1984). Chemical interactions in the marine world. *Chem. Brit.* **20**, 118–225.

Pasteels, J. M., Gregoire, J. C. and Rowell-Rahier, M. (1983). The chemical ecology of defence in arthropods. *A. Rev. Entomol.* **28**, 263–289.

Prestwich, G. D. (1983). Chemical systematics of termite exocrine secretions. *A. Rev. Ecol. Syst.* **14**, 287–311.
Prestwich, G. D. (1986). Chemical defense and self-defense in termites. In: Atta-ur-Rahman (ed.), 'Natural Product Chemistry', pp. 318–329. Springer, Berlin.
Ritter, F. J. (ed.) (1979). 'Chemical Ecology: Odour Communication in Animals'. Elsevier, Amsterdam.
Rockstein, M. (ed.) (1978). 'Biochemistry of Insects', 649 pp. Academic Press, New York.
Rothschild, M. (1973). Secondary plant substances and warning coloration in insects. In: van Emden, H. F. (ed.), 'Insect–Plant Relationships', pp. 59–83. Oxford Univ. Press.
Schildknecht, H. (1971). Evolutionary peaks in the defensive chemistry of insects. *Endeavour* **30**, 136–141.
Stoddart, D. M. (1980a). 'The Ecology of Vertebrate Olfaction', 234 pp. Chapman and Hall, London.
Stoddart, D. M. (ed.) (1980b). 'Olfaction in Mammals', 363 pp. Academic Press, London.
Stoddart, D. M. (1990). 'The Scented Ape'. Cambridge University Press.
Tursch, B., Braekman, J. C. and Daloze, D. (1976). Arthropod alkaloids. *Experientia* **32**, 401–407.
Wilson, E. O. (1972). Chemical communication within animal species. In: Sondheimer, E. and Simeone, J. B. (eds.), 'Chemical Ecology', pp. 133–156. Academic Press, New York.

Literature References

Albone, E. S. and Perry, G. C. (1976). *J. Chem. Ecol.* **2**, 101–111.
Amoore, J. E. and Forrester, L. J. (1976). *J. Chem. Ecol.* **2**, 49–56.
Bird, S. and Gower, D. B. (1983). *Experientia* **39**, 790.
Bird, S., Gower, D. B., Sharma, P. and House, F. R. (1985). *Experientia* **41**, 1134.
Bjorkman, C. and Larsson, S. (1991). *Ecological Entomol.* **16**, 283–289.
Bjostad, L. B., Lynn, C. E., Du, T. W. and Roelofs, W. L. (1984). *J. Chem. Ecol.* **10**, 1309–1324.
Blum, M. S., Jones, T. H., Holldobler, B., Fales, H. M. and Jaouni, T. (1980). *Naturwissensch.* **67**, 144–145.
Bowers, W. S. and Bodenstein, W. G. (1971). *Nature, Lond.* **232**, 259–261.
Bradshaw, J. W. S., Baker, R. and Howse, P. E. (1979) *Physiol. Entomol.* **4**, 15–46.
Bronson, F. H. (1979). *Q. Rev. Biol.* **54**, 265.
Brownlee, R. G., Silverstein, R. M., Muller-Schwarze, D. and Singer, A. G. (1969). *Nature, Lond.* **221**, 284–285.
Carlson, D.A. and Schlein, Y. (1991). *J. Chem. Ecol.* **17**, 267–284.
Carrière, Y., Millar, J. G., McNeill, J. N., Miller, D. and Underhill, E. W. (1988). *J. Chem. Ecol.* **14**, 947–956.
Carté, B. and Faulkner, D. J. (1983). *J. Org. Chem.* **48**, 2314.
Claus, R. and Hoppen, H. O. (1979). *Experientia* **35**, 1674–1675.
Collins, A. M., Rindever, T. E., Daly, H. V., Harbo, J. B. and Pesante, D. (1989). *J. Chem. Ecol.* **15**, 1747–1756.
Crisp, D. J. and Meadows, P. S. (1962). *Proc. R. Soc.* **156B**, 500–520.
Cromwell, B. T. and Richardson, M. (1956). *Phytochemistry* **5**, 735–746.
Crump, D. R. and Moors, P. J. (1985). *J. Chem. Ecol.* **11**, 1037–1044
Crump, D., Silverstein, R. M., Williams, H. J. and Fitzgerald, T. D. (1987). *J. Chem. Ecol.* **13**, 397–402.

Dawson, G. W., Griffiths, D. C., Janes, N. F., Mudd, A., Pickett, J. A., Wadhams, L. J. and Woodcock, C. M. (1987). *Nature* **325**, 614-616.

Duffey, S. S. and Blum, M. S. (1977). *Insect Biochem.* **7**, 57–66.

Dulka, J. G., Stacey, N. E., Sorensen, P. W. and Van der Kraak, G. J. (1987). *Nature* **325**, 251–253.

Eisner, T. (1964). *Science, N.Y.* **146**, 1318–1320.

Eisner. T., Hendry, L. B., Peakall, D. B. and Meinwald, J. (1971). *Science, N.Y.* **172**, 277–279.

Eisner, T., Johnessee, J. S., Carvell, J., Hendry, L. B. and Meinwald, J. (1974). *Science, N.Y.* **184**, 996–999.

Eisner, T., Conner, W. E., Hicks, K., Dodge, K. R., Rosenberg, H. I., Jones, T. H., Cohen, M. and Meinwald, J. (1977). *Science, N.Y.* **196**, 1347–1349.

Eisner, T., Wiemer, D. F., Haynes, L. W. and Meinwald, J. (1978). *Proc. Natn. Acad. Sci. U.S.A.* **75**, 905–908.

Eisner, T., Nowicki, S., Goetz, M. and Meinwald, J. (1980). *Science, N.Y.* **208**, 1039–1041.

Evershed, R. P., Morgan, E. D. and Cammaerts, M. C. (1982). *Insect Biochem.* **12**, 383.

Fish, L. J. and Pattenden, G. (1975). *J. Insect Physiol.* **21**. 741–744.

Gaston, L. K., Kaae, R. S., Shorey, H. H. and Sellers, D. (1977) *Science, N.Y.* **196**, 904–905.

Gilbert, L. E. (1976). *Science, N.Y.* **193**, 419–420.

Hill, J. O., Parlik, E. J., Smith, G. L., Burghardt, G. M. and Coulson, P. B. (1976). *J. Chem. Ecol.* **2**, 239–253.

Howard, R. W., McDaniel, C. A. and Blomquist, G. J. (1980). *Science, N.Y.* **210**, 431–432.

Kunesch, G., Zagatti, P., Pourreau, A. and Cassini, R. (1987). *Naturwissensch.* **42c**, 657–659.

Lanier, G. N. (1979). *Bull. Entomol. Soc. Am.* **25**, 109–111.

Linnaeus, C. (1756). *Amoenitates Academicae* **3**, 200.

Lofstedt, C., Herrebourt, W. and Du, J. W. (1986). *Nature* **323**, 621–623.

Moore, B. P. (1967). *J. Aust. Entomol. Soc.* **6**, 36–38.

Morris, H. R., Taylor, G.W., Masento, M. S., Jermyn, K. A. and Kay, R. R. (1987). *Nature* **328**, 811–814.

Muller-Schwarze, D., Singer, A. G. and Silverstein, R. M. (1974). *Science, N.Y.* **183**, 860–862.

Pickett, J. A., Williams, I. H., Smith, M. C. and Martin, A. P. (1981). *J. Chem. Ecol.* **7**, 543–554.

Raina, A. K., Kingan, T. G. and Mattoo, A. K. (1992). *Science* **255**, 592–594.

Raupp, M. J., Milan, F. R., Barbosa, P. and Leonhardt, B. A. (1986). *Science, N.Y.* **232**, 1408–1409.

Raymer, J., Wiesler, M., Novotny, C., Asa, U., Seal, U. S. and Mech., L. D. (1985). *J. Chem. Ecol.* **11**, 593.

Robinson, S. W., Moser, J. C., Blum, M. S. and Amante, E. (1974). *Insectes Soc.* **21**, 87–94.

Schwende, F. J., Wiesler, D. and Novotny, M. (1984). *Experientia* **40**, 213–215.

Sinnwell, V., Schulz, S., Franke, W., Kittmann, R. and Schneider, D. (1985). *Tetrahedron Lett.* **26**, 1707–1710.

Starr, R. C. (1968). *Proc. Natn. Acad. Sci. U.S.A.* **59**, 1082–1088.

Stowe, M. K., Tumlinson, J. H. and Heath, R. R. (1987). *Science, N.Y.* **236**, 964–966.

Sullivan, T. P., Nordstrom, L. O. and Sullivan, D. S. (1985). *J. Chem. Ecol.* **11**, 903–920.

Thiessen, D. D., Regnier, F. E., Rice, M., Goodwin, M., Isaaks, N. and Lawson, N. (1974). *Science, N.Y.* **184**, 83–85.

Tumlinson, J. H., Silverstein, R. M., Moser, J. C., Brownlee, R. G. and Ruth, J. M. (1971). *Nature, Lond.* **234**, 348–349.

Tschinkel, W. R. (1975). *J. Insect Physiol.* **21**, 753–783.

9 Biochemical Interactions Between Higher Plants

I. Introduction

As part of the Darwinian struggle for survival, higher plants compete with each other for moisture, light and soil nutrients in the ecosystem. In the course of this struggle, they have developed various means of defence against their neighbours; whenever this defence is chemical in nature, it is referred to as allelopathy. Thus, allelopathy represents chemical competition between plants and the phenomenon may be regarded as yet another phase of chemical ecology, the interference caused by one higher plant on another in the natural environment.

Molisch (1937) was the first to define the word allelopathy, using it in the widest sense to refer to 'biochemical interactions between all types of plant' and including both deleterious and advantageous interactions. Rice (1984), in his monograph on the subject, uses a similar definition. Like Molisch, Rice regards it as an all-embracing term to cover most types of biochemical interactions, including those between higher plants and micro-organisms. By contrast, Muller (1970), one of the chief pioneers in the modern development of the topic, prefers to restrict the term allelopathy to higher plant–higher plant interactions and this restriction is kept here, biochemical interactions between higher and lower plants being reserved for Chapter 10. While it is convenient from many points of view to make such a distinction, it is worth pointing out that lower plants are indirectly involved in higher plant–higher plant interactions, in that the effectiveness of the chemical substances produced by one higher plant to influence another may depend on the speed with which soil micro-organisms are able to detoxify and further metabolize such compounds.

The chemicals concerned in higher plant interactions, called allelopathic substances or toxins, are typical secondary constituents and appear to be mainly low molecular weight compounds of relatively simple structure. Most of those that have been positively identified are either volatile terpenes or else phenolic compounds. Whittaker (1972), among others, has suggested that because of their chemical nature allelopathic substances may only be secondarily functional in plants, having arisen initially in plants in response to herbivore pressures. This theory assumes that the evolution of feeding deterrents has led to the occasional production of compounds that leak out of the plant, are excreted from the leaves, stems or roots and enter the environment. Such substances may thus have become accidentally caught up in the interaction of one higher plant with another and because of the beneficial effects in terms of reduced plant competition, such plants have continued to synthesize them.

Chemical defence mechanisms are most frequently invoked when one type of plant, e.g. a shrub or tree, competes with another type, e.g. a herb or grass, and some of the best examples of allelopathy have arisen from such studies. Competition between plants of the same general type, e.g. herbs, may also, however, include allelopathic effects (see Newman and Rovira, 1975). Allelopathic effects, in addition, may be exerted between individuals of the same species, particularly when lack of moisture or soil nutrients is limiting growth; the term 'auto-toxicity' is sometimes used in such cases. Competition for the standard biological variables is most acute in extreme climates and some of the first recorded examples of allelopathy were demonstrated between desert plants. However, the effect has also been observed in plants growing in a range of other habitats from open grasslands to humid rain forests, so that allelopathy may take place in almost any climate.

Historically, the taxonomist de Candolle (1832) was one of the first to record situations in which chemical interactions between higher plant species appeared to be taking place. He noted, for example, that thistles growing in cornfields had an injurious effect on oat plants and similarly that *Euphorbia* harmed the growth of flax. De Candolle also described experiments in which bean plants, dipped in water containing matter exuded from the roots of other individuals of the same species, languished and died. Many other miscellaneous observations of a similar nature accumulated in the botanical literature up to about 1925. In this year, Massey provided one of the first clear-cut demonstrations of allelopathy between trees and herbs, when he set up a series of experiments to show that the black walnut *Juglans nigra* produced chemicals which killed tomato and alfalfa plants grown in its vicinity.

The study of allelopathy was advanced during the 1939–1945 war period by the largely accidental discoveries, made by plant physiologists working on war projects, of allelopathic interactions in plants growing in the Californian desert, especially in the shrubs *Encelia farinosa* and *Parthenium argentatum*. However, it was not until the pioneering efforts of Muller and his colleagues (summarized in Muller and Chou, 1972) with their work on the Californian chaparral and of Rice (1984) on chemical factors in field succession that the concept of plant allelopathy became really established. Most of our information on allelopathy at present derives from a series of papers published over a 20-year period by Muller and his coworkers at Santa Barbara, California.

Even today, not all plant ecologists accept the concept of allelopathy as a significant factor in competition in plant communities. The extreme view against the idea of chemical competition between plants is expressed by Harper (1977) in a critical review of the available literature up to that time. There is clearly a problem in that conclusive proof of allelopathy occurring in any given ecological situation is extremely difficult to obtain and even after careful experimentation, as has been carried out by Muller and his associates, there are still many facets of the interaction which require further study. Having said this, one can still point to a large body of circumstantial evidence, most of which is mentioned in Rice's book, which supports the view that chemical interactions do occur between higher plants. Indeed, considering the enormous capacity of angiosperms to synthesize such a wide range of generalized highly toxic compounds, it would be very surprising if no such interactions ever occurred. A failure to appreciate the chemical versatility of higher plants, the plethora of chemical structures produced and their considerable physiological activities, may rest behind some of the criticisms of allelopathy expressed in the past.

Muller himself has always been very careful not to overstress the case for allelopathy occurring in plants. As he says (Muller and Chou, 1972): 'it is one of several basic ecological processes whose chemical cause is but another major factor in the environmental complex. The chemical variable takes no precedence over light, temperature, moisture and mineral nutrients . . . and shares with each of them a part in determining the plant environment. However, it also shares in the potential for becoming a limiting factor and thus exerting control'.

In the present short account of allelopathy, emphasis will be given to the examples where the allelopathic substances have been chemically characterized. Most of the major references on allelopathy can be found in Rice (1984). More recent developments in allelopathy and in the methodology of allelopathic experimentation are considered in Putnam and Tang (1986). General reviews worth consulting include Muller and Chou (1972), Whittaker (1972), Newman (1978), Stowe and Kil (1983) and Fischer (1991).

II. The Walnut Tree

The concept of allelopathy has been unconsciously recognized for many years in the observations of gardeners and farmers that while some plants thrive when grown together in close proximity others do not. One tree which has been known for a long time, indeed since the time of Pliny (AD 23–79), to exert an allelopathic effect on other species if they are grown near it is the walnut, *Juglans nigra*. The antagonistic effects of walnut have been recorded on such diverse plants as pine trees, potatoes and cereals. There are even reports that the walnut toxin will kill apple trees if they are planted too close (Schneiderhan, 1927). Most observations on walnut refer to the North American black walnut, *J. nigra*, grown as timber, but the effect may well be true for the European *Juglans regia*, grown for its nuts, and for other species in the genus.

The first direct evidence of the reasons underlying the fatal consequences of the walnut toxin on herbs was obtained by Massey (1925) who planted tomato and alfalfa plants in a region up to 27 m from the trunk of a walnut tree and found that many plants died as a

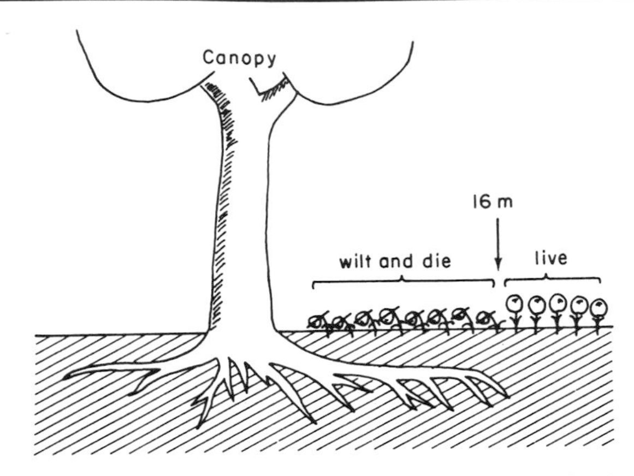

Figure 9.1 Effect of growing tomato plants in the vicinity of walnut trees

result (see Fig. 9.1). The position at which the tomato plants remained unaffected by the allelopathy (Fig. 9.1) coincided with the extent of root growth of the tree and Massey assumed at that time that the plants were killed by the exudation of toxin from the roots.

Later work by Bode (1958) indicated that this simple view of root exudation might be incorrect and that the toxic effects were actually due to leaching from the walnut leaves, stems and branches of a bound toxin, which underwent hydrolysis and oxidation in the soil with the release of the true toxin, which then killed any annual species growing in the vicinity. The area of toxicity was therefore determined by the leaf area of the tree and the ability of the leachate to saturate the surrounding soil. Our own preliminary experiments on this topic at Reading University suggest that leachates from *both* roots and leaves may actually be involved.

The bound toxin has been identified as the 4-glucoside of 1,4,5-trihydroxynaphthalene, which on hydrolysis and oxidation is converted to the naphthoquinone, juglone (Fig. 9.2). Juglone is a water-soluble yellow pigment and much of the characteristic brown staining of the hands caused by handling walnuts is due to the release of this compound. It is strictly limited in its occurrence to green parts of the tree and is lost from dead tissue and from the ripe nuts. Its considerable toxicity has been widely recognized. Many plants (e.g. tomato, alfalfa) are killed if juglone is injected into them via the petiole. It is also an important inhibitor of seed germination and may be conveniently bioassayed by

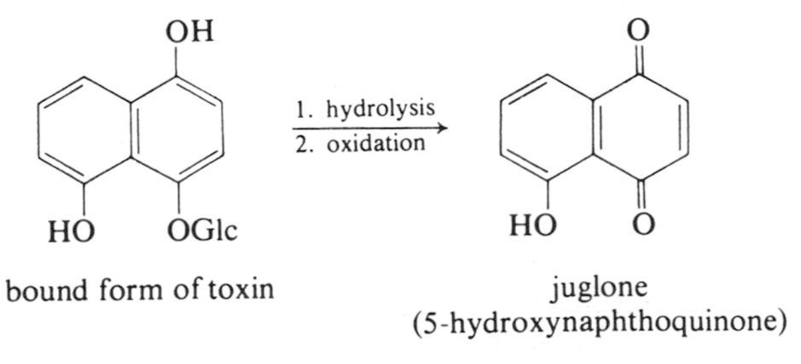

Figure 9.2 Release of juglone from bound form

such means. Thus, juglone at a concentration of 0.002% will completely prevent germination of lettuce seed treated with such a solution.

The walnut story is of especial interest because the toxin occurs in the plant in a safe, non-toxic bound form and it is only after leaching from the leaves and stems into the soils that it becomes active and can exert its effect. Clearly, to be a useful ecological agent, juglone must persist for a reasonable length of time in the soil around the tree and its concentration must presumably be regularly renewed by rainfall carrying down fresh supplies into the soil area.

Measurements of juglone concentrations in the soil under *J. nigra* have shown that while large amounts are, as expected, present in the top layers (0–8 cm), smaller amounts (1 μg/g of soil) are still detectable at a depth of 1.8 m. These amounts in the soil under mixed stands of *J. nigra* and black alder, *Alnus glutinosa*, are sufficient to account for the well-known mortality that this alder suffers if grown mixed with walnut trees (Ponder and Tadros, 1985).

The proportion of plants that are sensitive to the walnut toxin is still not certain. Reitveld (1983) tested the effect of juglone on germination and growth of herbs and trees and found that all species were sensitive to concentrations of 1×10^{-3} M; some, however, were more affected than others. The same differential effects have been observed in the field. While alders, broad-leaved herbs and ericaceous shrubs are excluded by juglone, it is known that *Rubus fruticosus* and *Poa pratensis* are well able to tolerate and grow under walnut trees (Brooks, 1951).

III. Desert Plants

In desert plants, it might be expected that the considerable competition for the limited water available in the soil would result in the development of many competitive effects including allelopathy which are exemplified in those plants able to survive in these harsh conditions. Indeed, evidence that allelopathy is one important factor comes from the fact that some, but not all, shrubs are found with bare patches of soil underneath their canopies and around them, with zones where annuals do not appear to flourish. One such plant, investigated by Went (1942), is the shrub *Encelia farinosa* (Compositae) which grows, for example, in the Mohave desert of Central California. By inhibiting the growth of annuals, it thus ensures for itself the available moisture within a metre or so of its site of growth.

Went (1942) suggested that the effect in *Encelia* was due to a root exudate having a toxic effect on the growth of annuals such as *Malacothrix* but Gray and Bonner (1948) were later able to isolate from the leaves a toxin that was not self-inhibitory but caused pronounced inhibition in many other plants. The substance was identified as 3-acetyl-6-methoxybenzaldehyde (Fig. 9.3), a simple benzene derivative with two carbonyl functions. While primarily produced in the leaves, this toxin is released when the leaves fall to the ground and decompose, and remains persistent in the soil, at least until it is washed away by heavy rainfall. Acetophenones of related structure to the *Encelia* toxin have been identified in a number of other composite semishrubs (Hegnauer, 1977) so it is possible that a similar isolating mechanism occurs in other species.

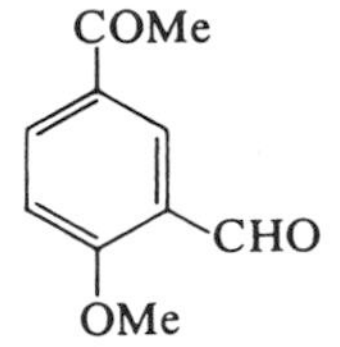

COMe | CHO | OMe

3-acetyl-6-methoxybenzaldehyde
Encelia farinosa

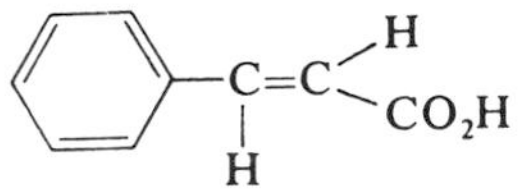

H | C=C | CO_2H | H

E-cinnamic acid
Parthenium argentatum

Figure 9.3 Toxins of desert shrubs

The role of toxins in desert shrub communities has also been examined in some detail by Muller and Muller (1953, 1956), who found that two other shrubs, *Franseria dumosa* (Compositae) and *Thamnosma montana* (Rutaceae), were capable of producing water-soluble toxins and yet did not exert any allelopathic effect on neighbouring annuals. Bioassays indicated a degree of toxicity, when tested with tomato seedlings, greater than in the case of *Encelia*. The role of toxins in controlling the development of an annual flora in the desert is thus a complex one, and other factors, including the build-up of organic litter in the soil around the shrubs, could account for the presence/absence of annuals. A further factor may be the ability of soil micro-organisms to differentially detoxify some allelopathic compounds and not others. The *Encelia* aldehyde may be resistant to breakdown while the toxins of other shrubs may be much more rapidly lost in the soil. While the *Franseria* toxin does not appear to have been identified, that of *Thamnosma* is a mixture of furanocoumarins (Bennett and Bonner, 1953) which could conceivably be turned over quickly by micro-organisms. Detailed information on the metabolism of these toxins in the soils, however, awaits further investigation.

One of the few examples of growth inhibition caused by a root, instead of a leaf, toxin is that of the rubber plant guayule, *Parthenium argentatum* (Compositae). Significantly in this case the substance produced in the root causes self-inhibition and does not appear to affect other species. The toxin was discovered during experiments devised to develop new plant sources of rubber. It was found that in regular plantations of *Parthenium* plants, individuals on the edge of the plot always grew better than those in the middle (Fig. 9.4). These differences could not be eliminated by extra watering or mineral application. Furthermore, roots of adjacent plants did not intermingle but grew separately. In addition, while seedlings never became established under larger *Parthenium* plants, they grew successfully under the canopy of other shrubs.

Subsequent experiments showed that a specific toxin was present in root exudates of *Parthenium* and this was identified as the simple aromatic compound, (*E*)-cinnamic acid (Fig. 9.3). The leachate from 20,000 roots eventually yielded 1.6 g of toxic material (Bonner and Galston, 1944). (*E*)-Cinnamic acid is toxic to *Parthenium* growth at a concentration of 0.0001%, whereas tomato seedlings are only affected if treated with a solution containing 100 times this concentration.

Cinnamic acid is effective in restricting growth when applied to the soil in pot cultures, but it does not persist very long so it must be continuously produced by the root in order to exert an allelopathic effect. It is not yet clear whether this compound has any

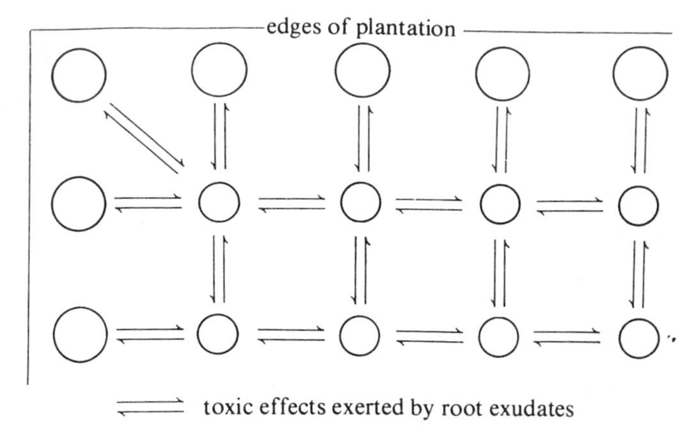

Figure 9.4 Effect of self-inhibition in plantations of *Parthenium argentatum*

importance in natural stands of *Parthenium argentatum* but presumably its production might ensure that the plants are so widely spaced that inhibition of growth is not normally observed. It could conceivably affect the growth of native competitors of *Parthenium* and thus have a dual role in reducing competition both from other members of the same species and also from plants of other species.

IV. Allelopathy in the Californian Chaparral

A. Volatile Terpenes and the Fire Cycle

The Californian chaparral is a vegetational area of relatively low Mediterranean rainfall, along the coastal strip of Southern California and adjacent to areas of natural, uncultivated grassland. One of the most striking natural phenomena of this shrubby grassland area is the zonation of herbs around the thickets of shrubs which dominate this flora. Two of the more important of these shrubs are the labiate *Salvia leucophylla* (sagebrush) and the composite *Artemisia californica*. Immediately surrounding each shrub or clump of shrubs there are bare zones of soil from one to two metres in width. Beyond the bare zones, there are areas of stunted growth where a few herbs show limited development. Finally, one reaches the grassland, where *Avena, Bromus* and *Festuca* species flourish and grow. These effects of zonation are most clearly revealed by aerial photographs (see Muller, 1966) but are also perfectly apparent in snapshots of the local vegetation (Fig. 9.5).

The explanation for this remarkable inhibitory effect of shrubs on surrounding herbs has been shown by Muller and his coworkers to be due to terpene toxins. Careful study of other ecological parameters showed clearly that biological factors were not responsible for these effects. Thus, shade, soil, drought, nutrient conditions, slope of ground, root competition, insect and animal predation and water-soluble components were all, in turn, ruled out. The possible role of animals, especially birds and rodents, in producing the bare zones was experimentally investigated by Muller (1970) and independently

Figure 9.5 Photograph of allelopathic effects of shrub versus herb in the Californian chaparral

by Bartholomew (1970), but no convincing evidence that they play a causal role could be established in numerous experiments.

The case in favour of volatile constituents of the shrubs, namely simple terpenes, being the major agents responsible for these allelopathic effects has been cogently argued by Muller (1970). This role for the terpenes was established from observations of their presence in all the various phases of the interaction. Thus terpenes (1) are richly present in the leaves; (2) are constantly 'turned over' by the shrubs and a vapour cloud of volatile essence hangs around the plants; (3) occur in the soil surrounding the plants; (4) remain in the dry soil until rains bring into activity soil micro-organisms which degrade them; (5) can be transported into plant cells through the waxy coatings of seeds or roots; and (6) have a significant effect on the germination of seeds of those annuals (e.g. *Avena fatua*) which grow in the adjacent grassland area.

The terpenes present in the two shrubs have been fully characterized and identified and identical compounds have been isolated from the appropriate soil samples. These same compounds have been tested and shown to inhibit plant growth and seed germination. Of the various terpenes of the sagebrush, *Salvia leucophylla*, the two most effective toxins are 1,8-cineole and camphor. Also present are α- and β-pinene and camphene (Fig. 9.6). *Artemisia californica* is remarkably similar in that its two most potent terpenes are again 1,8-cineole and camphor. Other volatile agents of *Artemisia* include artemisiaketone, α-thujone and isothujone (Halligan, 1975).

The chemical ecology of the Californian chaparral is complicated by the fact that the vegetation undergoes cyclical change as a result of natural fires which occur on average about every 25 years. The destruction of the shrubs by fire is followed by several years in

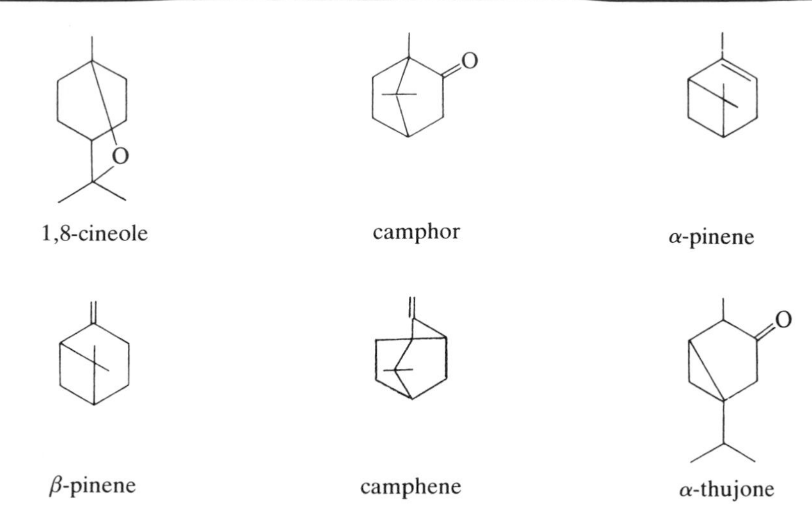

Figure 9.6 Formulae of terpenes concerned in plant allelopathy

which the annual herbs and grasses dominate the landscape. However, slowly but inexorably, the shrubs grow again and begin to exert their allelopathic effects. Finally, the characteristic tell-tale bare patches of soil develop around the shrubs some 6 or 7 years after the fire and remain until the next fire occurs. The cycle then repeats itself. The terpenes fit into this fire cycle perfectly, since they are hydrocarbon in nature and are rapidly volatilized and burnt off the soil during the period of the fire. As a result, the soil is no longer contaminated with them and annuals are able to grow and multiply for several years following a fire (Fig. 9.7). It is only when the shrub flora has developed sufficiently to synthesize terpenes in quantity that the influence of these toxins is again felt and the characteristic zonation appears.

If terpenes are so effective in reducing plant competition in sagebrush and *Artemisia*, one might ask the question whether other plant species make use of these toxins in a similar fashion. In fact, volatile mono- and sesquiterpenes are present in abundance in a wide range of angiosperms, particularly in many species of Myrtaceae, Labiatae and Rutaceae. They are also richly present in most gymnosperms. It is thus possible that these terpenes may be allelopathic substances in other plants, although their effects on surrounding vegetation may not be necessarily as dramatic as in the case of sagebrush. Indeed tests have been made on a number of chaparral shrubs and it is likely that similar limitations on annual species occur in these cases too. Plants in which terpenoids have been identified as effective allelopathic agents include *Eucalyptus globulus* (Baker, 1966), *E. camaldulensis* (Myrtaceae) (del Moral and Muller, 1970), *Artemisia absinthium* and *Sassafras albidum* (Lauraceae) (Gant and Clebsch, 1975).

Monoterpenes have also been shown to produce allelopathic effects in a Florida scrub community, where labiate shrubs such as *Calamintha ashei* and *Conradina canescens* exert a strong inhibition on the growth of native sandhill grasses. An aqueous wash of the leaves of *C. canescens* contains four monoterpenes—1,8-cineole, camphor, borneol and

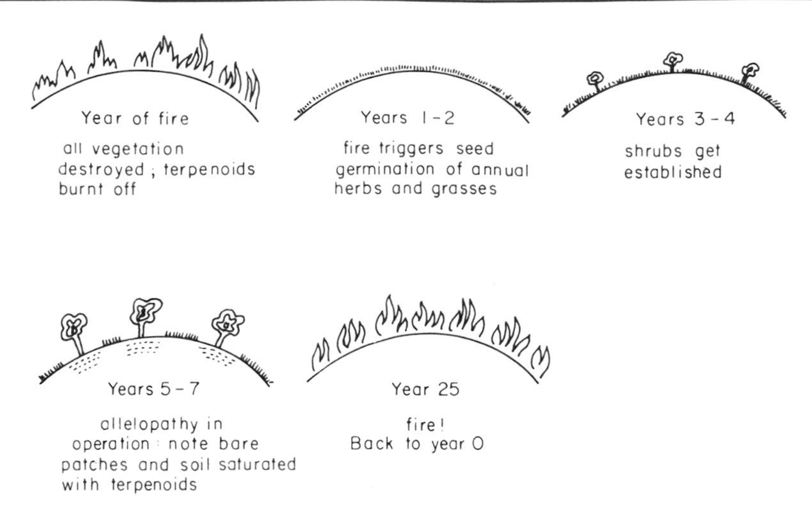

Figure 9.7 Fire cycle in the Californian chaparral vegetation

α-terpineol—together with the triterpenoid ursolic acid in quantity. Remarkably, these monoterpenes do have significant water solubility, in spite of their hydrocarbon nature, and in *in vitro* tests, saturated aqueous solutions totally inhibited the germination of test species. The role of ursolic acid, a natural detergent, in allelopathy is apparently to improve the penetration of the monoterpenes into the target seeds, causing them to fail to germinate (Fischer, 1991).

B. Water-soluble Inhibitors

Two other dominant shrubs in the Californian chaparral are chamise *Adenostoma fasciculatum* (Rosaceae) and *Arctostaphylos glandulosa* (Ericaceae). Both are widespread and exert allelopathic effects on herbs similar to those observed for sagebrush. *Adenostoma*, for example, grows in pure stands on dry exposed slopes; even though the soil is fully exposed to the sun and receives ample rain, no herbs grow in the vicinity of the lower slopes. Yet adjacent roadsides produce a heavy crop of annuals. If allelopathy is responsible for this situation, then toxins other than terpenoids must be responsible, since neither *Adenostoma* nor *Arctostaphylos* has any significant quantity of terpenes in the leaves.

Indeed, experiments of McPherson and Muller (1969) have shown that, in both these plants, substances responsible for the observed allelopathic effects are water-soluble and waterborne. As with the terpene toxins, there is a particular cycle of events which ensures that the chemical factors are effective in this particular ecological situation. One of the major climatic features of this vegetation is coastal fog, even in the summer months, and although rainfall is only moderate, there is, as a result of both forms of precipitation, a constant drip of moisture onto the leaves of these shrubs and hence onto the surrounding soil (del Moral and Muller, 1969). This can be sufficient to carry a regular supply of water-soluble inhibitor from leaf to the soil, where it remains in sufficient

Table 9.1 Phenolics identified as waterborne inhibitors in shrubs of the Californian chaparral

Class	Compound	*Adenostoma* Foliage	*Adenostoma* Soil	*Arctostaphylos* Foliage	*Arctostaphylos* Soil
Neutral phenols	Hydroquinone	+	–	+	–
	Phloridzin	+	–	–	–
	Umbelliferone	+	–	–	–
Hydroxybenzoic acids	*p*-Hydroxybenzoic	+	+	+	+
	Protocatechuic	–	–	+	–
	Vanillic	+	+	+	+
	Syringic	+	+	–	+
	Gallic	–	–	+	–
Hydroxycinnamic acids	Ferulic	+	+	+	+
	p-Coumaric	+	+	–	+
	o-Coumaric	–	–	–	+

Data from Muller and Chou (1972).

concentration to prevent the growth of any annual species that spreads its seeds near these shrubs.

The leaves of both *Adenostoma* and *Arctostaphylos* are relatively rich in water-soluble phenolic compounds and it was therefore no surprise when McPherson and Muller found that the inhibitors turned out to be varying mixtures of phenols and phenolic acids. A similar but not identical range of phenolics were isolated (a) by leaching the foliage of the shrubs and (b) by extracting the surrounding soil with alkaline ethanol. The major compounds obtained in these treatments are shown in Table 9.1. Differences in phenolic content between leachate and soil are probably due to the fact that some leachate compounds either become irreversibly bound to soil particles or else are metabolized by micro-organisms so rapidly that they are lost in the ecosystem.

Of the phenolic compounds isolated, those most effective in inhibiting the seed germination of grasses and other herbs are the hydroxybenzoic and hydroxycinnamic acids (Fig. 9.8) and it is just these compounds which are found in the soils as well as in the leachates. Furthermore, the fact that mixtures of related phenolic acids are present may be important since these acids can act synergistically on plant growth. It has been shown, for example, that vanillic and *p*-hydroxybenzoic acids are more effective inhibitors of seed germination in grain sorghum and radish when applied together than when applied separately (Einhellig and Rasmussen, 1978).

Phenolic substances are normally present in leaf tissue mainly in bound form but they are known to be turned over within the plant tissue so that the fact that the leachate contains significant amounts of free as well as bound phenolics is perfectly reasonable. There are, however, problems regarding the isolation of phenolics from soil. While undoubtedly the concentration of such phenols could be reinforced by this drip mechanism, it is also true that the same acidic phenols could arise in the soil by the microbial decomposition of leaf-litter of some antiquity. The rate of turnover of these compounds in the soil is also not completely known, so that there are technical problems concerning

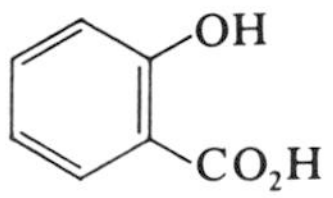

salicylic acid

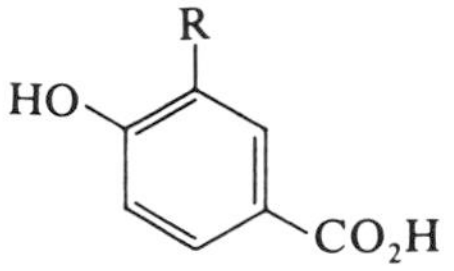

p-OH benzoic acid, R = H
vanillic acid, R = OMe

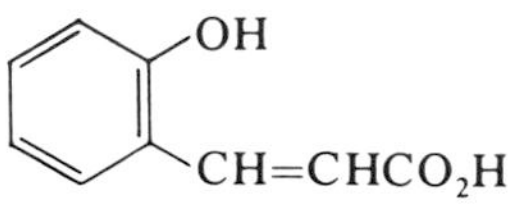

o-coumaric acid

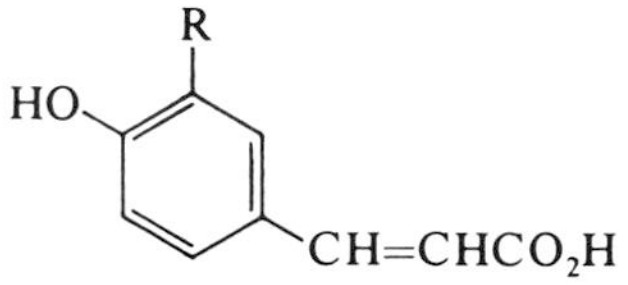

p-OH cinnamic, R = H
ferulic acid, R = OMe

Figure 9.8 Some water-soluble allelopathic agents in plants

the effectiveness of these phenolics as germination inhibitors. Nevertheless, the argument in favour of water-soluble toxins providing an ecological effect in these shrubs is clear. Further work is needed, however, on the identification of leachate constituents and their fate in the soils into which they are deposited.

The study of water-borne inhibitors has been extended to other habitats and it is possible that this may be a general phenomenon in any climate where the right humid conditions exist for regular leaching of organic material from plant leaves. In the subhumid deciduous forests of South Carolina, for example, two trees, *Quercus falcata* (Fagaceae) and *Liquidambar styraciflua* (Altingiaceae), have been observed to inhibit undergrowth within the dripline of the foliar crowns. In these conditions, there is a high rainfall, an abundance of mineral nutrients and the lack of shading phenomena; there is therefore, the likelihood of allelopathic effects operating. Indeed, in the case of *Quercus falcata,* salicylic acid has been isolated from the leaf leachate and shown to be toxic in bioassays (Muller and Chou, 1972). In tropical rain forests, other situations may be found. For example, Webb *et al.* (1967) explored the factors limiting the size of pure stands of *Grevillea robusta* growing in Queensland and found evidence that root exudates were self-inhibitory in this species. Such exudates were found to be highly toxic to seedlings of the same species and were presumably exerting such an effect in the natural environment.

Finally, there is one further plant which has been shown to exert allelopathic effects on other vegetation through water-soluble inhibitors, namely bracken. *Pteridium aquilinum* is a highly successful weed worldwide and it is so dominant that herbs are rarely ever found in bracken stands. As studied in Southern California (Gliessman and Muller, 1978), it has been found that phytotoxins leached by rain from dead, standing bracken fronds are largely responsible for this herb suppression. These extracts have a phenolic pattern similar to that of *Adenostoma* (Table 9.1) and caffeic and ferulic acids are probably major components. In this study, it was also possible to test the effects of herbivory by comparing mainland bracken stands, where animals (mainly rodents) were present, with stands on

Santa Cruz island, where small animals are essentially absent. While herbivory could not explain the suppression of most herbs, there is undoubtedly significant interaction with allelopathy. Some species like *Bromus rigidus* are not affected by the phytotoxins of bracken, but are kept out by animal grazing. Others like *Hypochoeris glabra* are fairly palatable but are also mildly inhibited by frond leachates. Yet others like *Clarkia purpurea* are so intensely inhibited by phytotoxins that they never develop within the stands to allow rodents to sample them.

Autotoxic effects are also apparent, since old stands of bracken degenerate after a number of years, go through a resting phase and then recolonize areas previously occupied. Such cyclic changes occur in plant populations generally and it is possible that allelopathy and auto-intoxication sequentially play important roles in shaping the ebb and flow of plant dominance with time.

V. Other Allelopathic Agents

A number of new classes of chemical have been identified as allelopathic agents in particular instances. The lignan nordihydroguaiaretic acid (Fig. 9.9), which occurs to the extent of 5–10% of the dry weight in the leaves of the creosote bush (*Larrea tridentata*), is thought to be responsible for the well-marked allelopathic effects of this shrub on surrounding vegetation. When it was applied to the seedlings of eight herbs which could compete with the creosote bush, the lignan dramatically reduced their growth (Elakovich and Stevens, 1985). Long chain fatty acids have been implicated for the first time in allelopathy of higher plants, being formed by the weed *Polygonum aviculare* for its allelopathic suppression of Bermuda grass (*Cynodon dactylon*). Nine acids were identified in the plant residues and seven in the soil around the plant. These were saturated and unsaturated acids, in the range C_{14} to C_{22}, and they were effective in inhibiting the growth of seedlings of this species of *Cynodon*; the acids were produced in conjunction with several phenolic inhibitors (Alsaadawi *et al.*, 1983).

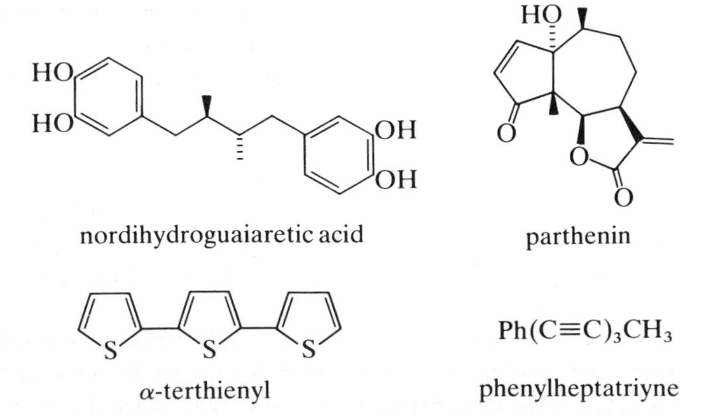

Figure 9.9 Some miscellaneous allelopathic agents in plants

While the allelopathic effects of the sulphur compound α-terthienyl and the polyacetylene phenylheptatriyne (Fig. 9.9), which are products of several species of Compositae, were being investigated it was found that they were more effective in the presence of sunlight or sources of ultraviolet light than in the dark. Their role as allelopathic agents may therefore be determined by their depth within the soil into which they are exuded. Extraction of the soil around the roots of *Tagetes erecta,* which produces α-terthienyl, yielded 0.4 ppm; this concentration is sufficient to inhibit the growth of seedlings of several test species (Campbell *et al.*, 1982).

Finally, sesquiterpene lactones, which also occur in plants of the Compositae, may be responsible for both inhibiting other plants and for autotoxic effects, although the evidence for this is still only suggestive (Fischer, 1986). Parthenin (see Fig. 9.9 for structure) is a major constituent of the aggressive weed *Parthenium hysterophorus*; it can be found in root exudates and leaf leachates and is inhibitory to growth in most but not all species tested. It is also autotoxic in seedlings and older plants (Picman and Picman, 1984).

VI. Ecological Importance of Allelopathy

The importance of chemical interactions between higher plants is undoubtedly underrated by ecologists but this is largely because of the scarcity of information relating to allelopathic effects in natural plant communities. One could argue that the chaparral vegetation studied by Muller and his colleagues has a number of special features associated with it, especially an unusual climate and the hazard of fire, which makes it impossible to extrapolate these results to the more equable conditions of temperate grasslands. Other hazards in interpreting the importance of particular chemicals in allelopathic situations are the physicochemical complexity of the soil, its ability or otherwise to bind organic molecules and also the presence in the soil of micro-organisms which can either convert a particular plant product to a more toxic agent or else destroy it so that it can no longer have any effect on higher plant tissue. Again, when testing crude extracts of one plant on the growth of another, it is necessary to see that the effects observed are not purely osmotic. This can be done by running controls with solutions of mannitol of equal osmotic potentials to the plant extracts (del Moral and Cates, 1971).

A series of allelopathic experiments which at least remove the problems of microbial involvement and osmosis from the situation have been carried out by Newman and Rovira (1975). These authors suggest that chemical interactions may occur commonly between herbs and grasses present in British meadows. Plants of eight species were grown in buckets of sand together with sterilized soil from the natural habitat. Leachates of these individual pots (and of control pots lacking plants) were then applied to all the other species in turn and the influence on growth rates examined after a period of some weeks.

The leachates from three species, *Holcus lanatus, Hypochoeris radicata* and *Trifolium repens,* consistently depressed the yield of the other five species. Of these, *Anthoxanthum odoratum* was the most sensitive, but this species was exceptional in that it grew faster in the presence of its own leachate than under any other conditions (including the control).

The other four species affected by leachates of the first three were *Cyonosurus cristatus*, *Lolium perenne*, *Plantago lanceolata* and *Rumex acetosa*. These results not only suggest that some species may compete with others in temperate grasslands by exerting allelopathic effects; equally they indicate that self-stimulation may be important in the attainment of dominance within a sward. These data are of further significance in that they indicate that allelopathic effects may occur in situations when zonation is absent and where no obvious chemical interaction would be suspected.

One ecological situation where allelopathic effects are likely to be more pronounced than in a stable pastureland is in field succession, where farming has removed the natural flora and has extensively altered the potentiality of the land in terms of mineral and organic matter content. Any such field, left fallow, will undergo a series of successive invasions of plant species before a more stable plant community becomes established. Rice (1984) has discussed in detail the possible role of allelopathy in such plant communities and its effect on the sequence of succession. He has listed five possible effects as follows: (1) the speeding up of the replacement of one species by another due to allelopathic self-toxicity (e.g. as in *Helianthus rigidus*); (2) the direct allelopathic suppression of the first species by the second via root or leaf exudates; (3) the slowing down of species replacement by the direct allelopathic effects of a dominant species on all potential invaders; (4) the indirect effect of a species through its decay products (shown by *Euphorbia supina*); and (5) allelopathic effects which determine what other species can invade the community and which cannot.

Clearly, many practical problems may arise in agriculture and horticulture from the effects of allelopathy. Thus weeds growing among crop plants, besides competing in the expected fashion for soil nutrients and light, may also exert a deleterious effect on the latter plants through the release of toxins. Again, the ploughing in, instead of burning, of cereal stubble causes the release of considerable quantities of phenolic acids during the breakdown of the cell walls and these could have an inhibitory effect on growth of certain crop plants grown subsequently. The quality of the microbial flora and its ability to further detoxify the phenolics are also important factors in such situations (Guenzi and McCalla, 1966). Other inhibitory products derived from cereal stalks following anaerobic breakdown of cellulose in the soil include volatile organic acids such as acetic (Lynch, 1976). Similarly, trees which produce toxic substances in the roots and which cause self-toxicity to the seedlings are at a disadvantage when replanted in the same soil and may not thrive. Such effects have been encountered in rosaceous fruit trees; in the case of the apple, the phenolic compound phloridzin is thought to be responsible for such effects (Börner, 1959). Autotoxic effects in coffee plantations are likewise thought to be due to the coffee purine alkaloid, caffeine (Waller *et al.*, 1986). Caffeine has been detected in soil and leaf litter in such plantations. Its antimicrobial activities probably limit its turnover by soil micro-organisms, explaining its persistence in these soils. Cases where allelopathic activities may interfere with crop rotation and with the establishment of nitrogen fixation in legumes are recorded by Rice (1984).

To end on a positive note, it may be mentioned that allelopathic agents may have some practical value in agriculture as replacement herbicides (Hathway, 1986). Caffeine, for example, besides being autotoxic in coffee plants, will inhibit seed germination in many weedy species such as *Amaranthus spinosus*. As it has no effect on the crop plant *Vigna mungo*, it could theoretically be employed as a selective herbicide in such a crop.

VII. Biochemistry of Host–Parasite Interactions

There are more than 3000 higher plant species which have the ability to form specialized intrusive organs, called haustoria, by which they are able to attach themselves to other higher plants and feed on them (Atsatt, 1983). They are thus parasitic on their host plant, depending on the host either completely or partly for their nutritional needs. Familiar examples in nature are the mistletoes (*Viscum* spp.) living on various trees, the twining dodders, *Cuscuta*, and the root parasites, *Orobanche* and *Striga*. Members of the latter two genera have become serious agricultural pests in certain parts of the world: *Orobanche crenata* on broad bean (*Vicia faba*) in Egypt and *Striga hermonthica* and *S. asiatica* on sorghum in Africa and Asia. Biochemically, plant parasites are of interest since the ability of their seeds to germinate and establish the haustorial bridge to the host depends entirely on chemical triggers, which exude from the roots of the unsuspecting host plant. There is considerable interest in determining what chemicals may pass between host and parasite, particularly those which may affect their secondary chemistry.

For successful parasitization, two distinct chemical messages are needed to pass out from the host plant root, one to stimulate seed germination of the parasite and the other to induce haustorial formation in the newly germinated seed, so that it can attach itself to the host. The first chemical message—the germination stimulant—may well be linked with the second, since if the parasite seed germinates too far away from the host root, it will not be in a position to attach itself. Furthermore, time is of the essence and unless attachment takes place within a few days, the seedling will die. Chemical studies of these germination stimulants and host-recognition substances are still at an early stage, but as a result of recent work, we know that typical secondary substances in root exudates are responsible for these signals in at least some cases.

The first germination stimulant to be characterized was the sesquiterpene strigol (Fig. 9.10) which was obtained from root exudates of cotton, a non-host plant in nature but one which stimulates *Striga* seed in the glasshouse. Strigol is highly active at a very low concentration. There is 50% stimulation of witchweed seed at a concentration of 1×10^{-6} M. The complete structure of strigol is not necessary for activity, since a simpler synthetic analogue (see Fig. 9.10) proved to be equally effective in triggering germination. This compound has some potential for controlling this weed, since it can be applied to the soil before the crop is planted. The *Striga* seeds then germinate and die in the absence of a suitable host (Johnson, 1980).

A natural stimulant for *Striga* seed has recently been isolated from the roots of *Sorghum bicolor*. It has been named sorgolactone (Hauck *et al.*, 1992) and has a structure closely similar to that of strigol (see Fig. 9.10). Its relationship with an unstable quinol stimulant obtained earlier from the same source (Chang *et al.*, 1986) is not yet clear. The second message of the *Striga*–*Sorghum* interaction—the haustorial inducer—has also been identified; it is 2,5-dimethoxy-*p*-benzoquinone (Chang and Lynn, 1986).

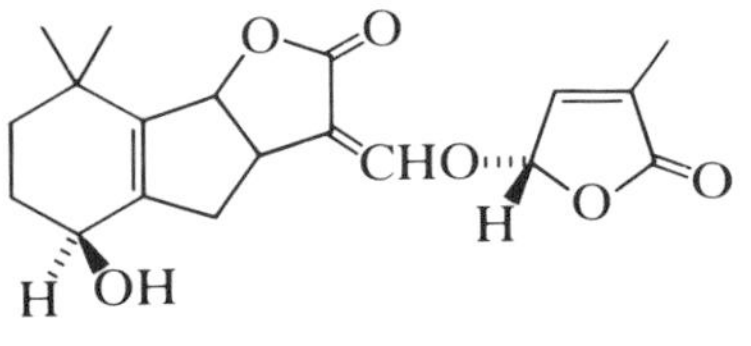

strigol
(from cotton)

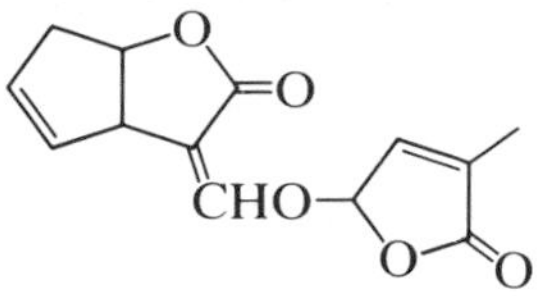

strigol analogue
(synthetic)

Figure 9.10 Germination stimulants for witchweed (*Striga*) seed

Other haustoria-inducing compounds have been described from two legume plants. Two were isolated from the root exudate of gum tragacanth (*Astragalus*) which stimulate haustorial formation in the parasite *Agalinis purpurea*. They are a simple dihydrostilbene xenognisin A and a related isoflavone, xenognisin B (Fig. 9.11). A third was isolated from the root of *Lespedeza*, another legume host of *Agalinis*, and identified as a triterpene soyasapogenol A (Chang and Lynn, 1986). Hence, several different classes of chemical apparently have the ability to trigger host recognition in this parasite. There is an interesting parallel with quite a different type of plant–parasite interaction, that of legume roots with the symbiotic nitrogen-fixing rhizobia. Here the host legume sends out a chemical signal to activate the nodulating genes of the bacterium, in order to initiate nodulation of the root. These chemical messages turn out to be flavones, compounds such as apigenin in the case of *Rhizobium leguminosarum* on pea (Firmin *et al.*, 1986). The parallel here is the fact that an isoflavone such as xenognisin B (see Fig. 9.11) has the ability to turn off or inhibit the nodulation process. There are therefore quite remarkable chemical similarities in the signals that emerge from legume roots and which may be detected by parasites as different as *Agalinis* and *Rhizobium*.

Legumes can also be infected by animal parasites. The soya bean, for example, can be attacked by a cyst nematode, which again depends on a root exudate factor for stimulating hatching of the eggs, which otherwise lie dormant in the soil. The hatching factor has been identified from this interaction and it is a pentanortriterpene, called glucinoeclipin (Fukuzawa *et al.*, 1985). This is yet another chemical messenger identified from legume roots.

While higher plant parasites are usually dependent on their host for primary metabolites (sugars, amino acids, etc.), this is not so in terms of secondary chemistry and most such parasites develop their own range of terpenoid and flavonoid constituents, which

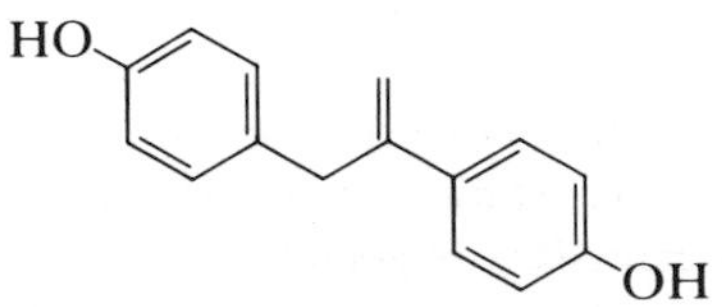

xenognisin A

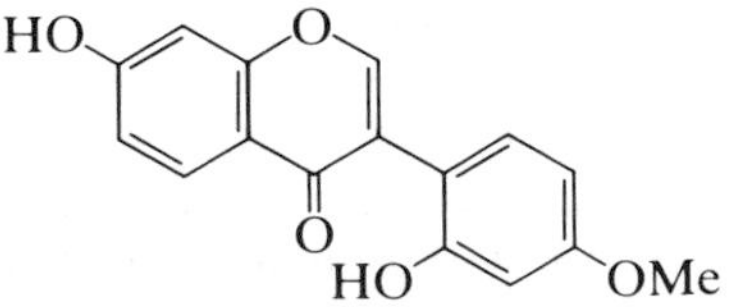

xenognisin B

Figure 9.11 Host-recognition factors from *Astragalus*

Table 9.2 Uptake of alkaloids by *Pedicularis* semiparasites from host plants

Pedicularis species	Host plant	Alkaloid	
		Class	Structure
P. bracteosa	*Picea engelmannii*	Piperidine	Pinidinol
	Senecio triangularis	Pyrrolizidine	Senecionine
P. crenulata	*Thermopsis montana*	Quinolizidine	Anagyrine
P. racemosa	*Lupinus argenteus*	Quinolizidine	Lupanine

are different from those of the host. This does not mean, however, that secondary substances do not, on occasion, pass from one plant to the other. This happens in the case of the semi-parasitic Indian paint brush genus *Castilleja* (Scrophulariaceae). Some of these plants were found unexpectedly to have pyrrolizidine or quinolizidine alkaloids in their tissues and the only rational explanation for the variable presence of these alkaloids was that they were host-derived and were passing through the haustorial connection from a pyrrolizidine- or quinolizidine-containing host, a *Senecio* or lupin species (Stermitz and Harris, 1987). Such a transfer is obviously beneficial to the parasite, since the alkaloids would supplement its own toxins (in this case, mainly iridoids) in its battle to survive herbivory (see Chapter 7).

Similar transfers of alkaloids occur with the semiparasite *Pedicularis,* which like the Indian paint brush genus is in the Scrophulariaceae. Different *Pedicularis* species attach themselves to legume or composite hosts and are found to contain some, but not all, of the alkaloids of the respective hosts (Table 9.2). One of these species, *P. bracteosa,* is unusual in being able to parasitize the Engelmann spruce, as well as *Senecio,* and the alkaloid pinidinol of the root, needle and bark of this spruce can be found in the tissues of the semiparasite (Schneider and Stermitz, 1990).

VIII. Conclusion

Sufficient experiments have now been carried out to indicate that allelopathic effects are exerted between one higher plant and another in a reasonable number of cases. The chemicals involved are generally secondary constituents, some of which are widely if not universally distributed in plants. It is therefore possible that allelopathy is a general phenomenon and could occur in many different ecological situations. How often allelopathy has a *controlling effect* in a particular case of plant competition is more difficult to predict and clearly all the biological aspects of the situation have to be fully investigated before the influence of chemical control can be considered as established.

The toxins involved in allelopathy generally occur in leaves and stems, although root constituents have a place in such interactions. How far compounds in other plant parts (e.g. flower, fruit) are deleterious to higher plant growth is not clear because these substances have rarely been studied as allelopathic agents. It is, however, well known that the seed coats of many plants contain chemicals which inhibit seed germination. These substances, besides contributing to the internal control of dormancy, may eventually leak out into the environment and have a deleterious effect on other species, before being degraded to harmless materials. Such possible allelopathic effects, however, need further study.

The method of release into the environment of the toxins can take one of several pathways. Extraction of phenolics from the living leaf is clearly effective, but depends on regular rainfall to accomplish the leaching. Volatilization from the leaf surface seems to be important in more arid climates and is the way that essential oils are released into the atmosphere and into the soil. Other volatile toxins, not yet fully implicated in allelopathy but which could be equally effective, are the mustard oils (glucosinolates) of the Cruciferae and the cyanogens of *Prunus, Trifolium* and many other plants.

Exudation of toxin from the root would seem to be the method *par excellence* for producing a harmful effect on a neighbouring species. However, of the situations studied, such exudations seem to occur in a minority of cases. However, a number of crop plants such as wheat, oats, guayule, cucumber and tomato are known to exude root toxins so that such root emanations may be important agriculturally. Finally, toxins may be released from plants during the decay of leaf material and these, besides producing self-toxicity, may be effective against a range of other species. Here, microbial activity may be partly or wholly responsible for the interactions observed, so that such effects pass out of the realm of allelopathy proper into that of microbial ecology.

The sorts of secondary compound so far implicated in allelopathy are mainly either terpenoids (mono- or sesquiterpenes) or phenolics (phenols, phenolic acids, cinnamic acids, hydroxyquinones). Other types of plant constituent have only been recorded on a few occasions (see Fig. 9.9). This is a limited range of natural products and it is perhaps surprising that alkaloids, the most important group of plant constituents from the viewpoint of toxicity to animals, have rarely been assigned a role in higher plant–higher plant interactions. This may be because of their low concentrations in plants or else to their possible rapid turnover in the soil. Another group of substances which could be important are the condensed tannins and their relatively slow turnover in the ecosystem would ensure the persistency of their effects. They have, of course, the disadvantage of relative immobility within the plant. Future work will no doubt show whether these or other as yet undescribed structures also have a role as toxins in the chemical competition between higher plants.

Finally, there is the question of the origin of allelopathic effects, already discussed briefly in the introduction. Newman (1978) has considered in detail whether or not natural selection has specifically favoured the development of allelopathy in plants. If this occurred, one would expect: (1) that plants on average are more toxic to other species than to members of their own species; (2) that competition results in enhanced toxin synthesis; and (3) that species which have grown together for many years are more tolerant to each others' exudates than those which have not. On testing these three hypotheses, it was found that the literature data show a random distribution, sometimes supporting but equally often opposing these suppositions. Newman concludes his analysis with the following words: 'this strongly suggests that there is not often specific selection for allelopathic ability or tolerance of allelopathy but that these plant properties are better viewed as the fortuitous outcome of characteristics of the plant whose primary reason for existence is not allelopathy. This is not to deny that allelopathy occurs, not to imply that it has negligible ecological significance, but rather to suggest that it is not always beneficial to the producer plant and that it should not be viewed in isolation from the other ecological implications of plant secondary metabolites.'

Bibliography

Books and Review Articles

Atsatt, P. R. (1983). Host–parasite interactions in higher plants. In: Lange, O. L., Nobel, P. S., Osmond, C. B. and Ziegler, H. (eds.), 'Encyclopedia of Plant Physiology, New Series', Vol. 12C, pp. 519–535. Springer, Berlin.

Gliessman, S. R. and Muller, C. H. (1977). *J. Chem. Ecol.* **4**, 337–362.

Harper, J. L. (1977). 'Population Biology of Plants', 892 pp. Academic Press, London.

Johnson, A. W. (1980). Plant germination factors. *Chem. Brit.* **16**, 82–85.

Molisch, H. (1937). 'Der Einfluss einer Pflanze auf die andere-Allelopathie'. Fischer, Jena.

Muller, C. H. (1970). Phytotoxins as plant habitat variables. *Recent Adv. Phytochem.* **3**, 106–121.

Muller, C. H. and Chou, C. H. (1972). Phytotoxins: an ecological phase of phytochemistry. In: Harborne, J. B. (ed.), 'Phytochemical Ecology', pp. 201–216. Academic Press, London.

Newman, E. I. (1978). Allelopathy: adaptation or accident? In: Harborne, J. B. (ed.), 'Biochemical Aspects of Plant and Animal Coevolution', pp. 327–342. Academic Press, London.

Putman, A. R. and Tang, C. S. (1986). 'The Science of Allelopathy', 332 pp. John Wiley, Chichester.

Rice, E. L. (1984). 'Allelopathy', 2nd edn., 422 pp. Academic Press, New York.

Stowe, L. G. and Kil, B. S. (1983). The role of toxins in plant–plant interactions. In: Keeler, R. F. and Tu, A. T. (eds.), 'Handbook of Natural Toxins', Vol. 1, pp. 707–741, Marcel Dekker, New York.

Whittaker, R. H. (1972). The biochemical ecology of higher plants. In: Sondheimer, E. and Simeone, J. B. (eds.), 'Chemical Ecology', pp. 43–70. Academic Press, New York.

Literature References

Alsaadawi, I. S., Rice, E. L. and Karns, T. K. B. (1983). *J. Chem. Ecol.* **9**, 761–774.

del Amo, S. and Anaya, A. L. (1978). *J. Chem. Ecol.* **4**, 305–314.

Baker, H. G. (1966). *Madrone, S. Francisco* **18**, 207–210.

Bartholomew, B. (1970). *Science, N. Y.* **170**, 1210–1212.

Bennett, E. and Bonner, J. (1953). *Am. J. Bot.* **40**, 29–33.

Bode, H. R. (1958). *Planta* **51**, 440–480.

Börner, H. (1959). *Contrib. Boyce Thompson Inst.* **20**, 39–56.

Bonner, J. and Galston, A. W. (1944). *Bot. Gazz.* **106**, 185–198.

Brooks, M. G. (1951). *West Va, Univ. Agr. Expt. Sta. Bull.* **347**, 1–31.

Campbell, G., Lambert, J. D. H., Arnason, T. and Towers, G. H. N. (1982). *J. Chem. Ecol.* **8**, 961–972.

de Candolle, M. A. P. (1832). 'Physiologie Vegetale', Vol. III. Bechet Jenne, Lib. Fac. Med., Paris.
Chang, M. and Lynn, D. G. (1986). *J. Chem. Ecol.* **12**, 561–579.
Chang, M., Netzly, D. H., Butler, L. G. and Lynn, D. G. (1986). *J. Am. Chem. Soc.* **108**, 7858–7860.
Einhellig, F. A. and Rasmussen, J. A. (1978). *J. Chem. Ecol.* **4**, 425–436.
Elakovich, S. D. and Stevens, K. L. (1985). *J. Chem. Ecol.* **11**, 27–33.
Fischer, N. H. (1986). In: Putnam, A. R. and Tang. C. S. (eds.), 'The Science of Allelopathy', pp. 203–218. John Wiley, Chichester.
Fischer, N. H. (1991) In: Harborne, J. B. and Barberan, F. A. T. (eds.), 'Ecological Chemistry and Biochemistry of Plant Terpenoids', pp. 377–398. Clarendon Press, Oxford.
Firmin, J. L., Wilson, K. E., Rossen, L. and Johnston, A. W. B. (1986). *Nature* **324**, 90–92.
Fukuzawa, A., Furnsaki, A., Ikura, M. and Masamune, T. (1985). *J. Chem. Soc. Chem. Commun.* 222–223.
Gant, R. E. and Clebsch, E. E. C. (1975) *Ecology* **56**, 425–436.
Gray, R. and Bonner, J. (1948). *Am. J. Bot.* **34**, 52–57.
Guenzi, W. D. and McCalla, T. M. (1966). *Agron. J.* **58**, 303–304.
Halligan, J. P. (1975). *Ecology* **56**, 999–1003.
Hathway, D. E. (1986). *Bot. Rev.* **61**, 435–486.
Hauck, C., Muller, S. and Schildknecht, H. (1992). *J. Plant Physiol.* **139**, 474–478.
Hegnauer, R. (1977). In: Heywood, V. H., Harborne, J. B. and Turner, B. L. (eds.), 'Biology and Chemistry of the Compositae', pp. 283–336. Academic Press, London.
Kobayashi, A., Morimoto, S., Shibata, Y., Yamashita, K. and Numata, M. (1980). *J. Chem. Ecol.* **6**, 119–131.
Lynch, J. M. (1976). *CRC Crit. Rev. Microbiol.* 67–107.
Massey, A. B. (1925). *Phytopathology* **15**, 773–784.
del Moral, R. and Cates, R. G. (1971). *Ecology* **52**, 1030–1037.
del Moral, R. and Muller, C. H. (1969). *Bull. Torrey Bot. Club* **96**, 467–475.
del Moral, R. and Muller, C. H. (1970). *Amer. Midl. Nat.* **83**, 254–282.
McPherson, J, K. and Muller, C. H. (1969). *Ecol. Monogr.* **39**, 177–198.
Muller, C. H. (1966). *Bull. Torrey Bot. Club* **93**, 332–351.
Muller, C. H. and Muller, W. H. (1953). *Am. J. Bot.* **40**, 53–60.
Muller, C. H. and Muller, W. H. (1956). *Am. J. Bot.* **43**, 354–361.
Newman, E. I. and Rovira, R. D. (1975). *J. Ecol.* **63**, 727–737.
Picman, J. and Picman, A. K. (1984). *Biochem. Syst. Ecol.* **12**, 287–292.
Ponder, F. and Tadros, S. H. (1985). *J. Chem. Ecol.* **11**, 937–942.
Reitveld, W. J. (1983). *J. Chem. Ecol.* **9**, 245–308.
Schneider, M. J. and Stermitz, F. R. (1990). *Phytochemistry* **29**, 1811–1814.
Schneiderhan, F. J. (1927). *Phytopathology* **17**, 529–540.
Stermitz, F. R. and Harris, G. H. (1987). *J. Chem. Ecol.* **13**, 1917–1926.
Waller, G. R., Kimari, D., Friedman, J. N. and Chou, C. H. (1986). In: Putnam, A. R. and Tang, C. S. (eds.), 'The Science of Allelopathy', pp. 243–270. John Wiley, Chichester.
Webb, L. J., Tracey, J, G. and Haydock, K. P. (1967). *J. appl. Ecol.* **4**, 13–25.
Went, F. W. (1942). *Bull. Torrey Bot. Club* **69**, 100–114.

10 Higher Plant–Lower Plant Interactions: Phytoalexins and Phytotoxins

I. Introduction

While the interaction between higher and lower plants can take many forms, it is the attack of the micro-organism on the higher plant leading to plant disease that is the major topic of the present chapter. The enormously damaging effect that microbial attack can have on the growth and development of plants is reflected in the common terms used in plant pathology to describe the different diseases. Diseases of the potato plant, for example, range from common scab, black leg, ring rot and skin spot to blight, gangrene and leaf roll. As these descriptive names imply, the symptoms are many and various; if unchecked, the end result of disease is usually the same whatever the invading micro-organism: death of the plant.

It should be emphasized, however, that susceptibility to disease, all too frequently apparent in cultivated plants, is really the exception rather than the rule. In fact, most higher plants, especially those growing in natural communities, are either resistant to microbial attack or coexist in a symbiotic relationship with the parasites without the production of any visible symptoms. Experience shows that even with crop plants, while the cultivars are often highly susceptible to a range of diseases, most of the related wild species are relatively immune. In fact, breeding for disease resistance in cultivars often involves the introduction of genetic material from the disease-free wild relatives.

It is the basis of this resistance to disease or microbial attack which has been of special interest to biochemists and plant pathologists in recent years and which will be considered in some detail in the present chapter. There is also a significant agricultural incentive in developing such studies. Practically all the chemicals used for crop protection have been discovered empirically rather than by design and this is true even of the systemic fungicides which control, for example, powdery mildew in cereals (Greenaway and Whatley, 1976). Knowledge of disease resistance factors may help in approaching the problem of disease control in crop plants on a more rational and scientific basis.

Attacking micro-organisms, in order to invade a plant, have to penetrate the surface layers. Obvious barriers to such invasion might be the presence of a waxy coating, numerous surface hairs or a thick cuticle. While much work has been done on the role of the leaf surface in disease, there is little concrete evidence to indicate that the surface layers provide any real protection from invasion. The main effect of the cuticular barriers is to reduce the speed of attack; successful invasion is nevertheless assured.

If there is no impenetrable physical barrier in plants to microbial invasion, one is forced to postulate the existence of a chemical barrier. Indeed, ideas of a chemical basis to disease resistance date back many years. Marshall Ward (1905) in his toxin theory was one of the first to suggest that there were compounds present in plants capable of inhibiting fungal growth. This theory states that: 'infection and resistance to infection depend on the power of a fungus to overcome the resistance of the cells of the host plant by means of enzymes or toxins; and reciprocally, on that of the cells of the host to form antibodies or toxins which repel the fungus protoplasm'. Among the many types of compound known to be present in plants, one group known as phenolics or polyphenols fit this role rather well. Phenolic compounds are universally distributed among higher plants and are often toxic to micro-organisms *in vitro* at physiological concentrations (10^{-4} to 10^{-6} M). It is, however, not clear, even today, whether such phenolic compounds have a causal role in protecting plants from disease *in vivo*.

A more dynamic view of phenolics and plant disease was developed by Offord (1940), who proposed that one particular class of phenolic—the tannins—were especially important as preinfective agents in higher plants. He suggested that 'the toxic action is initiated and conditioned by enzymes of the fungus; ultimate toxicity depends on the type of phenolic constituent formed by interaction of host and parasite and partly on the quantity and distribution of tannin'. Unfortunately for Offord's theory, experimental evidence that tannins are involved in disease resistance is very thin. While it can be shown that tannins prevent the mechanical transmission of plant viruses, they cannot similarly provide control against soil-borne or insect-transmitted disease attack.

The major breakthrough in the study of chemically based disease resistance mechanisms in higher plants came with the enunciation of the phytoalexin theory by Müller and Börger in 1941. This theory proposed that certain chemical substances were produced by plants *de novo* at the time of infection and they came into action as 'compounds warding off ('alexos') disease organisms from the plant ('phytos')' (hence the term phytoalexin). The theory developed from experiments carried out by the two authors on resistant factors in potato to the blight organism *Phytophthora infestans*: but they were unsuccessful themselves in actually demonstrating the existence of a discrete chemical substance in potato with the properties of a phytoalexin. It remained for Cruickshank and Perrin (1960) 20 years later to be the first to isolate and identify a phytoalexin, namely the substance pisatin from *Pisum sativum*. Since then, a large number of phytoalexins have been described from a range of plants and phytoalexin induction is now widely regarded as a major protective device in higher plants. The precise role of phytoalexins in disease situations is still being actively investigated and much remains to be learnt about this. Work on other modes of biochemical resistance have also proceeded recently and the existence of preinfectional chemical barriers in certain plants is also generally accepted.

While there are clearly chemical barriers to invasion, disease still occurs. Thus virulent strains of a given micro-organism are capable of breaching these defences. When this happens, a different set of biochemical reactions occur, set off by the invader. A variety of changes in the respiration rate and the primary metabolism occur in the host plant. At the same time, the micro-organism as it multiplies in the cells of the host may secrete substances called pathotoxins, with which it directly 'poisons' the host plant. These pathotoxins lead to the typical symptoms of disease susceptibility in the host, for example the wilting of the leaves as a result of biochemical interference with the plant's water supply. The host plant, in turn, attempts to detoxify the pathotoxins by conjugation or degradation and a battle ensues between the higher plant and the invading micro-organism for ultimate control of the situation. Thus, there is a clearly defined second stage in the biochemical interaction between plant and microbe when the primary defence has broken down and the plant is fighting for its survival and protection from the harmful metabolites synthesized by the microbe.

In the present chapter, biochemical defence mechanisms involving both pre- and post-infectional changes in the host plant will first be considered. Emphasis will be placed on phytoalexin formation. The second part of the chapter will outline the biochemical compounds produced in disease susceptibility and in the eventual mutilation or destruction of the host. For general reviews covering many of the topics discussed here, the reader is especially directed towards six books (Friend and Threlfall, 1976; Heitefuss and Williams, 1976; Horsfall and Cowling, 1980; Durbin, 1981; Bailey and Deverall, 1983; Callow, 1983) and the monograph on plant pathogenesis by Wheeler (1975).

II. Biochemical Basis of Disease Resistance

A. Preinfectional Compounds

As mentioned in the introduction, it is now clear that plants adopt a variety of biochemical defences to ward off microbial attack. Several attempts have been made to classify these different modes and there is some confusion in nomenclature regarding the chemicals involved. Ingham (1973) has attempted to provide a classification of these compounds, making a primary division between pre-infectional and post-infectional factors (Table 10.1). It should be remembered, however, that there is no sharp division between such factors since pre-infectional compounds can undergo significant post-infectional changes. While this system is thus to a certain degree an arbitrary one, it is convenient

Table 10.1 A classification of disease resistance factors in higher plants

Class	Description
Pre-infectional compounds	
1. Prohibitins	Metabolites which reduce or completely halt the *in vivo* development of micro-organisms
2. Inhibitins	Metabolites which undergo post-infectional increase in order to express full toxicity
Post-infectional compounds	
1. Post-inhibitins	Metabolites formed by the hydrolysis or oxidation of pre-existing non-toxic substrates
2. Phytoalexins	Metabolites formed *de novo* after invasion by gene derepression or activation of a latent enzyme system

Modified from Ingham (1973).

and will be used here as a framework in which to discuss some of the chemical constituents that have been implicated as disease resistance factors.

The occurrence in plants of substances which will inhibit the germination and/or growth of micro-organisms has been recognized for a long time. Indeed, many of the so-called secondary constituents, especially terpenoids and phenolics, have been demonstrated as having such effects in *in vitro* tests. One of the most striking accumulations of secondary compounds in the plant kingdom is in the heartwood of trees. An enormous profusion of chemical structures, with terpenoid, quinonoid and phenolic skeletons, have been isolated from such tissues, where they frequently occur in considerable quantity (Hillis, 1962). The majority of such tissues are unusually resistant to natural decay (Scheffer and Cowling, 1966) and there have been many suggestions that the role of these heartwood constituents is to provide disease resistance. Among the many groups of compound specifically linked with resistance to fungal attack are the hydroxystilbenes. One such compound is pinosylvin (for structure, see Fig. 10.1) which is widely distributed in *Pinus* and other Pinaceae. Unfortunately, evidence that these compounds are important to the plant is still almost entirely circumstantial, since the experimental difficulties in providing more direct evidence have largely prevented further progress in this field. The occurrence of hydroxystilbenes as pre-infectional compounds in tree heartwoods is supported, at least to some degree, by the identification of similar structures as phytoalexins in the leaves of several legumes (see p. 279).

One of the most often quoted examples in the plant pathological literature of preinfectional compounds apparently providing disease resistance in non-woody plants is the rather exceptional situation of onion bulbs suffering from onion smudge disease, *Colletotrichum circinans* (Walker and Stahmann, 1955). The dead outer scales of resistant onion varieties contain large quantities of protocatechuic acid and catechol (see Fig. 10.1), both of which are highly toxic to spores of *C. circinans*. Extracts of these scales reduce spore germination to below 2% whereas extracts of susceptible onion varieties which lack these agents in quantity allow a germination rate of over 90%. The occurrence of

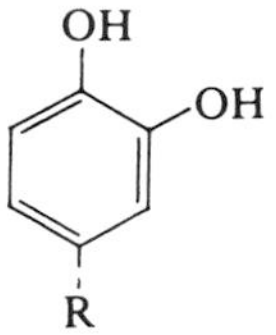

R = H, catechol
R = CO_2H, protocatechuic acid
(onion bulbs, *Allium cepa*)

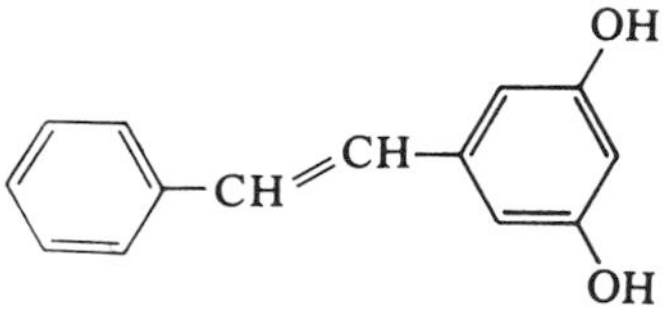

pinosylvin
(*Pinus* heartwood)

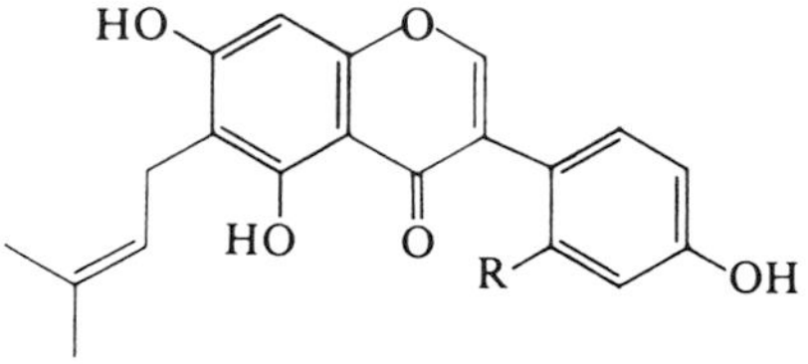

R = OH, luteone
R = H, 2′-deoxyluteone
(lupin leaves, *Lupinus* spp.)

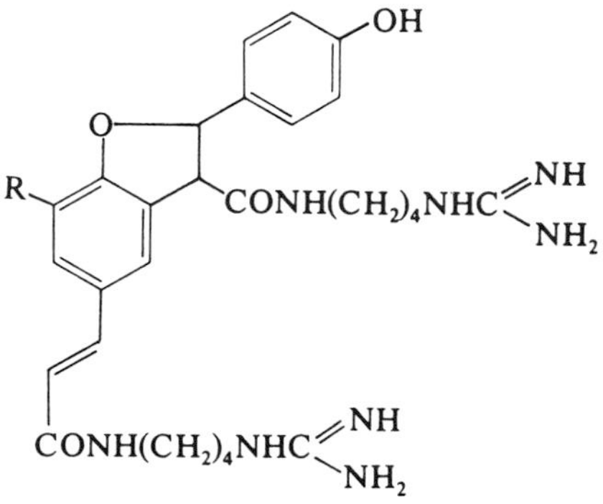

R = H, hordatine A
R = OMe, hordatine B
(barley seedlings, *Hordeum vulgare*)

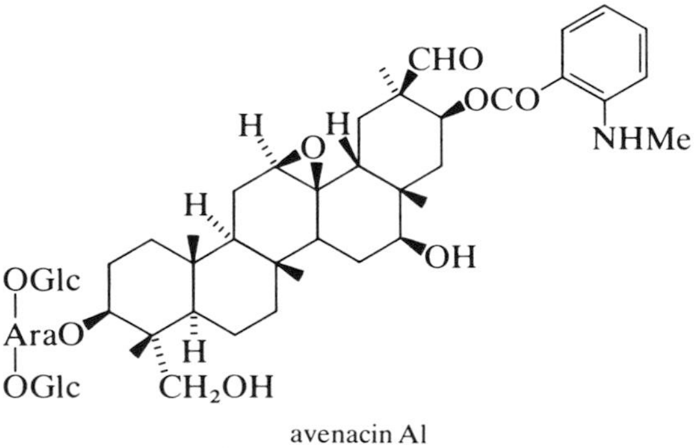

avenacin Al
(oat roots, *Avena sativa*)

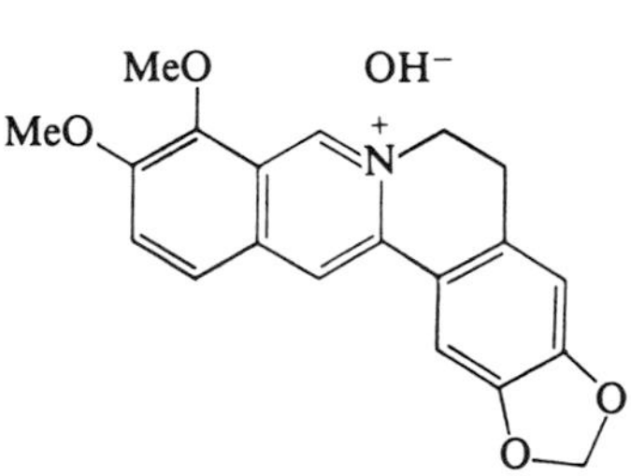

berberine
(*Mahonia trifoliata* roots)

Figure 10.1 Structures of some pre-infectional compounds present in plants

these two phenols in resistant varieties is correlated with anthocyanin colour in the scales, a correlation which is probably purely fortuitous in that other fungi are able to invade both coloured and colourless onion varieties indiscriminately. Anthocyanins based on cyanidin (i.e. with a catechol nucleus in their structures) have the potentiality of being fungitoxic and indeed cyanidin itself has been shown to inhibit germination of *Gloeosporium perennans*, a fungus causing apple rot (Hulme and Edney, 1960).

Evidence that preinfectional compounds also have a role in the disease resistance in living tissue, in fresh leaves, has been obtained from studies with lupins (Harborne *et al.*, 1976). Whilst examining leaves of white lupin *Lupinus albus* for phytoalexins using the drop diffusate technique (see p. 278) it was found that at least two fungitoxic compounds were present in significant concentration *both* on the leaf surface *and* within the leaf. The compounds were not induced by the microbial spores, since they were present in equal amounts in control droplets which lacked any fungal spores; these two compounds were identified as luteone, an isopentenylisoflavone, and its 2′-deoxy derivative (for structures, see Fig. 10.1). Fungitoxicity was evident from *in vitro* tests in which luteone showed an ED_{50} value of 35–40 $\mu g/ml$, when tested against the mycelial growth of *Helminthosporium carbonum*. The ED_{50} is a measure of fungitoxicity and refers to the median effective dose required for the inhibition of mycelial growth of a given fungus. In actual practice, rather higher concentrations will be required for complete (100%) inhibition. The presence of an isopentenyl group in the lupin isoflavones seems to be important for fungitoxicity since comparison of ED_{50} values shows that luteone is ten times more active on a molar basis than isoflavones lacking this terpenoid-based side chain attachment, e.g. biochanin A and formononetin.

Preinfectional compounds appear to occur throughout the genus *Lupinus*, since examination of 11 other species besides *L. albus* showed the presence on leaf surfaces of luteone or related isoflavones. Thus, under natural conditions, the isoflavone concentration in the leaf surface moisture film may produce an *in vivo* environment highly unfavourable for fungal development. The toxic effect is probably on fungal spore germination and/or germ tube development rather than on mycelial proliferation. This was evident from histological studies which indicated curled, distorted and highly branched germ tubes when *Helminthosporium carbonum* was allowed to invade lupin leaves; on leaves of its normal host plant, maize, the same organism produced normal, straight germ tubes.

Another group of flavonoids, this time methylated flavones, have been partly implicated in the resistance of citrus leaves to fungal attack. Substances such as nobiletin (5,6,7,8,3′,4′-hexamethoxyflavone) are highly fungitoxic and occur in leaves in sufficient amount to ward off attack. However, a quantitative investigation of the relationship between the concentration of nobiletin in the leaves and resistance to the fungus *Deuterophoma tracheiphila* showed no simple correlation (Piattelli and Impellizzeri, 1971). Nevertheless, nobiletin has many of the characteristics of a preinfectional agent and its high lipid solubility and contrasting lack of water solubility suggests that it is most likely to occur on the leaf surface rather than within it. The role of flavonoids as possible preinfectional agents in disease resistance is further discussed by Harborne (1987).

Yet another example of phenolic compounds being involved in disease resistance is the case of the hordatines A and B, which have a protective function in barley seedlings *Hordeum vulgare* attacked by *Helminthosporium sativum* (Stoessl, 1967); these compounds are of an unusual structure, being based on two *p*-coumaric acid residues linked to the amino acid arginine (see Fig. 10.1). An interesting complication in this instance is that the effectiveness of these fungitoxins decreases as the seedlings develop due to the accumulation in the tissues of Ca^{2+} and Mg^{2+} ions. *In vitro* tests have shown that the two hordatines are inactivated by divalent cations, presumably due to complex formation.

Evidence that structures other than phenols can act as prohibitins has come from work on root infections of cereals. Resistance in oats, *Avena sativa*, to the take-all fungus, *Gaeumannomyces graminis*, is due to a mixture of four closely related pentacyclic triterpene glycosides, of which avenacin A1 (Fig. 10.1) is typical (Crombie *et al.*, 1986). While growth of non-pathogenic strains of the fungus, e.g. *G. graminis* var. *tritici*, are completely suppressed by these compounds, the pathogenic *G. graminis* var. *avenae* is able to detoxify the avenacins by removing one or both of the glucose residues of the trisaccharide unit at the 3-position. Such deglucosylated compounds are no longer fungitoxic; hence the ability of oats to resist take-all disease is limited by the β-glucosidase activity that may be present in the pathogen. A similar system involving a different triterpenoid, avenacoside A, and the enzymic removal of a 26-glucose residue appears to operate in oat leaves when infected with the fungus *Drechslera avenacea* (Lüning and Schlösser, 1976). Another example of a non-phenolic prohibitin is that of the alkaloid berberine which is thought to provide resistance in roots of *Mahonia trifoliata* to invasion by the fungus *Phymatotrichum omnivorum* (Greathouse and Watkins, 1938).

The above triterpenoid and alkaloidal toxins are located inside the leaf or root, probably occurring in the cell vacuole. Other prohibitins, by contrast, may only occur outside the leaf, in special sites on the surface, thus providing a first line of defence against invading micro-organisms. This is true of the sesquiterpene lactone parthenolide, which is present in two-lobed glands on the surface of the leaf and seed of *Chrysanthemum parthenium*. Parthenolide appears to have a useful role in preventing microbial attack, since it is toxic to both Gram-positive bacteria and filamentous fungi (Blakeman and Atkinson, 1979). It is also true of sclareol and isosclareol, two diterpenes present on the leaf of *Nicotiana glutinosa* (Bailey *et al.*, 1974). These compounds are present in high concentration in liquid droplets visible under the microscope on the upper leaf surface. They effectively inhibit fungal growth *in vitro*, possibly because they are structural analogues of the gibberellins and thus interfere with normal hormonal development in the fungus, e.g. by increasing the amount of hyphal branching.

In summary then, there is evidence that a small number of plants have a disease resistance mechanism based in part on the presence of preformed fungitoxins which are sufficient in themselves in preventing microbial invasion. The compounds are mainly phenolic in nature but other types of chemical structures, especially triterpenoids and alkaloids, have also been implicated in this type of defence mechanism.

A second class of preinfectional compounds thought to be involved in disease resistance in plants are the inhibitins, metabolites which undergo post-infectional increases in order to express their toxic potential (Ingham, 1973). This group of compounds has been proposed on the basis of the general observation that infection in many plants by micro-organisms leads to the accumulation of various aromatic compounds, especially coumarin derivatives near the site of infection. One of the most striking examples of this is in blight-infected potatoes, where an intensely blue fluorescent zone can be seen in the tuber near the edge of the infection (Fig. 10.2). If Majestic potatoes are inoculated at one end with spores of *Phytophthora infestans*, incubated for 14 days at 18°C and then sliced

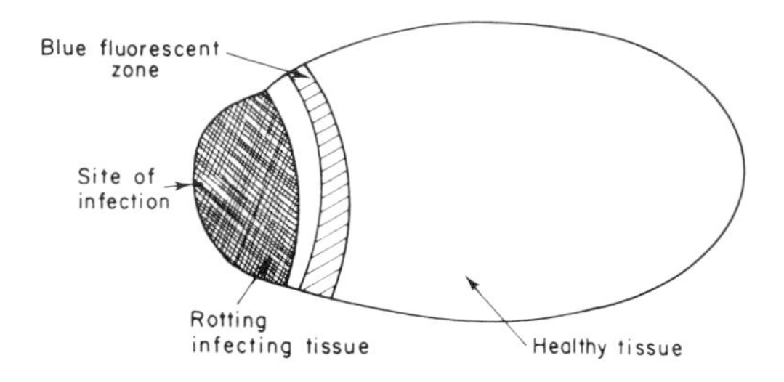

Figure 10.2 Formation of inhibitory zone of coumarin in potato infected with blight

longitudinally in half, the potato will appear as shown in Fig. 10.2. The intense blue fluorescent zone appears between the infected area and the healthy tissue and gives the appearance of being formed as a protective zone between the two types of tissue. Comparison of phenolic constituents in healthy tissue and in the blue fluorescent zone shows that in this zone, there are large increases in the synthesis of two particular phenolics. One, the fluorescent coumarin scopolin, increases 10–20-fold, whilst the other, chlorogenic acid, of weaker fluorescence, increases 2–3-fold (Hughes and Swain, 1960). Both coumarins and hydroxycinnamic acids such as chlorogenic are known to be significantly toxic when tested *in vitro* against a range of micro-organisms (cf. Jurd *et al.*, 1971).

Such changes in phenolic metabolites (for structures, see Fig. 10.3) may be purely incidental to the effect of infection on the normal metabolism in the tissue. The increased synthesis of these substances may not be directly related to disease resistance, but may simply be a symptom of the disease. It may, however, be mentioned that similar increases in scopolin, scopoletin or chlorogenic acid levels occur in other plants besides the potato, e.g. in tobacco *Nicotiana tabacum* and sweet potato *Ipomoea batatas*. One plant where an increase in as yet unidentified blue fluorescent compounds has been correlated with disease resistance is apple, when the leaves are infected with the scab organism *Venturia inaequalis* (Hunter *et al.*, 1968). Here, there are three compounds which inhibit conidial germination of *V. inaequalis* and which are present in healthy leaves but increase to inhibitory levels as a result of infection. On the assumption that the blue fluorescent components are hydroxycinnamic acid derivatives of some novel type, various synthetic acids (e.g. isobutyl *o*-coumarate) have successfully been applied as a spray on leaves of susceptible apple varieties to enhance their resistance to the scab disease (Kirkham and Hunter, 1964). However, in spite of this success, it is clear that much further work is needed to establish whether substances which increase in amount after infection really have a significant role in plant protection.

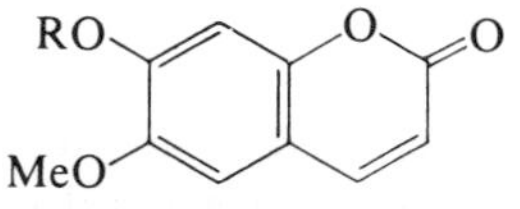

R = H, scopoletin
R = Glc, scopolin

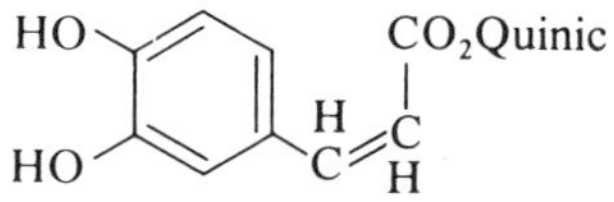

chlorogenic acid

Figure 10.3 Some phenolic inhibitins

The concept that catecholic phenols, such as chlorogenic acid (Fig. 10.3), do provide disease resistance to plants has been strengthened by the discovery that at least one pathogen has the ability to counteract their toxic effects. This is the anthracnose fungus *Colletotrichum graminicola*, a major disease on cereals throughout the world. The spores produce a water-soluble mucilage, the glycoprotein fraction of which has an exceptionally high affinity to bind condensed tannins and other phenols. Leachates from infected cereal leaves were shown to be rich in toxic phenols and these were immediately inactivated, through binding, in the presence of the fungal glycoprotein (Nicholson *et al.*, 1986). Thus this organism, through co-evolution, has developed a specific means by which it protects itself from the otherwise toxic phenols formed in plants in response to infection.

B. Post-infectional Compounds: Post-inhibitins

The concept of plants storing in an inactive form within their tissues toxins needed for their protection is now a familiar one. It applies to toxins produced to deter herbivores (Chapter 3) and toxins serving as allelopathic agents (Chapter 9). There is also evidence that disease resistance in plants is afforded by a similar mechanism. The toxins involved are described in Ingham's classification (Table 10.1) as post-inhibitins. They are present in healthy tissues as inactive glycosides, the active toxin being released by enzymic hydrolysis or oxidation following microbial invasion.

The simplest example of this type of mechanism involves the cyanogenic glycosides, which are not toxic as such, but which release prussic acid when hydrolysed by a specific β-glucosidase, the intermediate cyanohydrin formed breaking down spontaneously to aldehyde or ketone and HCN (see Fig. 10.4). The HCN so produced then serves to protect the plant from further fungal colonization. Release of cyanide occurs, for example, when leaves of birdsfoot trefoil, *Lotus corniculatus*, are invaded by the leaf pathogen *Stemphylium loti* (Millar and Higgins, 1970). In this example, the pathogen is relatively tolerant to HCN and can readily adapt to cyanide when cultured in its presence by producing the enzyme formamide hydrolyase, which detoxifies HCN by conversion to formamide $HCONH_2$. With *Stemphylium* infection, the HCN produced is relatively ineffective in halting fungal invasion. The value of HCN under field conditions is that a number of fungi with the potential to attack birdsfoot trefoil lack the ability to produce a detoxifying enzyme and hence are unable to become established within the tissue of this plant.

A parallel example to the cyanogenic glycosides are the glucosinolates of the Cruciferae, which have been shown to be involved in the resistance of wild and cultivated *Brassicas* to downy mildew (Greenhalgh and Mitchell, 1976). In these plants, tissue damage releases significant amounts of the volatile oil, allyl isothiocyanate, formed by enzymic hydrolysis of the glucoside sinigrin by myrosinase (Fig. 10.4). This isothiocyanate, which is one of the major flavour principles of cabbage and other vegetable crucifers, is highly toxic to the mildew pathogen *Peronospora parasitica*. Evidence that release of isothiocyanate is causally connected with the control of mildew infection is based on two findings. First, in cultivated *Brassica oleracea*, there is a correlation between isothiocyanate content and disease resistance: thus allyl isothiocyanate content, measured in

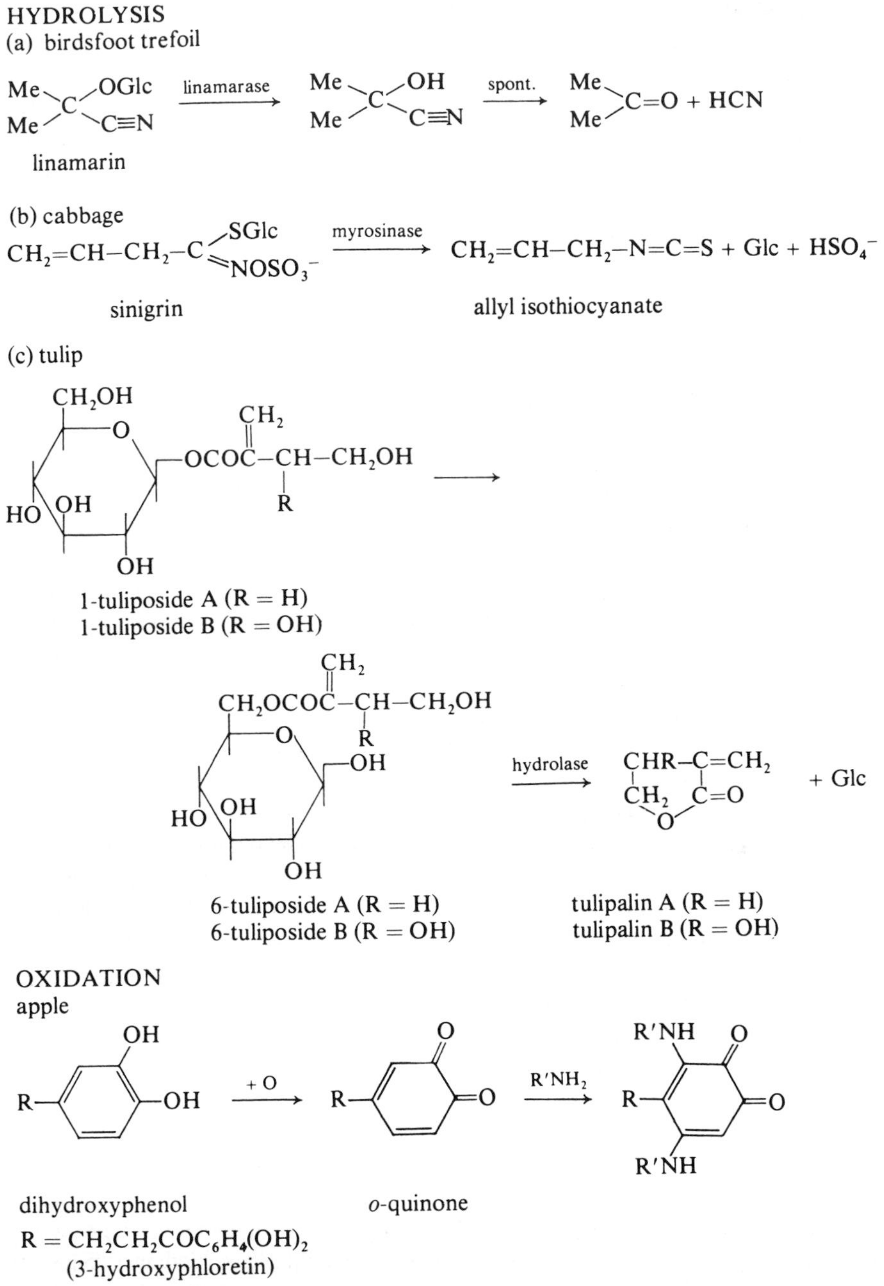

HYDROLYSIS

(a) birdsfoot trefoil

$$Me_2C(OGlc)C{\equiv}N \xrightarrow{\text{linamarase}} Me_2C(OH)C{\equiv}N \xrightarrow{\text{spont.}} Me_2C{=}O + HCN$$

linamarin

(b) cabbage

$$CH_2{=}CH{-}CH_2{-}C(SGlc){=}NOSO_3^- \xrightarrow{\text{myrosinase}} CH_2{=}CH{-}CH_2{-}N{=}C{=}S + Glc + HSO_4^-$$

sinigrin — allyl isothiocyanate

(c) tulip

CH_2OH, O, $OCOC(=CH_2){-}CH{-}CH_2OH$, R, HO, OH, OH →

1-tuliposide A (R = H)
1-tuliposide B (R = OH)

$CH_2OCOC(=CH_2){-}CH{-}CH_2OH$, R, O, OH, HO, OH, OH $\xrightarrow{\text{hydrolase}}$ $CHR{-}C{=}CH_2$, CH_2, $C{=}O$, O + Glc

6-tuliposide A (R = H)
6-tuliposide B (R = OH)

tulipalin A (R = H)
tulipalin B (R = OH)

OXIDATION

apple

R, OH, OH $\xrightarrow{+O}$ R, O, =O $\xrightarrow{R'NH_2}$ R'NH, O, R, =O, R'NH

dihydroxyphenol — *o*-quinone

$R = CH_2CH_2COC_6H_4(OH)_2$
(3-hydroxyphloretin)

Figure 10.4 Mechanisms of release of fungitoxins by hydrolysis or oxidation

μg/g dry wt, was found to be 630 in a resistant variety and between 450 and 21 in susceptible varieties. Second, in wild populations of cabbage, the highest proportion of resistant seedlings (up to 47%) occurs in those populations with the highest levels of flavour volatiles. Since wild populations in general have much higher isothiocyanate contents than cultivars, it appears that selective breeding for milder flavoured *Brassica* vegetables has been a contributory factor to the general lack of resistance to powdery mildew in modern varieties.

Another convincing example of post-inhibitins having importance in disease resistance is that of the 1-tuliposides A and B. These two glucosides occur in young tulip bulbs and provide them with resistance to the pathogen *Fusarium oxysporum* during most of the growing season. The two glucosides are themselves only mildly antibiotic but they rearrange to the isomeric 6-tuliposides A and B, which are inert but which in turn undergo enzymic hydrolysis followed by cyclization to yield the highly fungitoxic tulipalins A and B (see Fig. 10.4 for relevant structures). The fungitoxicity of the latter pair of substances is thought to be related to their ability to complex with SH groups and to inhibit the enzymic activities of the attacking fungus (Beijersbergen and Lemmers, 1970). It is interesting, in passing, that the same substances have allergenic properties in man and tuliposide A and its reaction products are responsible for the skin disease caused by excessive handling of tulip bulbs. The contribution of tuliposides A and B to disease resistance may be fairly widespread in the Liliaceae, since these glucosides have been identified in 25% of some 200 species surveyed; although mainly present in *Tulipa*, they also occur in many *Alstroemeria* species (Slob *et al.*, 1975).

The role of phenolics as pre-infectional metabolites has already been discussed in a previous section (p. 267). It is possible that certain of these phenolics may also have a role as post-inhibitins. Thus, one general mechanism by which disease resistance is enhanced is by the oxidation of pre-existing 3,4-dihydroxyphenols, which may or may not be first released by hydrolytic cleavage of esters or glucosides. The *o*-quinones so formed (Fig. 10.4) appear to be highly toxic; moreover, they are capable of undergoing condensation with amino compounds, including amino acids, to give even more toxic products. This type of mechanism has been invoked to explain the resistance of apple varieties to their various fungal pathogens. The pre-existing compound is phloridzin, the β-glucoside of phloretin. After hydrolysis of phloridzin, the phloretin formed is first oxidized to 3-hydroxyphloretin and this then undergoes further oxidation, mediated by the enzyme phenolase, to quinone. The process by which these various steps occur is still not clear. Nevertheless, in resistant leaves, it appears that the fungitoxic *o*-quinone is formed only when the invading micro-organism damages cellular membranes, allowing the plant enzyme phenolase to oxidize the natural substrate. The product of this oxidation then arrests further fungal development. By contrast, in susceptible varieties, cellular damage is minimized, the potential of the leaves to form fungitoxic quinone is suppressed and the fungus is then able to invade the host tissues without let or hindrance (Sijpesteijn, 1969). Whether such a system actually provides disease resistance in apples is still debatable, since Hunter (1975) found that in the case of apple scab, *Venturia inaequalis*, neither phloridzin nor its simple breakdown products provided any degree of resistance. However, the related quinones were not directly tested, because of their high instability *in vivo*.

One final example of post-inhibitins are sulphur amino acids present in onion tissues, which release pungent and lachrymatory sulphides after enzymic or hydrolytic attack. Diallyl disulphide $(CH_2{=}CH{-}CH_2{-}S)_2$, is one such toxin. These sulphides are undoubtedly protective in *Allium*, but like all defence mechanisms, are capable of being overturned. Indeed, Coley-Smith (1976) has shown that the soil-borne fungus *Sclerotium cepivorum* depends on the release of the sulphide from onion roots to trigger sclerotial germination and subsequent root infection. This finding can be used to the pathogen's disadvantage, since sclerotia will germinate in the presence of sulphide and absence of host. They will then die and healthy onions can be subsequently safely cultivated in sulphide-treated soil.

C. Post-infectional Compounds: Phytoalexins

The Phytoalexin Concept

The phytoalexin concept of disease resistance has undoubtedly led to one of the major developments in physiological plant pathology of the last 25 years and it has probably stimulated more research into the mechanisms of disease resistance in plants than any other single idea. Today, the biochemical aspects of these antifungal agents are well known and only a summary is needed here. It should, however, be emphasized that physiological, ultrastructural and pathological aspects of phytoalexin synthesis are not yet fully documented and although it is widely accepted that phytoalexins have a place in disease resistance, many aspects of their production and metabolism *in vivo* are not yet fully understood. Before discussing the present state of the phytoalexin field, it is worth restating the main tenets of the theory, as postulated by Müller and Börger in 1941 from their studies of the reaction of potato varieties to virulent and avirulent strains of *Phytophthora*. These tenets are as follows:

(1) A phytoalexin is a compound which inhibits the development of the fungus in hypersensitive tissues and is formed or activated *only when* the host plants come in contact with the parasite.
(2) The defence reaction occurs only in living cells.
(3) The inhibitory agent is a discrete chemical substance, a product of the host cell.
(4) The phytoalexin is non-specific in its toxicity towards fungi; however, fungal species may be differentially sensitive to it.
(5) The basic response in both resistant and susceptible cells is the same, the basis of differentiation between resistant and susceptible hosts being the speed of formation of the phytoalexin.
(6) The defence reaction is confined to the tissue colonized by the fungus and its immediate neighbourhood.
(7) The resistant state is not inherited; it is developed after the fungus has attempted infection. The sensitivity of the host cell which determines the speed of the host reaction is specific and genotypically determined.

As already mentioned, belief in the phytoalexin theory was not vindicated until 20 years later when Cruickshank and Perrin (1960) crystallized and chemically characterized the first phytoalexin; this was pisatin, a pterocarpan derivative produced by pods of *Pisum sativum* inoculated with conidia of the brown rot fungus, *Monilinia fructicola*. Cruickshank and Perrin (1964) were able to show that pisatin fulfulled all the criteria required by Müller and Börger's theory and this substance remains one of the most fully investigated phytoalexins known today. Subsequent studies soon established that other legumes, particularly *Phaseolus vulgaris*, produced similar pterocarpans (e.g. phaseollin) on fungal inoculation. At the same time reconsideration of compounds isolated from diseased plants in other families showed that phytoalexin production is a feature of the Convolvulaceae (ipomeamarone from infected sweet potato) and of the Orchidaceae (orchinol from orchid tubers). Efforts were then directed towards identifying the compounds formed in response to microbial attack in many other crop plants and a range of chemical structures were found to fit the phytoalexin concept. In particular, several substances, among them the sesquiterpenoid rishitin, were characterized in the potato-blight interaction originally investigated by Müller and Börger. The chemical structures of a representative sample of phytoalexins known today are illustrated in Fig. 10.5.

Our knowledge of phytoalexins has increased considerably since the original hypothesis was proposed and although most known phytoalexins fit remarkably well into the general scheme, some modifications of the tenets of the theory have to be considered in some instances. We now know that the interaction is more complex than originally envisaged, particularly since some fungi have the capacity to further metabolize and detoxify phytoalexins. Thus, the ability of a fungus to parasitize a particular plant is related at least in part to its ability to deal with the phytoalexins produced by the host. This further metabolism will be mentioned in a later section. First, however, some comments are required on factors inducing the phytoalexin response, on the structural variation among phytoalexins and on taxonomic and evolutionary aspects.

Factors Inducing the Phytoalexin Response

For a chemical substance to be regarded as a phytoalexin, its formation has to be induced experimentally in healthy plant tissue by inoculation or infection with micro-organisms. A simple procedure for testing the phytoalexin response of plants has been devised. This is the drop diffusate technique (Fig. 10.6), in which leaves are floated on water in light and droplets of a spore suspension of a non-pathogenic fungus are placed on the upper leaf surface (Higgins and Millar, 1968). A surface-active agent such as Tween-20 is added to prevent the droplets from spreading over the leaf surface. A second batch of leaves are set up at the same time, with aqueous Tween-20 droplets as a control. These are left for 48 h and then the droplets are collected. In the case of a positive response, the droplets are found to contain high levels of the phytoalexin largely uncontaminated with other plant constituents. Thus, in a very simple way, it is possible to obtain a relatively pure phytoalexin sample, which can then be used for further investigation.

What happens in this leaf-droplet system is that the fungal spores start germinating in the droplets within 1–2 h, and the resulting germ tubes then penetrate the host cells. As a

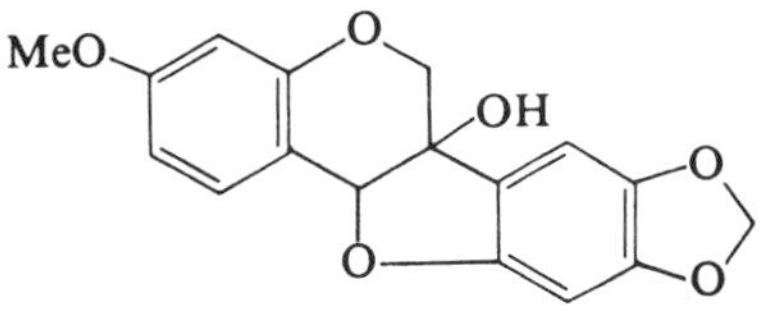

pisatin
(*Pisum sativum*, Leguminosae)

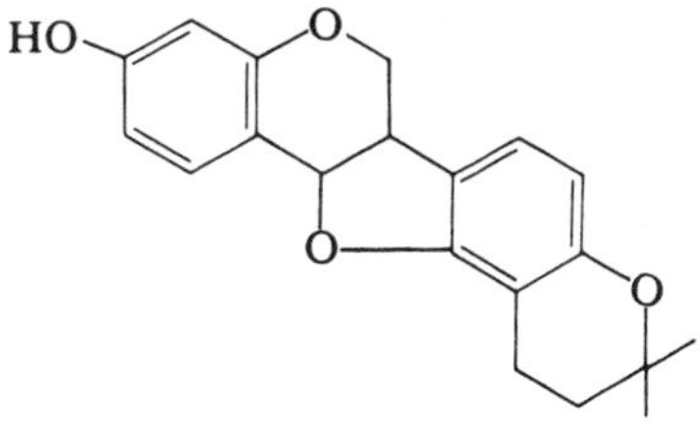

phaseollin
(*Phaseolus vulgaris*, Leguminosae)

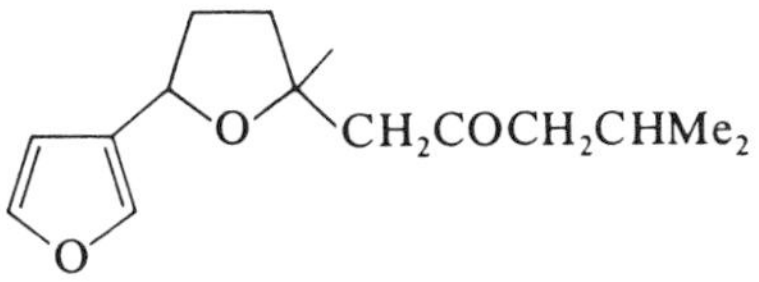

$CH_2COCH_2CHMe_2$

ipomeamarone
(*Ipomoea batatas*, Convolvulaceae)

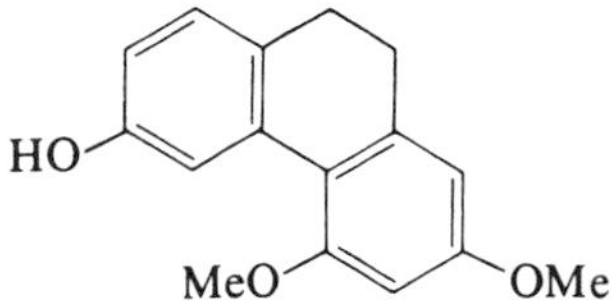

orchinol
(*Orchis militaris*, Orchidaceae)

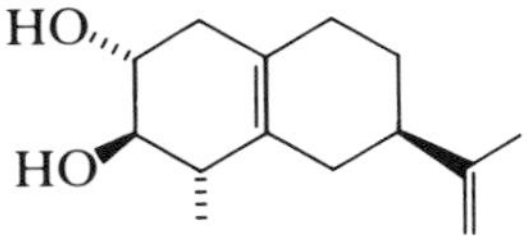

rishitin
(*Solanum tuberosum*, Solanaceae)

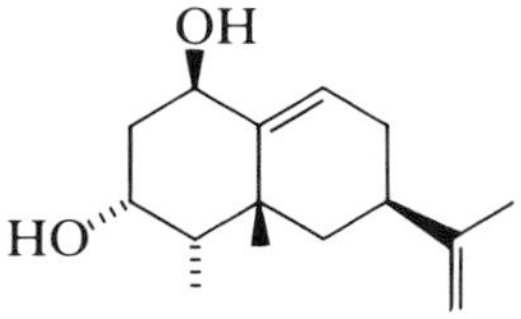

capsidiol
(*Capsicum frutescens*, Solanaceae)

$HOCH_2CHOHCH{=}CH(C{\equiv}C)_3CH{=}CHMe$

safynol
(*Carthamus tinctoria*, Compositae)

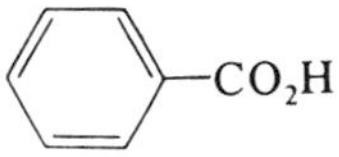

$-CO_2H$

benzoic acid
(*Malus pumila*, Rosaceae)

Figure 10.5 Structures of representative phytoalexins of higher plants

consequence of this 'trigger', the plant immediately responds by *de novo* synthesis of phytoalexin; such compounds, which are often detectable after several hours, reach maximum production at 48 to 72 h. The phytoalexins are themselves synthesized within the leaf but much material is 'pushed out' onto the leaf surface where the fungal invasion is occurring: hence, the accumulation of phytoalexin in the droplets. It has been argued (see Hargreaves *et al.*, 1976a) that the drop diffusate technique does not provide a complete picture of phytoalexin production, since certain substances may not diffuse from the leaf into the overlying droplets. However, tests have shown that for many species the same compounds are present in both the leaf and in the droplets, although there may be quantitative differences if more than one compound is produced. In some plants, it may be more

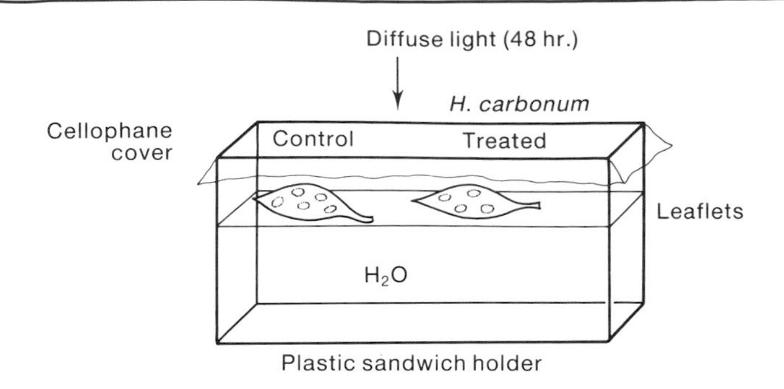

Figure 10.6 Drop diffusate technique for phytoalexin induction

convenient to challenge other parts than the leaf, and stem, hypocotyl and seed have all been used on occasion.

In order to confirm phytoalexin synthesis, the droplet extracts must be tested for fungitoxicity. This can conveniently be done by TLC bioassay (see Fig. 10.7), in which the two extracts (phytoalexin solution and water control) are developed on a TLC plate in a suitable solvent. After drying, the plate is sprayed with a fungal spore suspension (e.g. of *Cladosporium herbarum*) and incubated for about 5 days at 25–30°. During this time, the fungus grows over the whole plate except where fungitoxic zones are present; these areas appear as white spots on a grey background. The TLC plate after development can also be sprayed with a range of diagnostic reagents (e.g. diazotized *p*-nitroaniline solution—a test for phenols) in order to determine to which chemical class the phytoalexin belongs. The final stage in the investigation is to identify the compounds present, using the standard procedures of organic chemistry.

While phytoalexins are formed reproducibly, most frequently and in the highest yield when plants are invaded with fungi, they are also formed on occasion when plants are subjected to bacterial or viral invasion. However, they can also be formed under abiotic conditions, when plants are subjected to stress. Among such stress factors are UV irradiation, temperature shock, wounding and treatment with inorganic salts (e.g. aqueous mercuric chloride). It is even possible that phytoalexins occur in trace amounts in healthy tissue, although the evidence here is to some extent equivocal since a completely healthy unstressed plant is difficult to define precisely.

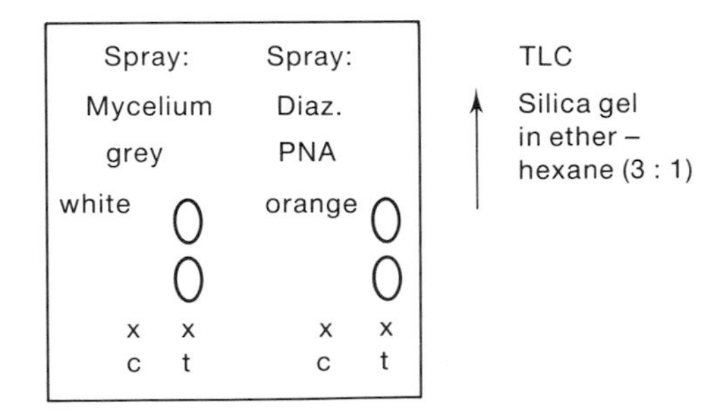

Figure 10.7 Bioassay and detection of phytoalexins in control (c) and treated (t) droplets

In conclusion, therefore, one has to accept the fact that phytoalexins are antimicrobial substances which can be produced in plants as part of a general repair and defence system and which are triggered off by a variety of agencies. In terms of disease resistance, it is not simply the ability to synthesize phytoalexin that is important, but the capacity to produce it in quantity, quickly enough in the right place and at the right time. This plants do when invaded by fungi and undoubtedly it is as antifungal agents that phytoalexins are most important from the ecological point of view.

Structural Variation Among Phytoalexins

Some of the structural variation encountered among the phytoalexins is illustrated in Fig. 10.5, which shows the typical phytoalexins produced by six plant families. It is clear from this that no simple relationship exists between chemical structure and fungitoxicity, since quite different structures can give high toxicity. The only general property that most, if not all, phytoalexins share is a degree of lipid solubility. Even the phenolic compounds (cf. pisatin, orchinol, phaseollin) have most of the polar hydroxyl groups masked by methylation or by methylenedioxy ring formation and are thus rendered lipid soluble. This property may be a necessary requirement for toxicity if the substance is going to interfere with, attack or block the membrane permeability of fungal cells.

Chemically, the simplest phytoalexin is benzoic acid, a phytoalexin formed in response to the storage rot of apples, *Nectria galligena* (Swinburne, 1973), and to the needle blight fungus of *Pinus radiata* (Franich *et al.*, 1986). Most other phytoalexins are more complex, being isoflavonoid (pisatin), terpenoid-derived (rishitin, ipomeamarone) or fatty acid derivatives (safynol). More recently, indole-derived sulphur-containing phytoalexins have been found in *Brassica juncea* and *Camalina sativa* (see Fig. 10.8), while several dibenzo-

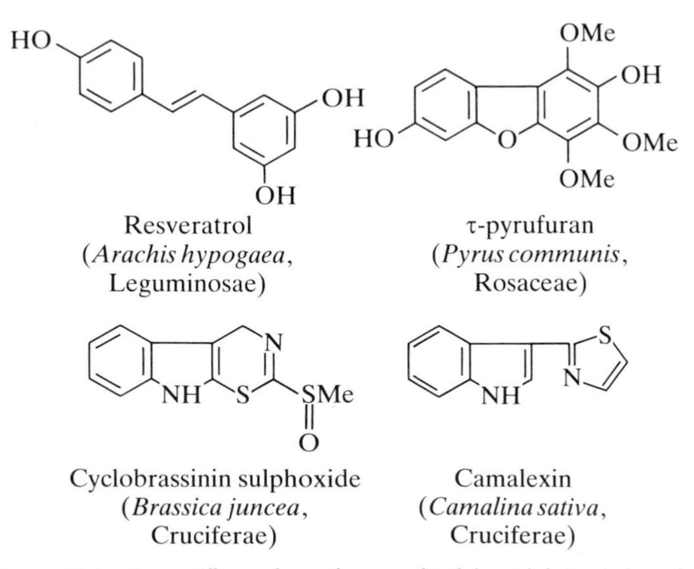

Figure 10.8 Some stilbene-, benzofuran- and indole acid-derived phytoalexins

furans, including τ-pyrufuran from the pear *Pyrus communis* have been identified in the infected sapwood of rosaceous trees. Further details of phytoalexin structures can be found in Harborne and Ingham (1978), Bailey and Mansfield (1982) and Harborne (1986).

The relationship between fungitoxicity and structure has been explored at the family level and some results are available with regard to the products of the Leguminosae (VanEtten, 1976). It is apparent, for example, that the different classes of isoflavonoid fall into a series of increasing toxicity: isoflavones–isoflavanones–pterocarpans–isoflavans. Fungitoxicity is also dependent to a lesser extent on the nature and number of substituents in either aromatic ring. While almost all legumes that have been studied produce isoflavonoids, an apparently different class of phytoalexin, hydroxystilbenes, have been detected in *Arachis* and *Trifolium* (Ingham, 1976a). However, such molecules (e.g. resveratrol), although biosynthetically distinct, are structurally very close to the isoflavonoid skeleton (Fig. 10.8). There is, indeed, a striking parallel here between fungitoxicity and oestrogenic activity. Thus legume isoflavones are well known as weak oestrogens (see Chapter 4) and one of the most active synthetic oestrogens is the reduced hydroxystilbene, diethyl-stilboestrol. As mentioned above, both classes—isoflavonoid and stilbene—are also fungitoxic. This suggests that antifungal activity in legumes may be related to the production of a molecule which is a steroid 'mimic' and can thus either interfere with the steroid nutrition of the fungus or more directly affect the membrane permeability of the fungal cell.

Taxonomic and Evolutionary Aspects

The known phytoalexins now number over 200 chemical structures, about half of which have been characterized from a single family, the Leguminosae (Ingham, 1981). A significant proportion of these structures are uniquely produced in this way and are not otherwise known as natural products. They have been characterized from at least 20 plant families (see Table 10.2), ranging from monocots to dicots and from trees to herbs.

In any given plant, several substances are likely to be produced *de novo* on fungal infection, their concentrations varying according to the experimental conditions used. Sometimes, there may be a major phytoalexin, with several minor components being formed as well. In lima bean, as many as 25 isoflavonoids have been obtained in the phytoalexin response (O'Neill *et al.*, 1983). Occasionally, chemically unrelated phytoalexins may be formed in the same plant. In the sweet pea, *Lathryus odoratus*, the main phytoalexin is the isoflavonoid pisatin, first obtained from *Pisum sativum* (Fig. 10.5), but two unrelated chromones are also produced in minor amounts (Robeson and Harborne, 1980).

There is a strong taxonomic element in phytoalexin production by plants and in some families, one particular type is likely to be formed in most species. This is true in the Solanaceae, where all plants so far tested have yielded sesquiterpenoids, such as rishitin and capsidiol (Fig. 10.5). Occasionally, several types may be formed in the same family (e.g. in Compositae or Leguminosae), so that it may be more difficult to predict what compound may be formed in a given plant. In the tribe Vicieae of Leguminosae, for example, two of

Table 10.2 Chemical variation in phytoalexins within the flowering plants

Family	Genus	Chemical type	Structural example
Monocotyledons			
Amaryllidaceae	*Narcissus*	Flavan	7-Hydroxyflavan
Costaceae	*Costus*	Pterocarpan	Glyceollin II
Gramineae	*Avena*	Benzoxazin-4-one	Avenalumin I
	Oryza	Diterpene	Momilactone A
Orchidaceae	*Orchis*	Phenanthrene	Orchinol
Dicotyledons			
Caryophyllaceae	*Dianthus*	Benzoxazin-4-one	Dianthalexin
Chenopodiaceae	*Beta*	Isoflavone	Betavulgarin
Compositae	*Carthamus*	Polyacetylene	Safynol
	Helianthus	Coumarin	Scopoletin
	Lactuca	Sesquiterpene lactone	Costunolide
Convolvulaceae	*Ipomoea*	Furanoterpene	Ipomeamarone
Euphorbiaceae	*Ricinus*	Diterpene	Casbene
Leguminosae	*Pisum*, etc.	Isoflavonoid	
		Isoflavone	Wighteone
		Isoflavanone	Kievitone
		Pterocarpan	Medicarpin
		Isoflavan	Vestitol
	Lathyrus	Chromone	Lathodoratin
	Arachis	Stilbene	Resveratrol
	Vigna	Benzofuran	Vignafuran
	Vicia	Furanoacetylene	Wyerone
Linaceae	*Linum*	Phenylpropanoid	Coniferyl alcohol
Malvaceae	*Gossypium*	Naphthaldehyde	Gossypol
		Naphthaldehyde	Vergosin
Moraceae	*Morus*	Benzofuran	Moracin-C
		Stilbene	Oxyresveratrol
	Broussonetia	Flavan	7-Hydroxy-4′-methoxy flavan
Rosaceae	*Eriobotrya*	Biphenyl	Aucuparin
	Malus	Phenolic acid	Benzoic acid
	Pyrus	Dibenzofuran	γ-Pyrufuran
Solanaceae	*Lycopersicon*	Polyacetylene	Falcarinol
	Solanum, etc.	Sesquiterpene	Rishitin
Tiliaceae	*Tilia*	Sesquiterpene	7-Hydroxy-calamenene
Ulmaceae	*Ulmus*	Terpenequinones	Mansonone A
Umbelliferae	*Daucus*	Polyacetylene	Falcarinol
		Dihydro-isocoumarin	6-Methoxy-mellein
	Pastinaca	Furanocoumarin	Xanthotoxin
Verbenaceae	*Avicennia*	Quinonoid	Naphthafuranone
Vitaceae	*Vitis*	Stilbene oligomers	α-Viniferin

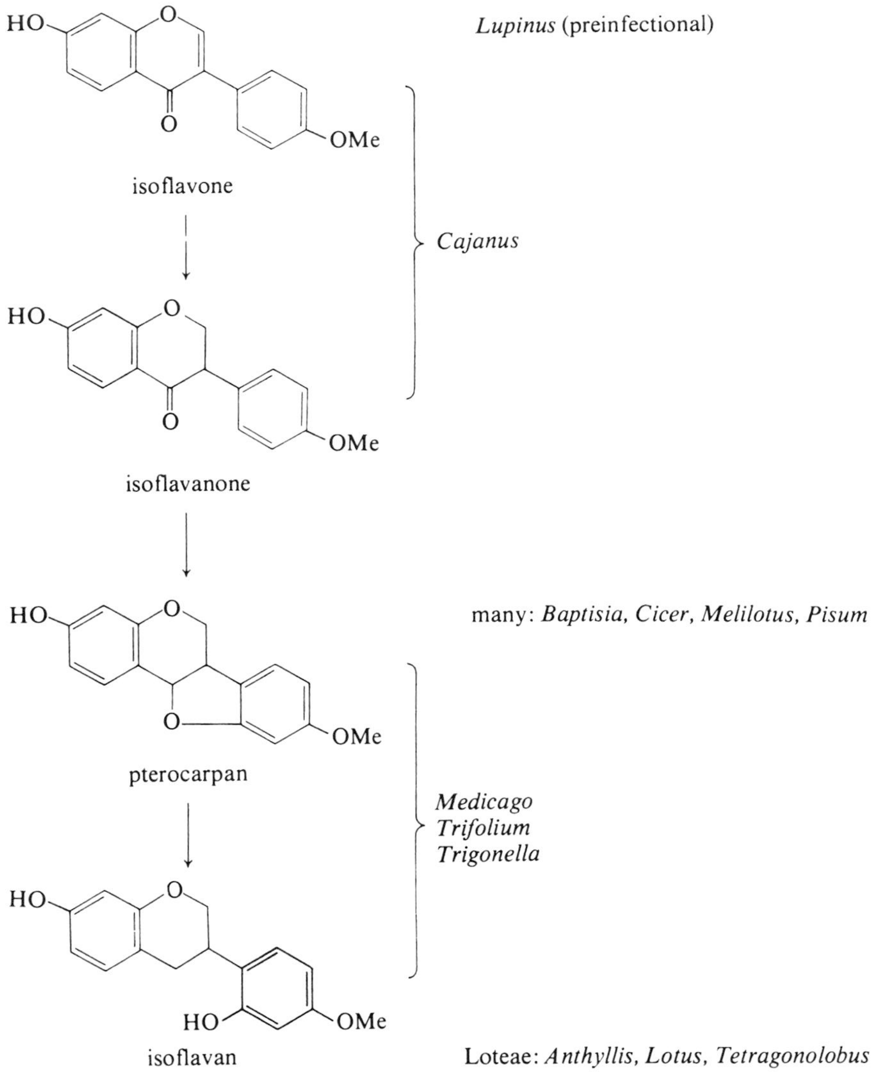

HO O O OMe

isoflavone

Lupinus (preinfectional)

Cajanus

HO O O OMe

isoflavanone

HO O O OMe

pterocarpan

many: *Baptisia, Cicer, Melilotus, Pisum*

Medicago
Trifolium
Trigonella

HO O HO OMe

isoflavan

Loteae: *Anthyllis, Lotus, Tetragonolobus*

Figure 10.9 Phyletic sequence of phytoalexins of increasing fungitoxicity

the four genera (*Lathyrus, Pisum*) expectedly produce the typical legume phytoalexin pisatin, but the other two (*Lens, Vicia*) produce furanoacetylenes instead.

In spite of some exceptions, it is possible to discern an evolutionary progression in phytoalexin synthesis within the Leguminosae (Harborne, 1977) (Fig. 10.9). Different structural types of isoflavonoid can be placed in an evolutionary series based on increasing biosynthetic complexity and increased fungitoxicity. The more highly evolved legumes (e.g. *Lotus*) tend to have the more fungitoxic phytoalexins. These results thus suggest that as plants co-evolve with their fungal parasites, they develop more and more effective phytoalexins to combat such invasion.

Further Metabolism of Phytoalexins

While plants are developing, by natural selection, more effective phytoalexins to ward off fungal attack, the fungi are themselves producing detoxification mechanisms which inactivate these antimicrobial agents. Certainly, there is increasing evidence that these compounds are further metabolized, principally by pathogenic fungi, to substances with decreased fungitoxicity. In the case of isoflavonoids, such detoxification appears to be an essential prerequisite for fungal pathogenicity in some interactions. It should be pointed out that non-pathogenic fungi may also have the enzymic machinery for detoxification.

A variety of mechanisms have been detected for detoxification. For isoflavonoids, two principal steps are hydroxylation of the nucleus and demethylation of methoxyl groups; both processes at once reduce lipid solubility and increase water solubility. The products of such reactions are immediately more susceptible to oxidative cleavage of the aromatic rings and undoubtedly the ultimate fate of these phytoalexins is to be broken down by well-described aromatic-cleavage pathways (Towers, 1964) to eventually yield CO_2. The first steps in the metabolism of medicarpin, the sweet clover phytoalexin, by *Botrytis cinerea*, are as shown in Fig. 10.10 (Ingham, 1976b). Loss in fungitoxicity is quite dramatic: thus, medicarpin is highly active (ED_{50} 25 μg/ml against mycelial growth of *Helminthosporium carbonum*) while its fungus-derived hydroxylation production (6a-hydroxymedicarpin) has only weak antifungal properties (ED_{50} > 100 μg/ml). A second

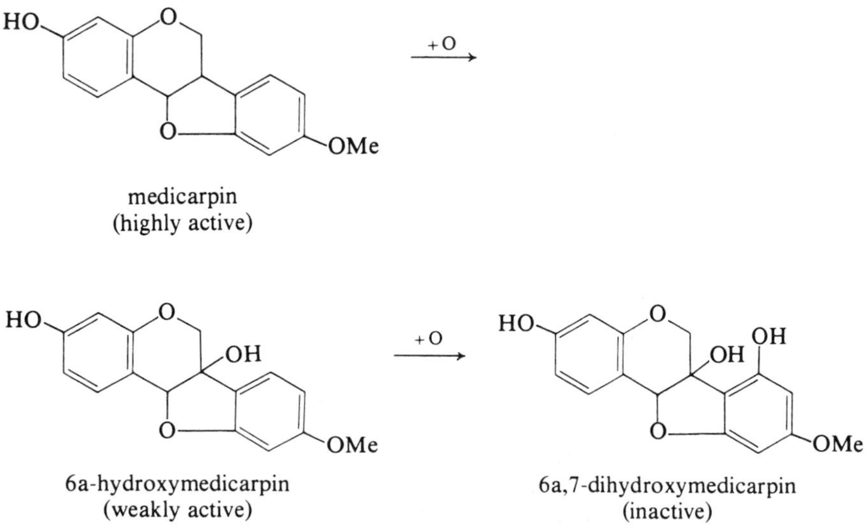

HO, O, O, OMe — medicarpin (highly active) $\xrightarrow{+O}$

HO, O, OH, O, OMe — 6a-hydroxymedicarpin (weakly active) $\xrightarrow{+O}$ HO, O, OH, OH, O, OMe — 6a,7-dihydroxymedicarpin (inactive)

Figure 10.10 Metabolism of medicarpin by *Botrytis* and *Colletotrichum*

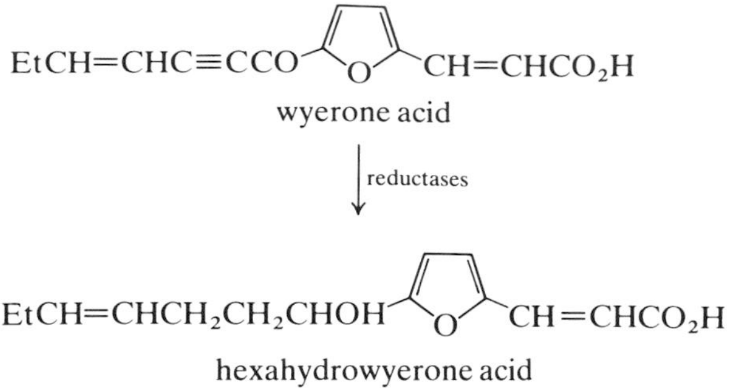

EtCH=CHC≡CCO–(furan)–$CH=CHCO_2H$

wyerone acid

↓ reductases

$EtCH=CHCH_2CH_2CHOH$–(furan)–$CH=CHCO_2H$

hexahydrowyerone acid

Figure 10.11 Detoxification of the phytoalexin wyerone acid by *Botrytis cinerea*

oxidation, which is carried out by *Colletotrichum coffeanum* but not by *Botrytis*, produces 6a,7-dihydroxymedicarpin (Fig. 10.10), which is virtually inactive as a fungitoxin.

Pisatin, the phytoalexin of peas, is likewise metabolized by demethylation to the much less fungitoxic 6a-hydroxymaackiain. The demethylase enzyme responsible for this detoxification has been surveyed in a number of pea pathogens and there is good evidence that the virulence of these organisms is dependent on the possession of a highly active demethylase system. In *Nectria haematococca*, for example, different isolates can vary 100-fold in the amount of demethylase enzyme present and it is only the highly active strains which are virulent on pea (Kistler and VanEtten, 1984).

Other types of phytoalexin also undergo detoxification *in vivo*. Wyerone, the phytoalexin of broad bean (*Vicia faba*), is reduced to the hexahydro derivative, which is no longer fungitoxic. Reduction of the keto group adjacent to the furan ring seems to be important in eliminating fungitoxic activity. Detoxification of wyerone (Fig. 10.11) occurs in *Botrytis cinerea*, the chocolate-spot fungus of broad bean, but not in *B. fabae*, a related species which is non-pathogenic on bean. The failure of *B. fabae* to successfully colonize the bean plant probably lies in its inability to detoxify the bean phytoalexins (Hargreaves *et al.*, 1976b).

In summary, it would seem that metabolic and enzymic studies have been important in confirming indirectly the role of phytoalexins as disease resistance factors. Only those micro-organisms which have developed the means of detoxification, via genetic selection, are likely to be pathogenic on their chosen host plants.

Elicitation of Phytoalexin Synthesis

Much interest has centred on the biochemical mechanism by which phytoalexin synthesis is induced by the fungus in the host plant. Factors of fungal origin which elicit the plant's response to invasion have been termed 'elicitors'. They have only been partially characterized chemically but in general they seem to be either protein, glycoprotein or polysaccharide (usually β (1→3)-glucan) in nature. Some of the better characterized elicitors are highly active and stimulate phytoalexin production in the host at 10^{-9} M, a concentration level at which plants respond to endogenous hormones (Darvill and Albersheim, 1984).

The site of production of the glucan elicitors seems to be at the fungal cell wall. It is possible that the elicitor is only released into the infection droplet after some metabolite in the host plant leaf leachate interacts with the fungus. At a later stage in the interaction, there is presumably a receptor site in the plant cell at which the elicitor is able to trigger off phytoalexin synthesis. A further complication is the fact that phytoalexin synthesis can occur in the absence of a potential pathogen, e.g. in the presence of mercuric salts (see p. 278), so that one has to postulate the existence within the plant itself of one or more constitutive elicitors, as distinct from those of fungal origin (Hargreaves, 1979).

The elicitors that have so far been described lack specificity in that they can be obtained from both compatible and incompatible races of a given fungus. Also, their production has not been correlated with the hypersensitive response, which is the pathological characteristic of true resistance. Phytoalexin induction by plant parasites at some stage involves recognition phenomena and it is likely that more specific elicitors, possibly lectin in nature, will eventually be characterized as mediating in these resistance interactions (see Sequeira, 1980).

The idea that elicitors must necessarily be high molecular weight recognition-type macromolecules was upset by the report that two long chain unsaturated fatty acids, eicosapentaenoic and arachidonic acids, present in crude mycelial extracts of *Phytophthora infestans* are able to elicit cell death and phytoalexin synthesis in potato tuber tissue (Bostock *et al.*, 1981). The discovery was unexpected also in view of the normal harmless nature of fatty acids in living cells. Indeed, in animal tissues, arachidonic acid is an essential fatty acid associated with vitamin F activity. These two acids, however, are specific to oomycete fungi, to which group *Phytophthora* belongs, and most other related acids tested for elicitor activity responded negatively. Since these two fatty acids are not normal higher plant metabolites, their effectiveness in eliciting phytoalexin synthesis may be due to the fact they are 'foreign compounds' as far as the potato is concerned.

III. Phytotoxins in Plant Disease

A. The Pathotoxin Concept

All the chemicals discussed so far in this chapter have been higher plant products, either present as such in host tissue or induced by microbial attack. In a pathogenic situation, the invading bacterium or fungus is able to establish itself within the host and when this happens, it starts producing its own secondary constituents. It is these microbial substances which are often responsible for disease symptoms in the host and it is their deleterious effect on the growth and metabolism of the higher plant which eventually leads to death. Their production and fate is closely related to the susceptibility of higher plants to disease, once the biochemical barriers of resistance have been broken down. Their synthesis represents the aggressive force of the micro-organism. Thus pathotoxins, as these harmful compounds are called, are the expression of the virulence of the pathogen.

A number of pathotoxins have been characterized in a variety of fungal and bacterial disease situations (Durbin, 1981). They can be either low molecular weight or high

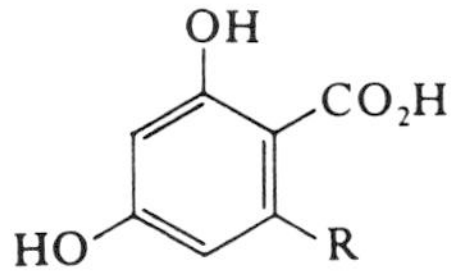

Ceratocystis toxins
R = CH_2COCH_3, $CHOHCOCH_3$ and $COCOCH_3$
(Dutch elm disease)

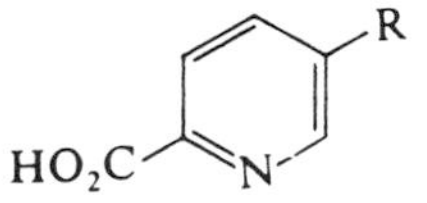

picolinic acid R = H
(*Pyricularia oryzae*, rice blast)
fusaric acid R = *n*-butyl
(*Fusarium oxysporum*, tomato wilt)

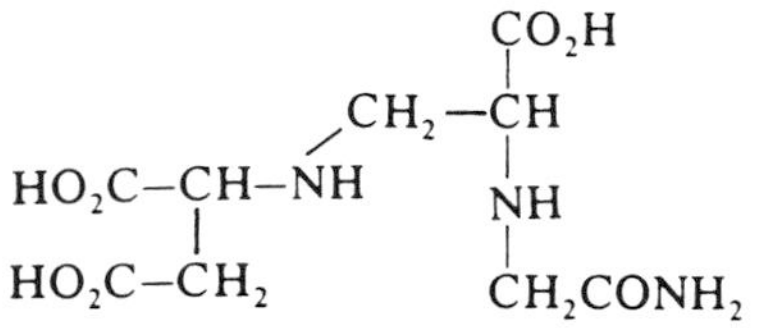

lycomarasmin
(tomato wilt)

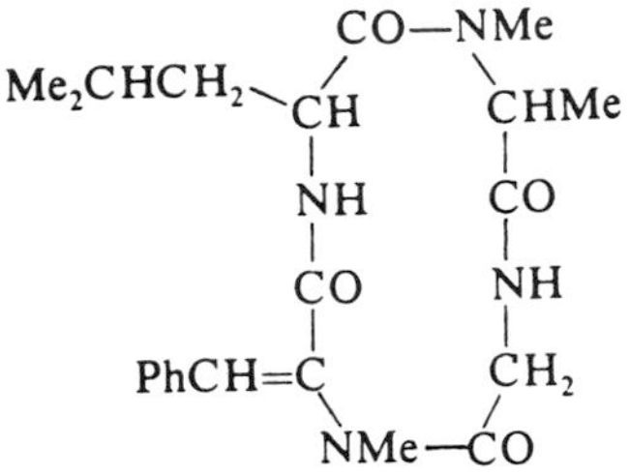

tentoxin
(*Alternaria tenuis*, cotton chlorosis)

Figure 10.12 Characteristic low molecular weight pathotoxins produced by fungal pathogens

molecular weight compounds. Low molecular weight pathotoxins include those which have an effect on growth or cause wilting, i.e. the so-called wilting factors. High molecular weight compounds include the peptides which are causal agents in plant necrosis and also the enzymes which bring about tissue maceration and loss of cellular cohesion in the host plant. Both low and high molecular weight substances may be produced by the same organism. For example, Dutch elm disease, which is disseminated by a bark beetle of the genus *Scolytus*, is due to the fungus *Ceratocystis ulmi*. Its toxins, which cause necrotic lesions in the leaves as well as wilting, are of two types: proteins or glycoproteins and also three low molecular weight phenolic metabolites (for structures, see Fig. 10.12) (Claydon *et al.*, 1974).

The simple hypothesis that a microbial toxin is directly responsible for the symptoms of a plant disease has only been demonstrated so far in a limited number of cases and our evidence that the vast majority of toxins are directly associated with particular diseases is still circumstantial. Among symptoms which can be directly attributed to toxin action are chlorosis, growth abnormalities, necrosis and wilting. Chlorosis, the destruction of the chloroplast and hence the loss of green colour in the leaf, may be due to the simple accumulation of ammonia in the tissues. For example, the phytotoxin of tobacco disease, i.e. the 'wild fire toxin' of the bacterium *Pseudomonas tabaci*, is a small peptide which may exert its effect by interfering with the enzyme glutamine synthetase. This enzyme is a key one in nitrogen metabolism and if it is inhibited, the ammonia produced by reduction of nitrate cannot be coupled to glutamic acid and hence accumulates with disastrous consequences (Sinden and Durbin, 1968).

Growth abnormalities are familiar symptoms of disease susceptibility and many are brought about by the synthesis of unusually high levels of one or other of the major growth hormones. The fungal disease of rice *Gibberella fujikuroi* causes elongation of the rice internodes, this being due to the gibberellins synthesized in large amount by the fungus once it is established in the rice plant. The discovery that this fungus has the ability to synthesize gibberellins provided an important breakthrough in plant physiology during the 1950s and led to the recognition of these compounds as higher plant hormones. Another example of hormonal synthesis is in the bacterial disease of pea plants *Corynebacterium fascians*, which causes fasciations due to the production by the bacterium of the cytokinin, N^6-(Δ^2-isopentenyl)adenine. These fasciations can be produced artificially by treating healthy plants with kinetin.

Crown gall tumours in dicotyledonous plants, due to infection by *Agrobacterium tumefaciens*, provide yet another example of disease symptoms being due to hormonal disturbances rather than to specific toxins. In this interaction, initial infection is triggered by the release from wounded plant cells of two simple phenolics, acetosyringone and α-hydroxyacetosyringone (Stachel *et al.*, 1985). These signal molecules activate a process of genetic transfer of information via Ti plasmids from the bacterium to the higher plant nucleus. At a later stage, large amounts of cytokinin and auxin are produced, which leads to gall formation. The *Agrobacterium*–crown gall system has been utilized by plant geneticists as a method of artificially introducing new genetic material (including genes for pest resistance) into higher plant cells (cf. Callow, 1983).

Another common disease symptom is necrosis, which is characterized in leaf tissue by dark coloured lesions, dry in consistency and of a leathery or brittle texture. Such lesions are caused by complex biochemical changes in the tissue, but the primary effect may be a fairly simple one in terms of a blockage in primary metabolism. Fire blight, *Erwinia amylovora*, in apple shoots causes the evolution of ammonia, the toxic effects of which are expressed as necrosis in the twigs (Lovrekovich *et al.*, 1970).

Finally, wilting, the fourth symptom mentioned above, is usually considered to be due to the production by the invading organism of polysaccharide gums, which act by mechanical plugging of the xylem tissue, thus directly restricting water uptake. There is evidence, however, that the mechanism of wilting may be more complex than this and in some cases may be the direct effect of a toxin in inducing water stress, e.g. by influencing the hormonal control of the stomatal apparatus of the leaf. Some of the agents responsible for causing wilting are considered in more detail in the following section.

B. Pyridine-based Pathotoxins

Wilt diseases are commonly caused by a variety of bacteria which attack such plants as cotton, pea, banana and tomato. The symptoms are strikingly similar in all these host plants, with wilting of the leaves and shoots due to insufficient water flow through the xylem. Desiccation in the leaves can become acute and the disease often leads to the death of the plant. The consequences of wilting in trees are only too familiar to those who have witnessed during the last decade the demise of the Dutch elm in the United Kingdom, due to the phenomenal spread of a virulent strain of Dutch elm disease. The effects of wilting on crop plants can be just as dramatic. From the biochemical viewpoint, the most widely studied interaction is that of *Fusarium oxysporum* with the cultivated tomato.

Two different low molecular weight toxins have been found in *Fusarium* cultures and both have been implicated as wilting agents in the tomato. They are the nitrogen-containing lycomarasmin and a pyridine derivative, fusaric acid (5-*n*-butylpicolinic acid) (Fig. 10.12). The role of lycomarasmin in pathogenicity is still not entirely clear because although readily formed in culture, it has yet to be detected unequivocally in infected plants. This may be simply due to its great lability in solution. An interesting point regarding the mode of action of these pathotoxins is that both lycomarasmin and fusaric acid are metal chelators. The former has strong chelating properties and its translocation and activity may be related to the water-soluble complex it forms with iron. The activity of fusaric acid is also related to metal ion content, since its production, at least in culture, is dependent on the presence of zinc and, unless sufficient is added, its synthesis is depressed.

The role of fusaric acid in causing wilting is reasonably well established. Thus it has been detected in plants after infection and is present in much higher concentrations in plants infected by virulent strains than those treated with avirulent strains. A given virulent strain of *Fusarium* produces up to 80 mg/l in *in vitro* culture and as much as 100 μg/l fresh wt fusaric acid has been found in badly infected tomato plants. As part of the co-evolution of *Fusarium* with the tomato, some varieties have developed the ability to resist attack. Such varieties apparently resist infection or the effects of infection because they are able to conjugate the fusaric acid with glycine, the resulting conjugate being inactive. This is not an all or nothing effect, since susceptible varieties are able to conjugate between 5–10% of the toxin in this way. However, the resistant varieties conjugate up to 25%, the greater efficiency in conjugating ability being apparently sufficient to avoid wilting.

Fusaric acid has been detected in other plants (cotton, flax and banana) besides tomato after inoculation with wilt pathogens and it seems to be fairly widely produced by *Fusarium* species. A comparison of fusaric acid with a range of synthetic pyridine derivatives has shown that the carboxyl group in the α-position to the nitrogen is essential for toxicity. The aliphatic side chain in the β-position is also important, since its presence improves water permeability (Kern, 1972). It may be noted, however, that the side chain is not necessary for toxicity *per se*, since the parent compound, picolinic acid, which lacks any β-substituent, causes necrosis in rice. Indeed, it is a major toxin of the agriculturally important disease 'blast of rice' which is caused by the fungus *Pyricularia oryzae*. Picolinic acid is so highly toxic to rice plants that only 0.5 ng are needed to produce a lesion in the leaf following injection.

Picolinic acid, like fusaric, is a metal chelator and, in the rice blast disease, it acts largely by scavenging vital iron and copper ions from within the plant tissue; its toxic effects can be reversed by supplying these metal ions back to the plant. Picolinic acid is detoxified by the host plant by conversion to the methyl ester and the *N*-methyl ether and resistant rice varieties have been shown to have a greater capacity for detoxification than susceptible forms. The pathogenicity of *Pyricularia* in rice is partly due to the synthesis by the fungus of a second toxin, piricularin, $C_{18}H_{14}N_2O_3$, a compound still only partly characterized. It is interesting that this second pathotoxin is actually toxic to the conidia of the parasite and is capable of preventing germination of the spores at a concentration of 0.25 ppm. This inhibitory effect is circumvented *in vivo* by the fact that piricularin exists as a complex with protein, the complex being non-toxic to the fungus but still highly lethal to the host plant.

In addition to the toxic phenols produced in Dutch elm disease and the pyridine α-carboxylic acids of tomato wilt and blast of rice, a range of other low molecular weight toxins have been implicated as causative agents in plant disease. Among them are a number of cyclic peptides (e.g. tentoxin of *Alternaria*), several naphthaquinones and many terpenoids, one being fusicoccin from *Fusicoccum amygdali*. This tricyclic diterpene glucoside exerts its toxic effects on the higher plant cell by interfering with plasmalemma function (Marré, 1980). One of the simplest phytotoxins to be described to date is 3-methylthiopropionic acid, $MeSCH_2CH_2CO_2H$, which is produced by the bacterium *Xanthomonas campestris* pv. *manihotis*, the necrotic blight organism of cassava leaves. Levels of 6 μg/g fresh wt were detected in necrotic leaves and this concentration produced typical symptoms when inoculated into healthy leaves (Perreaux *et al.*, 1986). Many other phytotoxins are much more complex than this and two—helminthosporoside and victorin—will be described in the next section.

C. Helminthosporoside and Victorin

Two phytotoxins deserve more detailed consideration here because of their selective activity and chemical complexity. The first is helminthosporoside (HS), which is responsible for the eye spot symptom of sugar-cane infected by *Helminthosporium sacchari*. It produces characteristic reddish-brown stripes, called runners, when injected into susceptible sugar-cane leaves (Strobel, 1974). It is of special interest because of its selective mode of action: it apparently binds to a single membrane protein in susceptible but not in resistant varieties. Resistant clones of sugar-cane contain a similar protein and binding will take place here but only if the tissue is first treated with detergent. The difference between resistance and susceptibility in this plant seems to be based, remarkably enough, on the availability or otherwise of a particular binding protein normally associated with α-galactoside transport across the chloroplast membrane. Binding of the toxin at this site interferes with the normal membrane function of ion transport, the tissue is disrupted and the symptoms develop.

Helminthosporoside was originally thought to be 2-hydroxycyclopropyl α-galactoside but re-examination has shown that it is a mixture of three isomeric sesquiterpene glycosides (Fig. 10.13). Digalactosyl residues are attached at either end of the molecule and the galactose residues are in the rare furanose form (Macko *et al.*, 1983).

What is especially interesting about the work on the HS toxin is the discovery of a series of related molecules in the filtrates from the fungal cultures. A number of penta- and hexaglycosides have been detected, three of them being derivatives of the toxins in which an additional α-glucosyl residue is attached at each end to the terminal galactoses. These are 'latent' forms of the toxin, themselves non-toxic but yielding a toxin

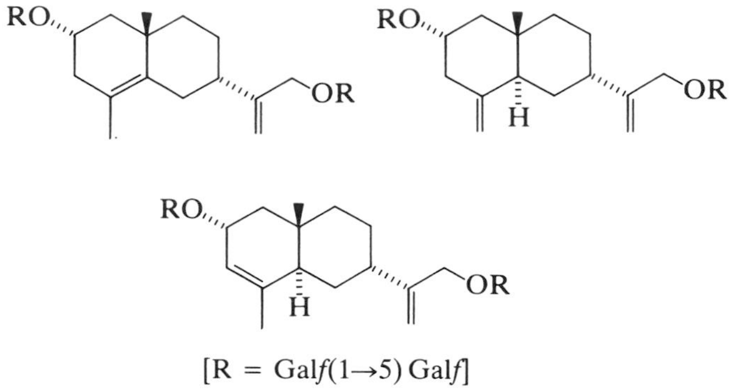

[R = Galf(1→5) Galf]

Figure 10.13 Structures of the three sesquiterpenoid components of helminthosporoside toxin

if they are treated with an α-glucosidase. Additionally, the culture filtrates show the presence of up to 21 non-toxic sesquiterpene glycosides, called toxoids, which are lower homologues from which one or more of the galactose units of the toxins are absent. The lack of toxicity in all of these related molecules indicates that there is a high specificity in the structure of the actual toxins and supports the idea that HS toxin produces its disease symptoms by binding to a receptor site in susceptible cells. This is also indicated by the fact that some of the toxoids are capable of acting as competitive inhibitors of HS toxin. Thus toxoid III (which contains three galactose residues) will give 90% protection from the virulence of HS toxin to susceptible plants of sugar cane if it is applied with the toxin in the molar ratio of 24 : 1 (Livingston and Scheffer, 1983).

The special feature of victorin, the pathotoxin of the oat disease *Cochliobolus victoriae*, is that it is the most potent and selective toxin known so far. Its potency and differential toxicity are illustrated by the fact that the crude filtrate of *C. victoriae* containing it has to be diluted down to one to ten million before it ceases showing a disease symptom in a susceptible oat plant. By contrast, a dilution of 1 : 25 is the limit where a symptom can be induced in a resistant variety. The selectivity of victorin, which is the dilution end point for susceptible plants over that for resistant plants, is thus 400,000. Other similar medium molecular weight toxins have been isolated from diseased tissues of maize,

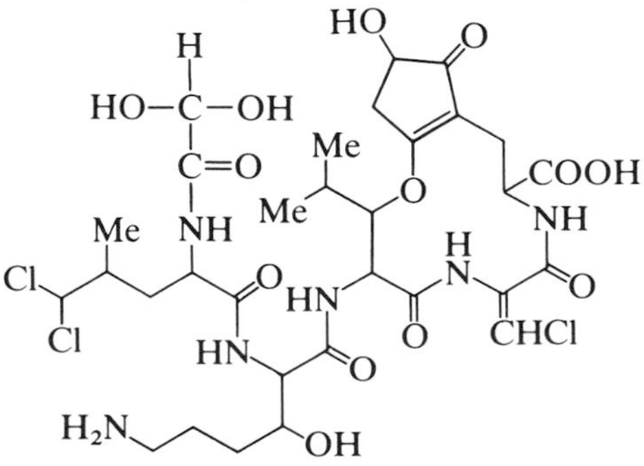

Figure 10.14 Structure of the phytotoxin victorin of *Cochliobolus victoriae*

sorghum and peas and these have selectivities of between 25 and 300. Victorin is a unique molecule in showing such a striking differential toxicity in cereal plants. Its toxicity seems to be related in the initial stages of infection to a disruption of cell permeability (Wheeler and Luke, 1963). Chemically, victorin has a peptide-like structure (Fig. 10.14) based on several unusual amino acids, including dichloroleucine, β-hydroxylysine and victalanine (Wolpert *et al.*, 1985).

D. Macromolecular Toxins

Less is known of the detailed structure of macromolecular phytotoxins. Two different toxins, i.e. a peptidorhamnomannan and a pure protein (ceratoulmin), have been claimed as being responsible for the wilting symptoms that are caused by *Ceratocystis ulmi*, which is the fungus that is responsible for Dutch elm disease (Burdekin, 1983). A glycoprotein phytotoxin has been partly purified from the rust fungus *Rhynchosporium secalis*. It contains the four sugars mannose, rhamnose, galactose and glucosamine in the ratio 13.6 : 1 : 1 : 1, and these are linked to the polypeptide backbone through threonine and serine residues (Mazars *et al.*, 1984). Another glycoprotein, named malseccin, is responsible for the symptoms in lemon leaves infected by *Phoma tracheiophila*. Its structure is based on mannose, galactose and glucose as well as most of the standard protein amino acids (Nachmias *et al.*, 1979).

Fungal polysaccharides are also capable of producing wilting symptoms in plants, but it is not yet clear whether they are necessarily the causative agents of disease. Glucans with $\beta 1 \rightarrow 3$ and $\beta 1 \rightarrow 6$ linkages have, for example, been isolated from several *Phytopthora* species and are capable of causing wilting in *Eucalyptus* seedlings (Woodward *et al.*, 1980).

No account of the biochemical effects of pathogenic micro-organisms on higher plants would be complete without at least a mention of the role of microbial enzymes on the disruption and destruction of host tissues following infection. This subject—the degradation of higher plant cell walls by parasites—has been extensively investigated by R. K. S. Wood and his students, among others, and much information is available in Wood (1967) and in later reviews (e.g. Friend and Threlfall, 1976). The degradative enzymes have been particularly studied in soft rots of storage tissues such as potatoes, carrots and citrus fruits. Organisms responsible include *Botrytis cinerea, Rhizoctonia solani, Sclerotinia fructigena* and various *Aspergillus* and *Penicillium* species.

In brief, then, these enzymes produced by the attacking organism cause loss of coherence in the invaded tissues. As a result, the cells are separated, a process called maceration, and the naked protoplasts released soon die. The effects can be seen in any rotting fruit or vegetable where the rotted tissue loses all its resistance to mechanical damage. The enzymes implicated are the cellulases, hemicellulases and various pectic enzymes (pectinesterases, polygalacturonases, (*E*)-eliminases, etc.). The pectinases are especially destructive since they break down the linkages between the cellulose microfibrils and the components of the cell wall matrix. The cellulose and other components (including glycoprotein) so released are then further broken down by cellulases, hemicellulases and proteases. The activities of these various enzymes are determined by the pH of the tissue and the presence of calcium ions.

It is clear from these experiments that the pathogenic fungi have highly effective enzymes for destroying plant tissues, once they become well established in the host plant. The materials released are ultimately used by the fungal parasite in the maintenance of its intermediary metabolism.

E. Other Effects of Phytotoxins

Phytotoxins are, by definition, microbial metabolites which are capable of producing one or more disease symptoms in the host plant. Although they are normally thought of as expressing the virulence of the pathogen in its ability to damage plant tissues, they could possibly have other functions. For example, they might protect an infected plant from further infection by a second fungus. This seems to be so in the case of *Epichloe typhina*, which is a pathogenic fungus of the grass *Phleum pratense*. The fungus produces three sesquiterpenes which are apparently fungitoxic. These could be responsible for the fact that infected *Phleum* plants are resistant to further infection by *Cladosporium phlei* (Yoshihara *et al.*, 1985).

In producing a toxin, a fungus runs the risk of autotoxicity, and there is evidence in the case of piricularin, which is a phytotoxin of rice blast (*Pyricularia oryzae*), that it is complexed with protein to avoid autotoxic effects (see Section IIIB). Autotoxic or self-inhibiting agents may, of course, be of value at certain stages of the fungal life-cycle. This appears to be true of gloeosporone, which is a germination self-inhibitor that has been isolated from the conidia of *Colletotrichum gloeosporioides* (a disease organism of many tropical plants). In restricting the germination of spores, this compound presumably enables them to survive longer and hence improves their chances of parasitizing a suitable host (Meyer *et al.*, 1983).

In soil-borne fungi, one fungus might produce a toxin which allows it to compete successfully with other soil-dwelling organisms. This appears to be true of *Laetisaria arvalis*, which has the ability to stop the growth of *Pythium ultimum*. The toxin has been identified as laetisaric acid, or 8-hydroxylinoleic acid (Bowers *et al.*, 1986), This allelopathic effect of one fungus on another has been used to man's advantage. *Pythium* is a major root pathogen of many crops and treating agricultural fields with *Laetisaria* is known to be an effective control.

IV. Conclusion

In the interaction between higher plants and micro-organisms leading to disease, there is evidence of co-evolution in both the resistance mechanisms developed by the higher plants and also in the various ways that the virulent micro-organism can damage the host. This is summarized in Table 10.3. In the case of disease resistance, there are at least three mechanisms: (1) accumulation of a fungitoxic compound at the surface of the plant where invasion is most likely, especially on the upper leaf surface; (2) after infective stimulation, the release of toxin already present in the tissue in bound form; (3) *de novo* synthesis of tailor-made fungitoxins, the phytoalexins. Similarly, in disease susceptibility,

Table 10.3 Summary of biochemical mechanisms of disease resistance and susceptibility in plants

	Higher plant	Micro-organism
Resistance (hypersensitive reaction)	1. Pre-infectional toxin 2. Post-infectional toxin formed from pre-existing substrate 3. Post-infectional toxin formed *de novo* (phyto-alexin)	Invasion arrested or toxin metabolized to harmless products
Susceptibility (disease symptoms)	Biodegraded or conjugated	1. Pathotoxins (low MW or peptides) 2. Growth hormone synthesis 3. Primary metabolite accumulation 4. Enzyme degradation

micro-organisms employ a variety of approaches to destroy their host. These are: (a) pathotoxin production; (b) overproduction of a growth hormone; (c) interference with primary metabolism; and (d) synthesis of hydrolytic enzymes which destroy the coherence of the organism.

These interactions are dynamic in that possibilities for changing the balance in favour of one or other party are always present. The micro-organism can evolve by developing phytoalexin detoxification mechanisms. Similarly, higher plants can retaliate to invading parasites by degrading or conjugating the pathotoxins formed in their tissues.

In any interaction that has been investigated closely, considerable chemical complexity has become apparent. Thus, investigations of the broad bean–*Botrytis* interaction indicate that at least seven different phytoalexins are formed (Hargreaves *et al.*, 1976a). Again, in the pea–*Fusarium solani* interaction, some phytoalexins are formed immediately after infection; others are produced several days after the primary infection (Pueppke and VanEtten, 1976). Further, in the potato–*Phytophthora infestans* interaction, recent research (Friend, 1976) indicates that defence through phytoalexin synthesis is less important than defence through the esterification of cell wall polysaccharide by phenolic acids. Similar chemical complexity exists among the pathotoxins of fungi and bacteria. In Dutch elm disease, a range of toxins (see p. 286) are produced by the invading fungus. In most other cases examined, two or more compounds, often of widely differing structures, have been implicated in pathogenicity.

There is, of course, considerable biological complexity in plant disease and physiological factors can be immensely important in determining whether the plant suffers infection and in determining the severity of the attack. As Wheeler (1975) puts it: 'pathogenesis can be viewed as a battle between a plant and a pathogen which is refereed by the environment'. Of many environmental factors, temperature and humidity are undoubtedly particularly important. A further complexity may be introduced by the presence of other micro-organisms into the system. These may be present either in the soil or in the aerial regions (the phyllospere) around the plant. They may be beneficial to the plant and antagonistic to the micro-organism. For example, damage to tomato plants caused by the fungus *Sclerotium rolfsii* is alleviated if a second fungus *Trichoderma harzianum* is added to the soil. Biological control is thus established by the second fungus overgrowing and killing the first (Wells *et al.*, 1972).

In spite of the many biological factors which may often be decisive in determining resistance or susceptibility, it is still true that the actual weapons of the fight are biochemical in nature. Further studies of the biochemistry of phytoalexins and pathotoxins can only enrich our present imperfect understanding of resistance mechanisms in higher plants. Potentially, there are enormous practical benefits to be gained by applying the results so gained to crop protection. Phytoalexins, for example, are antifungal by definition and they could be applied as fungicidal chemicals to disease control. Preliminary experiments with the sesquiterpenoid capsidiol, the phytoalexin of peppers, have shown that, when sprayed on in solution at concentrations 5×10^{-4} M, it will control late blight (*Phytophthora*) attack on tomatoes (Ward *et al.*, 1975).

Since phytoalexin production is a fairly general resistance mechanism, these substances offer interesting possibilities for cross-protection of crop plants. Thus the phytoalexin of one plant (in the above case, pepper) might be much more effective in controlling a disease of a second plant (e.g. tomato) than the pathogens it normally encounters. This has been achieved in tobacco, *Nicotiana tabacum*, by genetically engineering into it a gene for stilbene synthesis from the grapevine *Vitis vinifera*. The resulting transgenic plant produced the new stilbene phytoalexin resveratrol on infection and also had improved resistance to the fungus *Botrytis cinerea* (Hain *et al.*, 1993). Another way of employing phytoalexins in crop protection is by activating their synthesis in the field by abiotic means. This has been done with rice plants; application of a systemic fungicide induces massive phytoalexin synthesis and hence resistance to rice blast disease, *Pyricularia oryzae* (Cartwright *et al.*, 1977). Finally, synthetic analogues of phytoalexins might be developed fruitfully to produce fungicidal agents which are persistent and are not as readily biodegradable as the natural molecules.

Bibliography

Books and Review Articles

Bailey, J. A. and Deverall, B. J. (eds.) (1983). 'The Dynamics of Host Defence', 233 pp. Academic Press, Sydney.
Bailey, J. A. and Mansfield, J. W. (eds.) (1982). 'Phytoalexins', 334 pp. Blackie, Glasgow.
Burdekin, D. A. (ed.) (1983). 'Research on Dutch Elm Disease in Europe'. HMSO, London.
Callow, J. A. (ed.) (1983). 'Biochemical Plant Pathology', pp. 484. John Wiley, Chichester.
Cruickshank, I. A. M. and Perrin, D. R. (1964). Pathological function of phenolic compounds in plants. In: Harborne, J. B. (ed.), 'Biochemistry of Phenolic Compounds', pp. 511–544. Academic Press, London.

Durbin, R. D. (ed.) (1981). 'Toxins in Plant Disease', 515 pp. Academic Press, New York.
Friend, J. and Threlfall, D. R. (eds.) (1976). 'Biochemical Aspects of Plant–Parasite Relationships', 354 pp. Academic Press, London.
Harborne, J. B. (1977). Chemosystematics and coevolution. *Pure appl. Chem.* **49**, 1403–1421.
Harborne, J. B. (1986). The role of phytoalexins in natural plant resistance. In: Green, M. B. and Hedin, P. A. (eds.), 'Natural Resistance of Plants to Pests', pp. 22–35. American Chemical Society, Washington, DC.
Harborne, J. B. (1987). Natural fungitoxins. *Proc. Phytochem. Soc. Eur.* **26**, 195–211.
Harborne, J. B. and Ingham, J. L. (1978). Biochemical aspects of the coevolution of higher plants with their fungal parasites. In: Harborne, J. B. (ed.), 'Biochemical Aspects of Plant and Animal Coevolution', pp. 343–405. Academic Press, London.
Heitefuss, R. and Williams, P. H. (eds.) (1976). 'Physiological Plant Pathology', Vol. 4 in the New Series Encyclopedia of Plant Physiology, 890 pp. Springer-Verlag, Berlin.
Hillis, W. E. (ed.) (1962). 'Wood Extractives'. Academic Press, New York.
Horsfall, J. B. and Cowling, E. B. (eds.) (1980). 'Plant Disease Vol. 5. How Plants Defend Themselves', 543 pp. Academic Press, New York.
Ingham, J. L. (1973). Disease resistance in plants: the concept of preinfectional and post-infectional resistance. *Phytopath. Z.* **78**, 314–335.
Strobel, G. A. (1974). Phytotoxins produced by plant parasites. *A. Rev. Plant Physiol.* **25**, 541–566.
Swinburne, T. R. (1973). The resistance of immature Bramley's seedling apples to rotting by *Nectria galligena*. In: Byrde, R. J. W. arid Cutting, C. V. (eds.), 'Fungal Pathogenicity and the Plants Response', pp. 365–382. Academic Press, London.
Wheeler, H. (1975). 'Plant Pathogenesis', 106 pp. Springer Verlag, Berlin.
Wood, R. K. S. (1967). 'Physiological Plant Pathology', 570 pp. Blackwell Scientific, Oxford.

Literature References

Bailey, J. A., Vincent, G. G. and Burden, R. S. (1974). *J. gen. Microbiol.* **85**, 57–74.
Beijersbergen, J. C. M. and Lemmers, C. B. G. (1970). *Acta Bot. Neerl.* **19**, 114.
Blakeman, J. P. and Atkinson, P. (1979). *Physiol. Plant Path.* **15**, 183–192.
Bostock, R. M., Kuc, J. A. and Laine, R. A. (1981). *Science, N.Y.* **212**, 67–69.
Bowers, W. S., Hock, H. C., Evans, C. H. and Katayama, M. (1986). *Science, N.Y.* **232**, 105.
Cartwright, D., Langcake, P., Price, R. J., Leworthy, D. P. and Ride, J. P. (1977). *Nature, Lond.* **267**, 511–513.
Claydon, N., Grove, J. F. and Hosken, M. (1974). *Phytochemistry* **13**, 2567–2572.
Coley-Smith, J. R (1976). In: Friend, J. and Threlfall, D. R. (eds.), 'Biochemical Aspects of Plant–Parasite Relationships', pp. 11–24. Academic Press, London.
Crombie, W. M. L., Crombie, L., Green, J. B. and Lucas, J. A. (1986). *Phytochemistry* **25**, 2075–2083.
Cruickshank, I. A. M. and Perrin, D. R. (1960). *Nature, Lond.* **187**, 799–800.
Darvill, A. G. and Albersheim, P. (1984). *A. Rev. Plant Physiol.* **25**, 243–275.
Franisch, R. A., Carson, M. J. and Carson, S. D. (1986). *Physiol. Molec. Plant Path.* **28**, 267–268.
Friend, J. (1976). In: Friend, J. and Threlfall, D. R., 'Biochemical Aspects of Plant–Parasite Relationships', pp. 291–304. Academic Press, London.

Greathouse, G. A. and Watkins, G. M. (1938). *Am. J. Bot.* **25**, 743–748.
Greenaway, W. and Whatley, F. R. (1976). In: Smith, H. (ed.), 'Commentaries in Plant Science', pp. 249–262. Pergamon Press, Oxford.
Greenhalgh, J. R. and Mitchell, N. D. (1976). *New Phytol.* **77**, 391–398.
Hain, R., Reif, H. J., Krause, E., Langebartels, R., Kindl, H., Vornam, B., Wiese, W., Schmetzer, E., Schreier, P. H., Stocker, R. H. and Stenzel, K. (1993). *Nature* **361**, 153–156.
Harborne, J. B., Ingham, J. L., King, L. and Payne, M. (1976). *Phytochemistry* **15**, 1485–1488.
Hargreaves, J. A. (1979). *Physiol. Plant Path.* **15**, 279–287.
Hargreaves, J. A., Mansfield, J. W. and Coxon, K. R. (1976a). *Nature* **262**, 318–319.
Hargreaves, J. A., Mansfield, J. W., Coxon, D. T. and Price, K. R. (1976b). *Phytochemistry* **15**, 1119–1121.
Higgins, V. J. and Millar, R. L. (1968). *Phytopathology* **58**, 1377–1383.
Hughes, J. C. and Swain, T. (1960). *Phytopathology* **50**, 398–400.
Hulme, A. C. and Edney, K. L. (1960). In: Pridham, J. B. (ed.) 'Phenolics in Plants in Health and Disease', pp. 87–94. Pergamon Press, Oxford.
Hunter, L. D. (1975). *Phytochemistry* **14**, 1519–1522.
Hunter, L. D., Kirkham, D. S. and Hignett, R. C. (1968). *J. gen Microbiol.* **53**, 61–67.
Ingham, J. L. (1976a). *Phytochemistry* **15**, 1791–1793.
Ingham, J. L. (1976b). *Phytochemistry* **15**, 1489–1496.
Ingham, J. L. (1981). In: Polhill, R. M. and Raven, P. H. (eds.), 'Advances in Legume Systematics', pp. 599–626. HMSO, London.
Jurd, L., Corse, J., King, A. D., Bayne, H. and Mihara, K. (1971). *Phytochemistry* **10**, 2971–2974.
Kern, H. (1972). In: Wood, R. K. S., Ballio, A. and Graniti, A. (eds.), 'Phytotoxins in Plant Disease', pp. 35–48. Academic Press, London.
Kirkham, D. S. and Hunter, L. D. (1964). *Nature, Lond.* **201**, 638.
Kistler, H. C. and VanEtten, H. D. (1984). *J. gen. Microbiol.* **130**, 2545–2603.
Livingston, R. S. and Scheffer, R. P. (1983) *Plant Physiol.* **72**, 530.
Lovrekovich, L., Lovrekovich, H. and Goodman, R. N. (1970). *Can. J. Bot.* **48**, 999–1000.
Luke, H. H. and Gracen, V. E. (1972). In: Kadis, S., Ciegler, A. and Ajl, S. J. (eds.), 'Microbial Toxins', Vol. 8, pp. 139–168. Academic Press, London.
Lüning, H. U. and Schlösser, E. (1976). *Z. Pflanzenkrankheiten Pflanzenschutz* **83**, 317–327.
Macko, V., Acklin, W., Hildenbrand, C., Weibel, F. and Arigoni, D. (1983). *Experientia* **39**, 343.
Marré, E. (1980). *Prog. Phytochem.* **6**, 253–284.
Mazars, C., Hapner, K. D. and Strobel, G. A. (1984). *Experientia* **46**, 1244.
Meyer, W. L., Lax, A. R., Templeton, G. E. and Brannon, M. J. (1983). *Tetrahedron Lett.* **24**, 5059.
Millar, R. L. and Higgins, V. J. (1970). *Phytopathology* **60**, 104–110.
Müller, K. O. and Börger, H. (1941). *Arb. biol. Abt. (Ansl. Reichstanst.), Berl.* **23**, 189–231.
Nachmais, A., Borasch, I., Buckner, V., Solel, Z. and Strobel, G. A. (1979). *Physiol. Plant Path.* **14**, 135–140.
Nicholson, R. L. Butler, L. G. and Asquith, T. N. (1986). *Phytopathology* **76**, 1315–1318.
Offord, H. R. (1940). *Bull. US Bur. Ent.* E-518.
O'Neill, M. J., Adesanya, S. A. and Roberts, M. F. (1983). *Z. Naturforsch.* **38c**, 693.
Perreaux, D., Maraito, H. and Meyer, J. A. (1986). *Physiol. Molec. Plant Path.* **28**, 323–328.

Piattelli, M. and Impellizzeri, G. (1971). *Phytochemistry* **10**, 2657–2660.
Pueppke, S. G. and VanEtten, H. D. (1976). *Physiol. Plant Path.* **8**, 51–61.
Robeson, D. and Harborne, J. B. (1980). *Phytochemistry*, **19**, 2359–2366.
Scheffer, T. C. and Cowling, E. B. (1966). *A. Rev. Phytopath.* **4**, 147–170.
Sequeira, L. (1980). In: Horsfall, J. G. and Cowling, E. B. 'Plant Disease. Vol. 5, How Plants Defend Themselves', pp. 179–200. Academic Press, New York.
Sijpesteijn, A. K. (1969). *Meded. Landbouwhogesch. (Gent)* **34**, 379–391.
Sinden, S. L. and Durbin, R. D. (1968). *Nature, Lond.* **219**, 379–380.
Slob, A., Jekel, B., Jong, B. and Schlatmann, E. (1975). *Phytochemistry* **14**, 1997–2006.
Spencer, P. A. and Towers, G. H. N. (1988). *Phytochemistry* **27**, 2781–2785.
Stachel, S. E., Messene, E., van Motagu, M. and Zambryski, P. (1985). *Nature* **318**, 624–629.
Stoessl, A. (1967). *Can. J. Chem.* **45**, 1745–1760.
Towers, G. H. N. (1964). In Harborne, J. B. (ed.), 'Biochemistry of Phenolic Compounds', pp. 249–294. Academic Press, London.
VanEtten, H. F. (1976). *Phytochemistry* **15**, 655–659.
Walker, J. C. and Stahmann, M. A. (1955) *A. Rev. Plant Physiol.* **6**, 351–366.
Ward, E. W. B., Unwin, C. H. and Stoessl, A. (1975). *Phytopathology* **65**, 168–169.
Ward, H. M. (1905). *Ann. Bot.* **19**, 1–54.
Wells, H. D., Bell, D. K. and Jaworski, C. A. (1972). *Phytopathology* **62**, 442–447.
Wheeler, H. and Luke, H. H. (1963). *A. Rev. Microbiol.* **17**, 223–242.
Wolpert, T. J., Macko, V., Acklin, W., Jann, B., Seibl, J., Meili, J. and Arigoni, D. (1985). *Experientia* **41**, 1370–1371.
Woodward, J. R., Keane, P. S. and Stone, B. A. (1980). *Physiol. Plant Path.* **16**, 205–212.
Yoshihara, T., Togiya, S., Koshimo, H. and Sakamura, S. (1985). *Tetrahedron Lett.* **26**, 5551.

SUBJECT INDEX

INDEX OF PLANT NAMES

D

E

F

G

H

I

J

K

L

INDEX OF ANIMAL SPECIES